MECHANICAL MACHINE DESIGN-I

AF647020

DR. SADHU SINGH
Formely, Professor & Head, Mechanical Engineering Department
&
Dean, Faculty of Engineering & Technology
Goving Ballabh Pant University of Agriculture & Technology,
Pantnagar-263 145 (Uttarakhand)
Formely, Director (Colleges), Punjab Technical University, Jalandhar

S.K. KATARIA & SONS®
Publisher of Engineering Books
4885/109, Prakash Mahal, Dr. Subhash Bhargav Lane,
Opposite Delhi Medical Association, Daryaganj, New Delhi–110002 (INDIA)
Phone: +91-11-23243489, 43551243; *Mobile:* +91-9871775858
e-mail: katariabook@gmail.com, katariabooks@yahoo.com
Website: www.skkatariaandsons.com
(e-books, subscription and bulk buying also available on our website)

Published By:

S.K. Kataria & Sons®
4885/109, Prakash Mahal, Dr. Subhash Bhargav Lane,
Opposite Delhi Medical Association, Daryaganj, New Delhi-110002 (INDIA)
Phone: +91-11-23243489, 43551243; *Mobile:* +91-9871775858
e-mail: katariabook@gmail.com, katariabooks@yahoo.com
Website: www.skkatariaandsons.com

First Edition : 2011; Reprint: 2023

Price: ₹ 425/-

Laser Typesetting by : Stride Computers (Delhi) *Printed at* : New A.S. Offset Press Sahibabad (U.P.)

PREFACE

Mechanical Machine Design - I (ME – 303F) has been introduced from the session 2011 – 12 by the Maharishi Dayanand University, Rohtak. It is offered in the 5th semester to B.Tech./B.E. Mechanical Engineering students. The present book has been written strictly according to the new syllabus.

The book contains ten chapters on Design philosophy; Selection of materials; Manufacturing considerations in design; Strength considerations in design; Mechanical joints; Keys, couplings and flywheel; Screwed joints and power screws; Shafts; Belt, chain and rope drives; and Clutches and brakes. A large number of examples have been solved to illustrate the solution methods. An added feature of the book is that previous questions from MDU examinations have been solved in each chapter. A large number of problems have been given at the end of each chapter. As per the instructions issued by the MDU, the design data can be taken from any Design Data Book.

The author is highly thankful to Mr. Sanjeev Kataria and Mr. Sharda Singh for their help and cooperation during the preparation of the manuscript of the book and otherwise.

The author expresses his appreciations for his family members for their tolerance and cooperation. My grand children Kanupreet and Amitoej have been a source of great encouragement during the writing of manuscript and proof reading.

Suggestions for the improvement of the book shall be welcome. The author shall appreciate receiving information from readers regarding any errors or misprints left inadvertently.

—AUTHOR

CONTENTS

1. DESIGN PHILOSOPHY 1-17

1.1 Definitions 1

1.2 Mechanical Engineering Design 2

1.2.1 Morphology of Mechanical Engineering Design 2

1.2.2 Procedure for Morphological Analysis 3

1.3 Brain-Storming 4

1.3.1 Scheme for Brain-Storming Session 4

1.4 Skills of a Designer 4

1.5 The Design Process 5

1.5.1 Design Process : Shigley Model 5

1.5.2 Earle Model for Design Process 6

1.6 Aesthetic Considerations in Design 10

1.7 Ergonomic Considerations in Design 11

1.8 Role of Standardisation and Standards 12

1.8.1 Advantages of Standardisation 12

1.8.2 Standards 12

1.9 Interchangeability 13

1.10 Preferred Numbers 13

1.10.1 Derived Series 14

1.10.2 Preferred Value 14

1.10.3 Characteristics of Preferred Series 14

1.11 Requirements of a Good Design 16

Exercises 17

2. SELECTION OF MATERIALS 18-42

2.1 Introduction 18

2.2 Selection of Engineering Materials 18

2.3 Stress-Strain Diagram for Mild Steel 20

2.3.1 True stress-Stain Diagrams 21

2.3.2 Stress-Strain Diagrams for Other Materials 21

2.3.3 Linear and Non-Linear Materials 22

2.3.4 Idealized Stress-Strain Diagrams 23

2.3.5 Viscoelastic Materials 23

2.4 Factor of Safety 25

2.5 Classification of Engineering Materials 25

2.6 Cast Irons 26

2.6.1 Advantages of Cast Irons 26

2.6.2 Applications of Cast Irons 26

2.6.3 BIS system of Designation of Cast Irons 26

2.6.4 Effect of Alloying Elements on Cast Iron 27

2.6.5 Effect of Impurities on Cast Iron 28

2.7 Steels 28

2.7.1 Classification of Steels 28

2.7.2 Designation of Steel 28

2.7.3 Applications of steels 30

2.7.4 Effect of Alloying Elements on Steel 32

2.7.5 Reasons for use of Alloy Steels in Machines Design 32

2.7.6 Special Alloy Steels 32

2.8 Heat Treatment Methods 34

2.9 Non-Ferrous Metals and Alloys 35

2.9.1 Copper and Copper Alloys 35

2.9.2 Aluminium and Aluminium Alloys 37

2.10 Non-Metallic Materials 37

2.11 Composite Materials 39

2.12 Weighted Factor Method of Material Selection 40

Exercises 42

3. MANUFACTURING CONSIDERATIONS IN DESIGN 43-55

3.1 Assembly Considerations in Design 43

3.2 Support and Retainment of Rotating Parts 44

3.3 Frames and Housings 45

3.4 Castings 45

3.4.1 Iron Castings 45

3.4.2 Steel Casting 47

3.4.3 Die Casting 47

3.5 Welding 49

3.6 Fits and Tolerances 50
3.6.1 Fits 50
3.6.2 Types of Fits 50
3.6.3 Terminology 51
3.6.4 Standard Tolerances 52
3.6.5 Hole Basis and Shaft Basis for Fits 53
3.6.6 Selection of Fits 53
3.6.7 Dimensioning of Tolerances Rules 54
Exercises 55

4. STRENGTH CONSIDERATIONS IN DESIGN 56-99

Part-1 : Static Loading

4.1 Introduction 56
4.2 Review of Stresses in Machine Members 57
4.3 Simple and Principal Strains 60
4.4 Relationship between Elastic Constants 61
4.5 Working Stress 61
4.6 Factor of Safety 61
4.7 Modes of Failure 63
4.8 Theories of Failure 64
4.8.1 Selection of a Failure Theory 68

Part-II : Fluctuating Loading

4.9 Introduction 73
4.10 Stress-Concentration 73
4.10.1 Causes of Stress-Concentration 74
4.10.2 Determination of Stress-Concentration Factors 74
4.10.3 Mitigation of Stress-Concentration 75
4.11 Completely Reversed and Fluctuating Stresses 75
4.12 Fatigue Failures 76
4.13 Fatigue Considerations in Design 77
4.14 Endurance Limit and Wohler's Curve 77
4.15 Fatigue Stress-Concentration Factor 78
4.16 Notch Sensitivity 78
4.17 Factors Affecting Endurance Limit 80
4.18 Design for Finite (or Limited) and Infinite (or Unlimited) Life 83
4.19 Working Stresses 88
4.20 Combined Fluctuating Stresses 90
Exercises 98

5. MECHANICAL JOINTS 100-161

5.1 Introduction 100

Part-I : Riveted Joints

5.2 Introduction 100
5.2.1 Types of Riveted Joints 100
5.2.2 Terminology 103
5.2.3 Modes of Failure of a Riveted Joint 103
5.2.4 Design of Boller Joints 105
5.2.5 Structural Riveted Joint-Lozenge Joint 111
5.2.6 Eccentrically Loaded Riveted Joint 115

Part-II : Welded Joints

5.3 Introduction 121
5.3.1 Types of Welded Joints 122
5.3.2 Streses in Welded Joints 122
5.3.3 Design of Welded Joints 123
5.3.4 Axially Loaded Unsymmetrical Welded Sections 127
5.3.5 Eccentrically Loaded Welded Joints 128
5.3.6 Welded Joint Subjected to Fluctuating Forces 131

Part-III : Cotter and Knuckle Joints

5.4 Introduction 136
5.5 Spigot and Socket Cotter-Joint 136
5.6 Gib and Cotter Joint 139
5.6.1 Strap End of Connecting Rod 139
5.6.2 For Square Rods 141
5.7 Knuckle Joint 143
Exercises 160

6. KEYS, COUPLINGS AND FLYWHEEL 162-208

6.1 Keys 162
6.1.1 Types of Keys 162
6.1.2 Design of Square and Flat Keys 164
6.2 Design of Splines 165
6.3 Couplings 168
6.3.1 Types of Couplings 168
6.3.2 Sleeve or Muff Coupling 168
6.3.3 Clamp or Compression Coupling 171
6.3.4 Forged End Flange Coupling 173
6.3.5 Rigid Protected Type Flange Coupling 174
6.4 Bush Type Flexible Coupling 178
6.5 Flywheel 183
6.5.1 Coefficient of Fluctuation of Speed 183
6.5.2 Coefficient of Fluctuation of Energy 184
6.5.3 Energy Stored in a Flywheel 185
6.5.4 Types of Flywheels 186
6.5.5 Design of Belt-Pulley Flywheel 187
6.5.6 Stresses in Flywheel Rim 188

6.5.7 Stresses in Flywheel Arms 189
6.5.8 Design of Flywheel Arms 191
6.5.9 Design of Shaft, Hub and Key 191
Exercises 207

7. SCREWED JOINTS AND POWER SCREWS 209-246

7.1 Screwed Joints 209
7.1.1 Terminology 209
7.1.2 Forms of Screw Threads 210
7.1.3 Designation of Threaded Fasteners 211
7.1.4 Standard Dimensions of Screw Threads 211
7.1.5 Stresses in Screwed Joints 212
7.1.6 External Load with Preload 215
7.1.7 Eccentric Loading on Bolts 216
7.2 Power Screws 222
7.2.1 Types of Screw Threads 222
7.2.2 Force Analysis of Square Threads 223
7.2.3 Force Analysis of Trapezoidal Threads 225
7.2.4 Collar Friction 226
7.2.5 Stresses in Screw 227
7.2.6 Design of Screw Jack 235
Exercises 244

8. SHAFTS 247-258

8.1 Introduction 247
8.2 Materials for Shafts 247
8.3 Theories of Strength for Shafts 248
8.4 Design of Shaft against Static Loading 248
8.4.1 ASME Code for the Design of Transmission Shafting 251
Exercises 258

9. BELT, CHAIN AND ROPE DRIVES 259-304

9.1 Introduction 259
9.2 Flat Belt Drive 259
9.2.1 Factors for Selection of Belt Drive 259
9.2.2 Factors on which Coefficient of Friction Depends 260
9.2.3 Design of Flat Belts 260
9.2.4 Pulleys for Flat Belts 263
9.2.5 Shaft and Key Size 265
9.2.6 Diameter of Motor Pulley 266
9.2.7 Selection of Flat-Belts from Manufacturer's Catalogue 266
9.2.8 Creep in Belts 267
9.3 V-Belts Drive 269
9.3.1 Characteristics of V-Belt Drive 269
9.3.2 Design of V-Belt Drive 270
9.4 Chain Drive 273
9.4.1 Introduction 273
9.4.2 Advantages and Disadvantages of Chain Drive Over Belt and RopeDrives 274
9.4.3 Types of Chains 274
9.4.4 Design of Transmission Roller Chain 275
9.4.5 Chordal (or Polygonal) Effect in Chain 280
9.5 Ropes 283
9.5.1 Selecting Fibre Ropes 283
9.5.2 Steel Wire Ropes 284
9.5.3 Selection of a Wire Rope 284
9.5.4 Stresses in Hoisting Ropes 285
9.5.5 Wire Rope Sheaves 288
9.5.6 Steel Wire Rope Drums 288
9.5.7 Stresses in Rope Drums 289
Exercises 302

10. CLUTCHES AND BRAKES 305-340

10.1 Clutches 305
10.1.1 Types of Clutches 305
10.1.2 Design Considerations 306
10.1.3 Properties of Material for the Lining of a Clutch 306
10.1.4 Disc or Plate Clutch 306
10.1.5 Multiple Disc Clutch 307
10.1.6 Dry and Wet Clutches 308
10.1.7 Cone Clutch 308
10.1.8 Centrifugal Clutch 316
10.2 Brakes 320
10.2.1 Types of Mechanical Brakes 320
10.2.2 Factors on which Capacity of a Brake Depends 320
10.2.3 Single Block Brake 320
10.2.4 Simple Band Brake 321
10.2.5 Differential Band Brake 322
10.2.6 Thermal Considerations in Brake Designing 328
Exercises 339

PAPER 341-342

INDEX 343-344

Chapter

1 Design Philosophy

Inside This Chapter

1.1 Definitions
1.2 Mechanical Engineering Design
1.3 Brain-Storming
1.4 Skills of a Designer
1.5 The Design Process
1.6 Aesthetic Considerations in Design
1.7 Ergonomic Considerations in Design
1.8 Role of Standardisation and Standards
1.9 Interchangeability
1.10 Preferred Numbers
1.11 Requirements of a Good Design

1.1 DEFINITIONS

Engineering : It may be defined as the profession in which a knowledge of the mathematical and physical sciences gained by study, experience and practice is applied with judgement to develop ways to utilize economically, the materials and forces of nature for the progressive well being of mankind.

Design : To design is to innovate and to create.

Designing : The process of designing is to suggest or outline ways to put together man made things, or to suggest modifications in man-made things to satisfy optimally (under the given constraints) some specified human needs.

Engineering Design : It has been defined in many ways by various scientists, but the simplest way to define engineering design is ;

"An iterative decision making process to conceive and implement optimum systems to solve societies problems and needs."

Engineering design being practical in nature must be concerned with physical realizability, or economic and financial feasibility.

Machine : It may be defined as a combination of kinematic links having unique relationship of motion so as to result in useful work being done in a controlled manner.

Machine Design : It may be defined as a process of achieving a plan for the construction of a machine that works safely, reliably, and well.

Optimal Design : A design which is best for the given objective function under the specified constraints.

Industrial Design : It is a design process in which main considerations are given to aesthetics, economics and production aspects of design.

Empirical Design : It is based upon existing standard design practice and past experience.

1.2 MECHANICAL ENGINEERING DESIGN

Mechanical Engineering Design is defined as the use of scientific principles, technical information and imagination in the description of a mechanical structure, a machine or mechanical system to perform prescribed functions with the maximum economy and efficiency. The description of a mechanical system is in the form of hard copy of drawings.

1.2.1 Morphology of Mechanical Engineering Design

The most general design problem can be stated as :

- **Devise or design :** A product, system or process to accomplish the given task or fulfill the given need optimally.
- **Subject to :** Certain constraints.

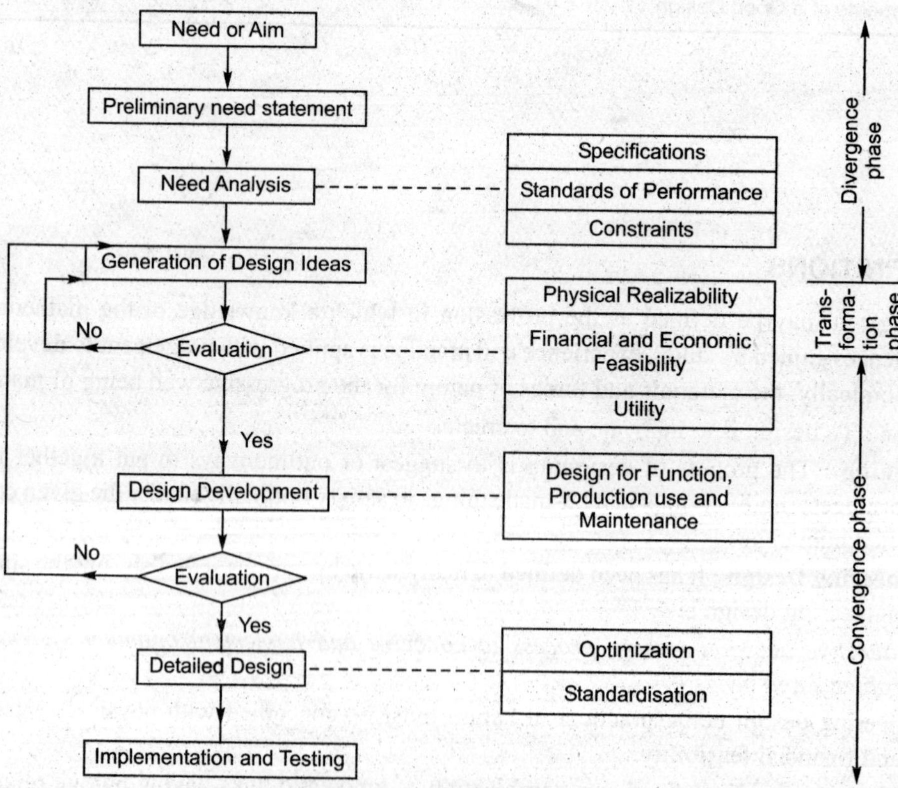

Fig. 1.1 Morphology of design.

The simplest sequence for problem solving activities is :

1. Analysis 2. Synthesis, and 3. Evaluation.

The collection of time phases of the various steps of mechanical engineering design is called the **morphology of design,** as shown in Fig. 1.1. The various phases are :

1. Feasibility Study

The three phases of design in chronological order are :

(i) Divergence phase

(ii) Transformation phase, and

(iii) Convergence phase.

(i) Divergence phase : The term divergence refers to the act of extending the boundary of a design situation. This is usually the exploration phase where the designer is to get as much understanding of the problem as possible. The typical divergence methods are : (a) Literature survey (b) Questionnaire (c) Strategy switching, and (d) Brain-storming, etc.

The analysis or need consists of exploring the design situation, establishing the specifications of the system desired, deciding on the standards against which the performance of the design can be measured and stating the constraints within which one has to work. The optimality criteria has also to be fixed.

(ii) Transformation phase : It is creative phase wherein the designer makes use of all his experience, innovative capabilities, insights and genius to conceive of a number of plausible alternate solutions to the problem.

(iii) Convergence phase : It starts with narrowing down the field of plausible solutions to the most promising one on the basis of physical realizability, anticipated utility and financial feasibility.

2. Preliminary Design

In this phase, the proposed design concepts are synthesized to achieve a pre-specified level of performance.

3. Evaluation of Preliminary Design

The evaluation of preliminary design is based upon reliability, wear and consequent deterioration in quality and performance.

4. Presentation of Detailed Design

In the detailed design, the detailed dimensions, tolerances, finishes, etc. are specified.

5. Implementation and Testing

Finally the design is implemented, a prototype prepared and tested.

1.2.2 Procedure for Morphological Analysis

The following procedure may be adopted for morphological analysis :

1. Recognize the essential parameters acceptable for design.
2. Determine the number of possible ways for achieving each of the parameters.

3. Set up a matrix in the left column of which are listed all the parameters and in the rows against each of these parameters are listed the possible methods of achieving them.
4. Study all combinations of sub–solutions for feasibility and select the best on the basis of some valid criterion.

1.3 BRAIN-STORMING

It is the best technique for idea generation. It is a group activity in which a group of designers sit down to generate the greatest number of alternative ideas and solutions to a problem. The following rules should be observed during a brain-storming session :

- **First Rule :** Criticism and judgement, whether favourable or unfavourable are not allowed.
- **Second Rule :** Generate a lot of ideas.
- **Third Rule :** Think wild.

1.3.1 Scheme for Brain-Storming Session

The following scheme is quite useful for conducting a brain-storming session.

1. Select a group of designers (around 5) to generate ideas.
2. State the problem as simply as possible.
3. Give about ten minutes to each person to write on a piece of paper as many solution ideas as possible.
4. Allow each person to read idea in turn and ask the remaining persons to suggest any additional ideas to improve or modify the idea being suggested.
5. Do not allow to criticize or discourage whether an idea is practical or not. Only shortcomings and improvements may be entertained.
6. Encourage humour to release the boredom of members and to build mutual trust.
7. At the end, find a logical classification of all the ideas generated.
8. Evaluate the design concepts generated and select one which you feel to be the best for further development.

1.4 SKILLS OF A DESIGNER

The basic skills of a designer are :

Inventiveness : It is the ability to think of or discover valuable, useful ideas or concepts for things or processes to accomplish given objectives.

Ability of engineering analysis : It is the capability of the designer to analyse a given component, system, or process using engineering or scientific principles in order to arrive quickly at meaningful answers.

- Knowledge and training in an engineering science speciality.
- Inter-disciplinary ability.
- Mathematical and computational skills.
- Thorough knowledge of use of digital computers.
- Knowledge of manufacturing processes.

- Ability to take appropriate decisions.
- Communication skills.

1.5 THE DESIGN PROCESS

The design is the art of devising an original solution to a problem by a combination of principles, resources, and products in design. The design process is the pattern of activities that is followed by the designer in arriving at the solution of a technological problem. The design process is an iterative procedure. It is a step-by-step procedure from some statement of need through identification of the problem, a search for solutions and development of the chosen solution to manufacture, test and use.

1.5.1 Design Process : Shigley Model

The design process involves six steps as suggested by Shigley and shown in Fig. 1.2.

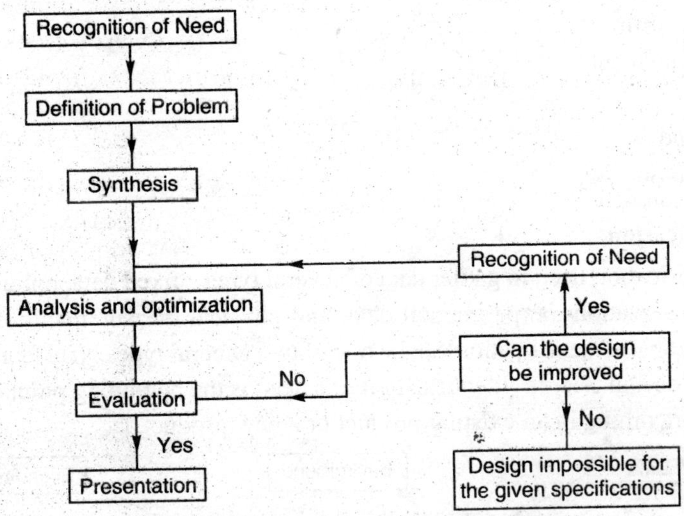

Fig. 1.2 The design process (Shigley model).

1. **Recognition of Need :** It involves the realization by some one that a problem exists for which some feasible solution is to be found. This might be identification of some defect in a current machine designed by an engineer or the perception of a new product marketing opportunity by a salesman.
2. **Definition of Problem :** It involves a through specification of the item to be designed. This specification will generally include functional and physical characteristics, cost, quality, performance, etc.
3. **Synthesis :** During the synthesis phase of design, various preliminary ideas are developed through research of similar products of designs in use.
4. **Analysis and Optimization :** The resulting preliminary designs are then subjected to appropriate analysis to determine their suitability for the specified design constraints. If these designs fail to satisfy the constraints, they are then redesigned or modified on the basis of the information gained from the analysis. This iterative process continues until the proposed designs meet the specification, or until the designer is convinced that the design is not

feasible within the specified constraints. The components, sub-assemblies or sub-systems are then synthesized into the final overall system in a similar iterative manner.

5. **Evaluation :** The assessment of evaluation of the design against the specifications established during the problem definition phase is then carried out. This often requires the fabrication and testing of prototype model to evaluate operating performance, quality, reliability, etc.
6. **Presentation :** The final phase in the design process is the presentation of design. This includes documentation of design through drawings, material specifications, assembly lists, and so on.

1.5.2 Earle Model for Design Process

The steps in the design process proposed by Earle are as follows :

1. Problem identification
2. Preliminary ideas
3. Problem refinement
4. Analysis
5. Decision, and
6. Implementation.

1. Problem Identification

Problem identification need to gather data of several types : fixed data, opinion surveys, historical records, personal observations, experimental data, and physical measurements and characteristics, as shown in Fig. 1.3. Problem identification can be one of two general types : (i) identification of a need, or (ii) identification of design criteria. Identification of a need is the beginning point of the process. It may be a defect, or a shortcoming in an existing product or a new product.

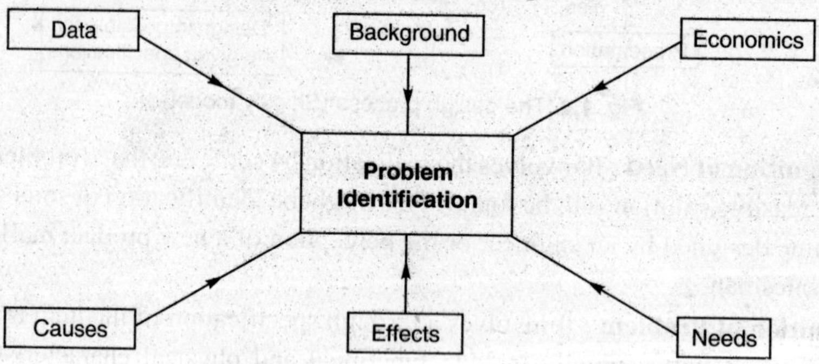

Fig. 1.3 Problem identification needs.

Identification of design criteria is that part of the problem where the designer conducts an in-depth investigation of the specifications that must be met by a new design. Types of problem identification are shown in Fig. 1.4.

The problem identification requires the designer to analyze the requirements, limitations, and other background information without becoming involved with the solution to the problem. The following steps should be used in problem identification (Fig. 1.5).

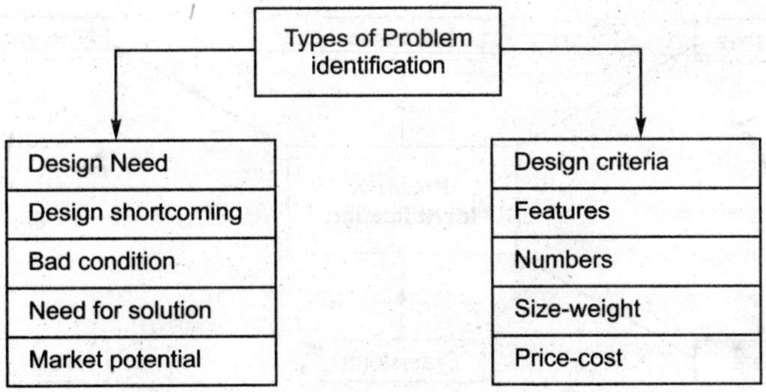

Fig. 1.4 Types of problem identification.

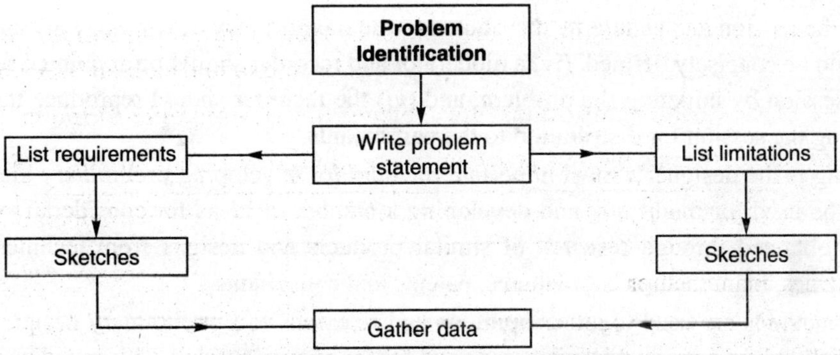

Fig. 1.5 Steps in problem identification.

(a) **Problem statement :** Write down the problem statement to begin the thinking process. The statement should be complete and comprehensive, but concise.

(b) **Problem requirements :** List the positive requirements that must be achieved in the design.

(c) **Problem limitations :** List negative factors that confine the problem to specified limitations.

(d) **Sketches :** Make sketches of physical characteristics of the problem. Add notes and dimensions that would make these sketches more understandable.

(e) **Gather data :** The data might be population trends, of related designs, physical characteristics, sales records, and market studies. This data should be graphed for easy interpretation.

2. Preliminary Ideas

Preliminary ideas is the generation of as many ideas for solution as possible (Fig. 1.6). These ideas should be sufficiently broad to allow for unique solutions that could revolutionize present methods. All ideas should be recorded in written form with sketches. A systematic approach should be used to gather preliminary ideas for the design problem. The following sequence of steps is suggested : (i) hold brainstorming session, (ii) prepare sketches and notes, (iii) research existing designs, and (iv) conduct surveys.

To 'brainstorm' is defined as the practice of a conference technique by which a group attempts to find a solution for a specific problem by amassing all the ideas spontaneously contributed by its members. The rules for brainstorming are : (i) no criticism, (ii) think wild, (iii) quantify ideas, and (iv) seek combination and improvement. The organization steps of a brainstorming session are : (i) select a panel of about 12 members with and without knowledge of the subject, (ii) prepare a one page note

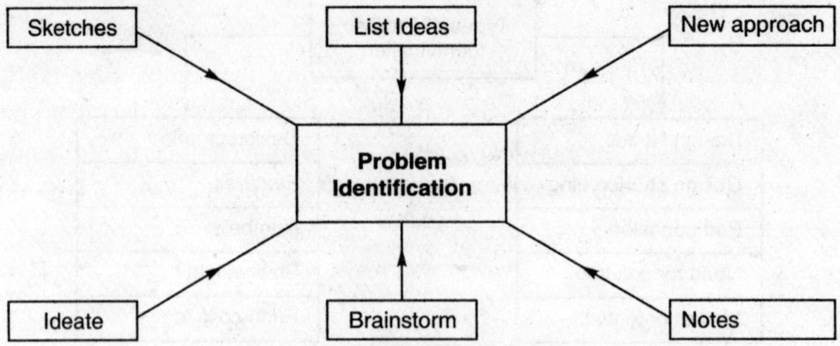

Fig. 1.6 Generation of ideas.

pertaining to the session and should be distributed to the members about two days in advance, (iii) the problem should be concisely defined, (iv) a moderator and recorder should be appointed for the session, (v) hold the session by initiating the problem, and (vi) the recorder should reproduce the list of ideas gathered during the session for distribution to the participants.

Sketching is the designer's most important medium for developing preliminary ideas. Computer graphics can be used for modifying and developing a number of ideas for consideration. Preliminary ideas can be obtained through research of similar products and designs from technical magazines, general magazines, manufacturer's brochures, patents and consultants.

Survey methods are used together opinions and reactions to a preliminary design or completed design, especially when a product is being designed for the general market. This could be accomplished by personal interview, telephone interview and mail questionnaire.

3. Design Refinement

Several of the better preliminary ideas are selected for further refinement to determine their true merits. Rough sketches are converted to scale drawings that will permit space analysis, critical measurements, and the calculation of ideas and volumes affecting the design. Consideration is given to spatial relationships, angles between planes, lengths of structural members, intersections of surfaces and planes. The sequence for refinement of ideas is shown in Fig. 1.7. The determination of physical properties of the proposed solutions is the important concern of the designer. Descriptive geometry can be applied for this purpose. Computer graphics is a powerful tool the can be used to refine the preliminary idea.

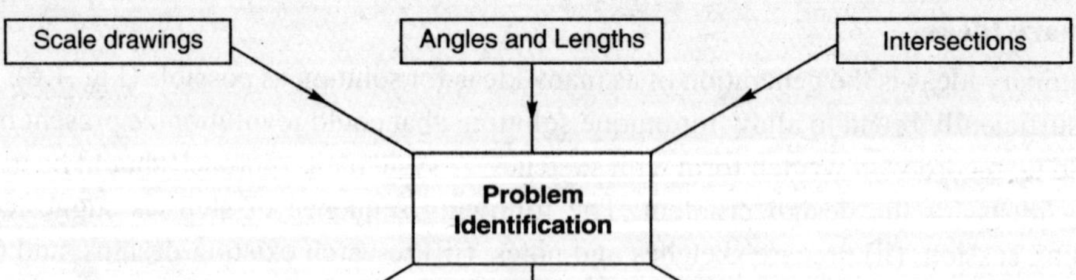

Fig. 1.7 Refinement of ideas.

4. Analysis

In involves the evaluation of the best designs to determine the comparative merits of each with respect to cost, strength, function, and market-appeal. The analysis phase of design is shown in Fig. 1.8. The general areas of analysis are : (a) functional analysis, (b) human engineering (c) market-and product analysis, (d) specification analysis, (e) strength analysis, (f) economic analysis, and (g) model analysis.

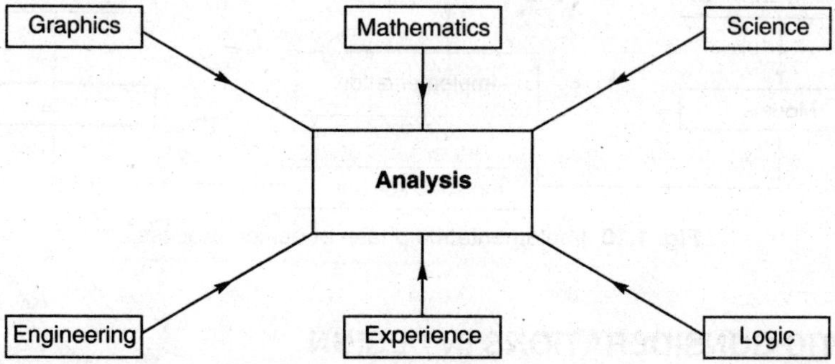

Fig. 1.8 Analysis phase of design.

Engineering graphics and descriptive geometry are valuable tools for analysis. The analysis of a design's functional characteristics can be performed using mathematics, graphics, and engineering disciplines. A product must be analysed to determine its acceptance by the market before it is released for production. Areas of product analysis are : (a) potential market evaluation, (b) market outlets, (c) advertising methods, and (d) sales features. The physical specifications of a product must be analysed to finalise the design, *e.g.,* sizes, ranges and shipping specifications. Much of the engineerng is devoted to the analysis of the strength of designs to support dead loads, withstand shocks, and to endure repetitive usages of a variety of motions ranging from slow to fast. Before a product is released for production, a cost analysis must be performed to determine the item's production cost and the margin of profit that can be realized from it.

Models are effective aids in analysing a design in final stages of its development. The types of models are : (a) conceptual models, (b) mock-up models, (c) prototype models, and (d) system layout models.

5. Decision

At this stage a single design is accepted as the solution of the design problem. Graphics is a primary means of presenting the proposed design for a decision. The graphs must compare costs of manufacturing, weights, operational characteristics, and other data that would be considered in arriving at the final decision. Fig. 1.9 shows the decision phase of the design process.

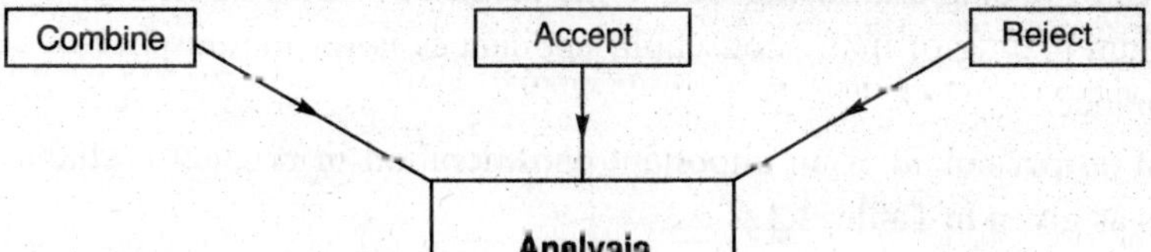

Fig. 1.9 Decision phase of design process.

6. Implementation

It is the presentation of the final design concept in a workable form, primarily as working drawings and specifications that are used as the actual instruments for fabrication of a product. Implementation phase of the design process is shown in Fig. 1.10.

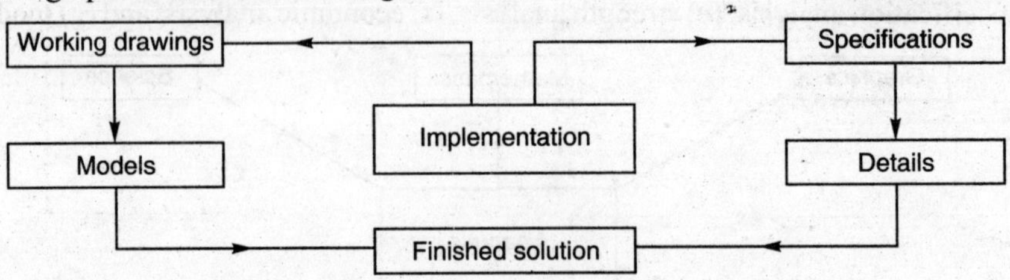

Fig. 1.10 Implementation phase of design process.

1.6 AESTHETIC CONSIDERATIONS IN DESIGN

The products are manufactured for a specific purpose and function to meet the requirements of the customers. The contact between the product and the user arises due to the dire necessity of this functional requirement. However, when there is a choice available to the customer of a product having the same qualities of efficiency, durability and cost, the customer is naturally attracted towards the most appealing product. The external appearance is an important feature, which gives grace and luster to the product and dominates the market. This is called aesthetic sense in design. Aesthetic considerations are very important for consumer durable like automobiles, household appliances and audio-visual equipment.

The aesthetic considerations in design have given rise to a separate discipline, known as industrial design. The job of an industrial designer is to create new forms and shapes which are aestherically pleasing to the mind.

There are five basic forms—step, stream, taper, shear and sculpture. The step form is similar to the shape of a 'skyscraper' or multistorey building. This involves shapes with a vertical rather than a horizontal accent. The stream or streamline form is seen in automobiles and aeroplane structure to reduce the air drag resistance. The taper form consists of tapered blocks interlocked with tapered plinths or cylinders. The shear from has a square outlook, which is ideally suitable for free-standing engineering products. The sculpture form consists of ellipsoids, paraboloids and hyperboloids. The sculpture and stream form are suitable for automobile, while step and shear forms for stationary products.

There is a relationship between functional requirement and appearance of a product. In many cases, functional requirements result in shapes which are anesthetically pleasing. The evolution of streamlined shapes of boeing and racing cars is the result of studies in aerodynamics for reducing air drag. The chromium plating of many household appliances is for the purpose of corrosion resistance rather than appearance.

Selection of proper colour is an important consideration in product aesthetics. The meaning of various colours is as given in Table. 1.1.

Table 1.1 : Aestheric meaning of colours

Colour	*Aesthetic meaning*
Red	Danger-Hazard-Hot
Orange	Possible danger
Yellow	Caution
Green	Safety
Blue	Caution-Cold
Grey	Dull

Other factors affecting aesthetic consideration are :

1. Rigidity and resilience.
2. Tolerance,
3. Surface finish.
4. Motion of individual components.
5. Materials.
6. Manufacturing methods.
7. Vibration and noise.

1.7 ERGONOMIC CONSIDERATIONS IN DESIGN

The word 'ergonomics' is derived from two Greek words : ergon = work and nomos = natural laws. Thus Ergonomics means the natural laws of work. It is the relationship between man and machine and the application of anatomical, physiological principles to solve the problems arising from man-machine relationship. It is also called human engineering.

- **Anatomy :** Study of body dimensions and relations for work design.
- **Physiology :** Study of man and his working environment.
- **Psychology :** Study of adaptive behavior and skills of people.

Ergonomica includes the following factors for study from design considerations :

1. Anatomical factors in design.
2. Control of physical work environment.
3. Design of man-machine system.
4. Design of controls and displays.
5. Accidents, fatigue and safety.
6. Work place design.

The fundamental concept in human engineering is the system. A system is composed of humans, machines and other things that interact to accomplish some goal which these same components could not produce independently. The man machine system is a closed loop system as shown in Fig. 1.11.

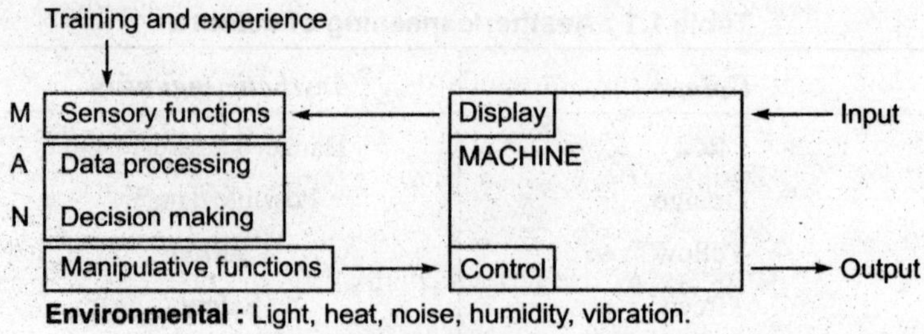

Fig. 1.11 Man–machine system.

1.8 ROLE OF STANDARDISATION AND STANDARDS

Standardisation aims at establishing mandatory or obligatory norms or standards to which different types, grades, parameters (like length, diameter, etc.) quality characteristics, test methods, rules of marking, packing, storage of finished items, raw materials and semi-finished articles should conform. The main purpose of standardisation is to minimise variety so that the number of models may be limited to a definite quantity. Standard items are of much use to a designer to achieve design for manufacture.

Bureau of Indian Standards (BIS) has standardised a number of items. For machine designers the items standardised are of the following categories :

1. Rules for preparing engineering drawings.
2. Symbols and notations for mechanical quantities and their units.
3. Engineering materials, their chemical composition, mechanical properties, method of heat treatment and method of mechanical testing.
4. Dimensions and preferred sizes of various machine components.
5. Fits and tolerances for various machine elements from assembly considerations.

1.8.1 Advantages of Standardisation

Standardization offers the following advantages :

1. Helps in obtaining design for manufacture.
2. Standardised products have better quality and are more reliable, giving higher service life.
3. Helps in having easy maintenance as standardised products are easily available for replacement.
4. The time and effort needed to manufacture a new machine is reduced remarkably as the standardised products can be easily obtained and assembled.
5. Interchangeability of components is possible due to standards, which reduces labour requirements and manufacturing costs.

1.8.2 Standards

There are three types of standards used in machine design, as described below :

1. International standards, such as the standards prepared by International Standards Organization (ISO).

2. National Standards, such as :

IS	: Bureau of Indian Standards, BIS
DIN	: German
GOST	: Russian
BS	: UK
ASME	: American Society of Mechanical Engineers
AISI	: American Iron and Steel Institute
SAE	: Society of Automotive Engineers
AGMA	: American Gear Manufacturer's Association
ASM	: American Society of Metals
ASTM	: American Society of Testing and Materials

3. Company Standards for use in a particular company or a group of sister concerns.

1.9 INTERCHANGEABILITY

Interchangeability of product means that any standardized component shall assemble correctly with the mating components, both being chosen at random. For this purpose, the parts should be as near as possible identical. Interchangeability ensures that the replacement parts can be drawn from the stock with the certainity that they will fit without alteration. Interchangeability of components is possible due to standardization.

1.10 PREFERRED NUMBERS

Preferred numbers form a general basis for standardizing and grading a series of similar dimensions, characteristics or articles. The preferred numbers are arranged in geometric series. The preferred numbers greatly reduce unnecessary variations in the sizes of an article, thus reducing the variety of production and of the material in the store.

The preferred numbers contain the powers of 10, such as 1, 10, 100, ... The intermediate numbers are obtained by dividing each decimal range (1 to 10, 10 to 100, ...) into 5, 10, 20, 40 and 80 geometrically equal steps. The series thus obtained are called the **Basic Series.**

Preferred numbers were first introduced by the French engineer Charles Renard in the 19th century. The geometric series are designated after his name by R5, R10, R20, R40 and R80. The progression ratio of the basic series are :

$$\phi_5 = \sqrt[5]{10} = 1.58$$

$$\phi_{10} = \sqrt[10]{10} = 1.26$$

$$\phi_{20} = \sqrt[20]{10} = 1.12$$

$$\phi_{40} = \sqrt[40]{10} = 1.06$$

$$\phi_{80} = \sqrt[80]{10} = 1.03$$

1.10.1 Derived Series

When the steps of a basic series are too close for a particular application, **Derived Series** many be obtained by taking every 2nd, 3rd, 4th or 5th term of a basic series. The derived series are designated as follows :

(a) R5.2 (1...1000000) : Derived series comparising every 2nd terms of R5 series limited by the term 1 and 1000000, both inclusive.

(b) R10.3 (.........80) : Derived series comprising every 3rd term of R10 series, unlimited in both directions an including the term 80.

(c) R20/4 (112 ......) : Derived series comprising every 4th term of R20 series and limited at the lower end by the term 112 inclusive.

(d) R40/5 (......60) : Derived series comprising every 5th term of R40 series and limited at the upper end by the term 60 inclusive.

1.10.2 Preferred Value

According to the accuracy of calculations the following may be defined :

- **Theoretical Value :** The exact value of a term of any of the series. This is of no practical use.
- **Calculated Value :** The value of a term, approximating its theoretical value, expressed to 5 significant digits. This is also of minor importance in practice.
- **Main or Preferred Value :** This is the rounded off value of a term as given in the standard and is used generally. The limits of rounding off are +1.26% and 1.09%.

Theoretical value	*Calculated value*	*Preferred value*
$(\sqrt[10]{10})^2$	3.1623	3.15
$(\sqrt[10]{10})^{11}$	12.589	12.5

Preferred numbers are used in standardizing quantities like lengths, weights, flow rates, force, pressure, allowable loads, revolutions, velocities, gear ratios, etc. Since standardization aims at minimizing variety, it is advisable to use R5 in preference to R10, R10 in preference to R20, R20 in preference to R40 and R40 in preference to R80. The preferred numbers are given in Table 1.2.

1.10.3 Characteristics of Preferred Series

The preferred series have the following special characteristics :

1. The product or quotient of any two terms of a series is always a term of that series.
2. The integral positive or negative power of any term of a series is always a term of that series.
3. The R series contain 10 and all +ve and –ve integral powers.
4. The number 3.15, which is nearly equal to π would be found among the preferred numbers of R10, R20 and R40. It follows that the length of circumference and the area of a circle, the diameter of which is a preferred number may also be expressed by preferred numbers.
5. The R40 series includes the numbers 3000, 1500, 1000, 750 and 375, which are the synchronous speeds of electrical machines for the standard frequency of 50 Hz.

Table 1.2 : Preferred Numbers

R5	*R10*	*R20*	*R40*
1.00	1.00	1.00	1.00
			1.06
			1.12
			1.18
	1.25	1.25	1.25
			1.32
		1.40	1.40
			1.50
1.60	1.60	1.60	1.60
			1.70
		1.80	1.80
			1.90
	2.00	2.00	2.00
		2.24	2.12
			2.24
			2.36
2.50	2.50	2.50	2.50
			2.65
		2.80	2.80
			3.00
	3.15	3.15	3.15
			3.35
		3.55	3.55
			3.75
4.00	4.00	4.00	4.00
			4.25
		4.50	4.50
			4.75
	5.00	5.00	5.00
			5.30
		5.60	5.60
			6.00
6.30	6.30	6.30	6.30
			6.70
		7.10	7.10
			7.50
	8.00	8.00	8.00
			8.50
		9.00	9.00
			9.50
10.00	10.00	10.00	10.00

1.11 REQUIREMENTS OF A GOOD DESIGN

A good design must fulfil the following essential requirements :

1. **Customers satisfaction :** The product should satisfy the consumer by fulfilling their need and expectations. In order to achieve this objective, the following points should be kept in view :
   (i) It should function properly.
   (ii) It should be of proper quality so as to achieve desired degree of accuracy and reliability.
   (iii) Easy to use or operate.
   (iv) Good aesthetic view.
   (v) It should be of a reasonable price.
2. **Adequate profit :** The product when manufactured and sold should give adequate profit. For this purpose it must satisfy the following requirements :
   (i) It should be able to be manufactured at a reasonable price so that it can compete other products in the market.
   (ii) It should have minimum number of parts.
   (iii) It should adopt latest technology so that manufacturing requires minimum cost per unit of production.

SOLVED QUESTIONS FROM UNIVERSITY EXAMINATIONS

■ **Example 1.1** Discuss preliminary design and detailed design. How to select best possible solution? [MDU, Dec. 2007]

Solution : Preliminary design : Refer to art. 1.5.2 preliminary ideas.

Detailed design : Refer to art. 1.5.2 implementation.

Best possible solution : Refer to art. 1.5.2 decision.

■ **Example 1.2** Discuss the influence of standardisation and manufacturing process on the design of a component. [MDU, Dec. 2008]

Solution : Refer to art. 1.8.

■ **Example 1.3** Explain the importance of ergonomic and aesthetic considerations in machine design. [MDU, Dec. 2008]

Solution : Refer to art 1.6 and art. 1.7

■ **Example 1.4** Discuss the importance of the following in machine design :

(i) Brain storming as an idea generation technique.
(ii) Environmental feasibility.
(iii) Ergonomic considerations. [MDU, May, 2009]

Solution :

(i) Refer to art. 1.3.
(ii) The design of machine should not pollute the environment. The things to be considered from this point of view are : sound level due to rotation of machine components, vapours generated

due to lubricants in rotating machinery bearing, noise generated due to meshing of gears, belts on pulleys, whirling of machine components, etc.

(iii) Refer to art. 1.6.

■ **Example 1.5** What is design philosophy? Discuss technical feasibility, economic and financial feasibility, social and environmental feasibility. Discuss in detail. [MDU, May, 2008]

Solution : Refer to art. 1.2

■ **Example 1.6** What are preferred number series? Explain their uses. [MDU, May, 2009]

Solution : Refer to art. 1.10.

■ **Example 1.7** Explain the following with examples :

(i) Manufacturing process considerations.

(ii) Standardisation. [MDU, Dec., 2009]

Solution : (i) Refer to art. 3.4.

(ii) Refer to art. 1.8.

EXERCISES

1. What do you mean by mechanical engineering design?
2. Explain the importance of aesthetic considerations in machine design.
3. Define standardisation. Explain various standards used in machine design.
4. Discuss the basic requirements for machine elements and machines.
5. What are the various stages in the design of a machine component.
6. Explain the design process according to Shigley model.
7. Explain the design process according to Earle model.
8. What is brain storming? What are its rules?
9. Carry out a brain storming session for the following problems :
   (a) Low cost housing for the urban middle class family.
   (b) A cheap family vehicle.
   (c) Hand drying in a hostel mess having 200 students.
10. Explain the following :
    (a) Standardisation (b) Preferred numbers.

❑❑❑

2

Chapter

Selection of Materials

Inside This Chapter

2.1 Introduction
2.2 Selection of Engineering Materials
2.3 Stress-Strain Diagram for Mild Steel
2.4 Factor of Safety
2.5 Classification of Engineering Materials
2.6 Cast Irons
2.7 Steels
2.8 Heat Treatment Methods
2.9 Non-Ferrous Metals and Alloys
2.10 Non-Metallic Materials
2.11 Composite Materials
2.12 Weighted Factor Method of Material Selection

2.1 INTRODUCTION

It is the responsibility of the design engineer to specify the material to be used for each element of the machine which he has designed. The material from which a part is to be made depends upon what the part is to do, how may parts are needed, and what material and processes are available to make it economically viable. The selection of a material depends primarily upon its properties and many more factors which a designer has to consider for a given situation.

2.2 SELECTION OF ENGINEERING MATERIALS

There are many factors which have to be considered while selecting a material. The most important to consider are those that are essential to the function of the part. Equal importance must be given to the properties, availability, manufacturing considerations, and economics aspects.

1. Properties

Selection of a material for a given application must be based on the following logical steps :

(i) Establishing the performance requirements that the material must meet.

(ii) Selecting the appropriate material for evaluation.

(iii) Selecting the material that best meets the performance requirements and economics constraints.

The properties of materials are :

(a) Mechanical

(i) Tensile/compressive strength
(ii) Shear strength
(iii) Endurance limit
(iv) Impact strength
(v) Creep strength
(vi) Stiffness

Fatigue : The reduced tendency of a material to offer resistance to the applied stresses due to fluctuating cyclic loads is called fatigue. This phenomenon leads to failure of the material much below the tensile strength.

Endurance Limit : The maximum stress below which a material can presumably withstands an infinite number of cycles of stress is called endurance limit or fatigue strength.

Creep : It may be defined as the slow and progressive deformation of a material with time under a constant stress.

Creep strength : It may be defined as the highest stress that a material can withstand for a specified length of time without excessive deformation.

Stiffness : It is defined as the load per unit deflection or deformation.

(b) Tribological

(i) Friction (ii) Wear, and (iii) Lubrication

(c) Physical

(i) Density (ii) Coefficient of expansion (iii) Melting point
(iv) Electrical properties (v) Magnetic properties (vi) Flammability

(d) Chemical

(i) Composition (ii) Bonding structure (iii) Environmental resistance

(e) Dimensional

(i) Flatness (ii) Surface finish (iii) Stability
(iv) Critical size

(f) Technological

(i) Machinability (ii) Weldability (iii) Castability
(iv) Forgability (v) Bendability (vi) Malleability
(vii) Ductility (viii) Hardenability

2. Availability

When initially considering the use of a material, the following points should be considered :

(i) Is the material on hand?
(ii) If answer is No; can we get it in one to two weeks, and so on?
(iii) If Yes, how much quantity is required?
(iv) What is the nature of the supplier? It is not advisable to use a proprietary material that is available only from one supplier.
(v) Is there any special method required for fabrication of raw material? In that case it is essential for the designer to establish a time line for procurement of materials.

3. Economics

Cost is used as a prime selection factor for materials. The cost of a material in a machine component consists of the initial cost, processing cost and the maintenance cost.

4. Manufacturing Considerations

The manufacturing processes, such as casting, forging, rolling, extrusion, welding and machining govern the selection of a material. Machinability of materials is an important consideration in their selection. The shape of the components also plays an important role in the selection of the manufacturing process. Casting method may be preferable over machining methods for complicated shapes.

2.3 STRESS-STRAIN DIAGRAM FOR MILD STEEL

The stress-stain diagram for a ductile material like mild steel is shown in Fig. 2.1. The curve starts from the origin *O*, showing thereby that there is no initial stress or stain in the specimen. Upto point *A*, Hooke's law is obeyed and stress is proportional to stain. Therefore, *OA* is a straight line. Point *A* is called the **limit of proportionality**. Upto point *B*, the material remains elastic, *i. e.*, on removal of the load, no permanent set is formed, *AB* is not a straight line. Point *B* is called the elastic **limit point**. Beyond point *B*, the material goes to the plastic stage until the **upper yield point** *C* is reached. At this point the cross-sectional area of the material starts decreasing and the stress decreases to a lower value to point *D*, called the **lower yield** point. Between *DE*, the specimen elongates by a considerable amount without any increase in stress. From point *E* onwards, the strain hardening phenomena becomes predominant and the

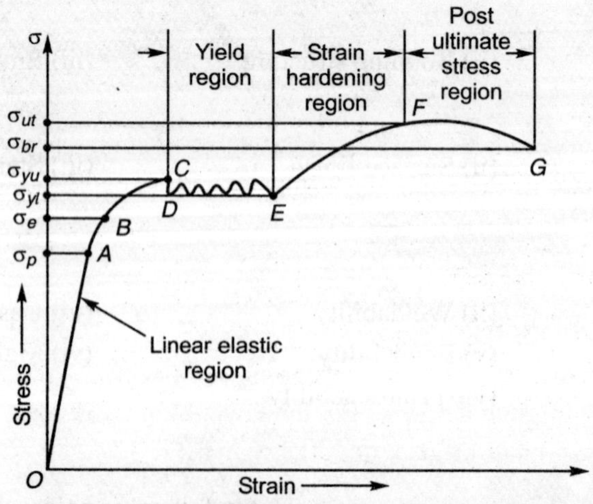

Fig. 2.1 Stress-strain diagram for ductile materials.

strength of the material increases thereby requiring more stress for deformation, until point F is reached. Point F is called the ultimate point and the corresponding stress is called the **ultimate strength**. At point F, necking of the material begins and the cross-sectional area decreases at a rapid rate. The apparent stress decreases but the actual or true stress goes on increasing until the specimen breaks at point G, called the point of fracture. The fracture of ductile material is of the cup and cone type.

The phenomena of yielding and necking is not exhibited by brittle materials. The ultimate strength is calculated at 0.2 percent of maximum strain.

The plot of the normal stress v's engineering strain can be divided into four well-defined regions, as shown in Fig. 2.1.

1. Linear elastic region.
2. Yield region.
3. Strain-hardening region.
4. Post-ultimate stress region.

2.3.1 True stress-Stain Diagrams

During plastic strain of a uniform specimen subjected to an axial tension (compression), the cross-sectional area gets smaller (larger) as the specimen elongates (shortens). The two stress-stain diagrams for ductile materials (such as mild steel), the compression and the tension true stress-strain diagrams practically coincide, where as the two engineering stress-strain diagrams drift apart as shown in Fig. 2.2.

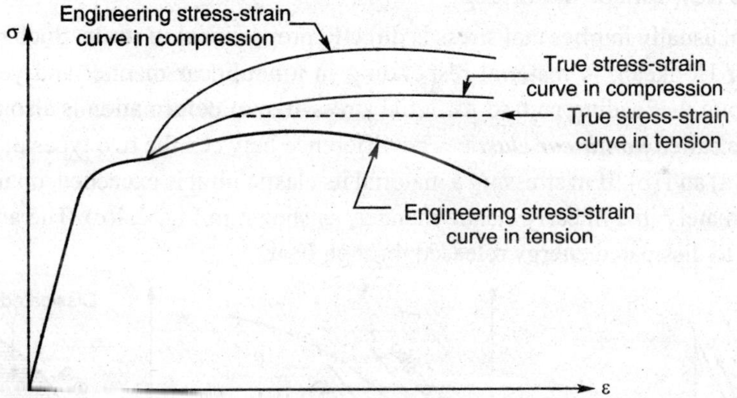

Fig. 2.2 Comparison of tension and compression stress-strain diagrams for ductile materials.

Comparison of the engineering and the true stress strain diagrams shows that in tension, since the cross-sectional areas decreases as the specimen elongates, the true stress is greater than engineering stress, where as in compression as the specimen shortens, the cross-sectional area increases and thus the true stress is less than the corresponding engineering stress.

2.3.2 Stress-Strain Diagrams for Other Materials

The engineering stress-strain diagrams for a few representative materials are shown in Fig. 2.3. For most engineering applications, to limit the deformation, the lower range of strains is particularly important. The large deformations of materials is of great importance in the analysis of metal forming processes like forging, drawing and rolling, etc.

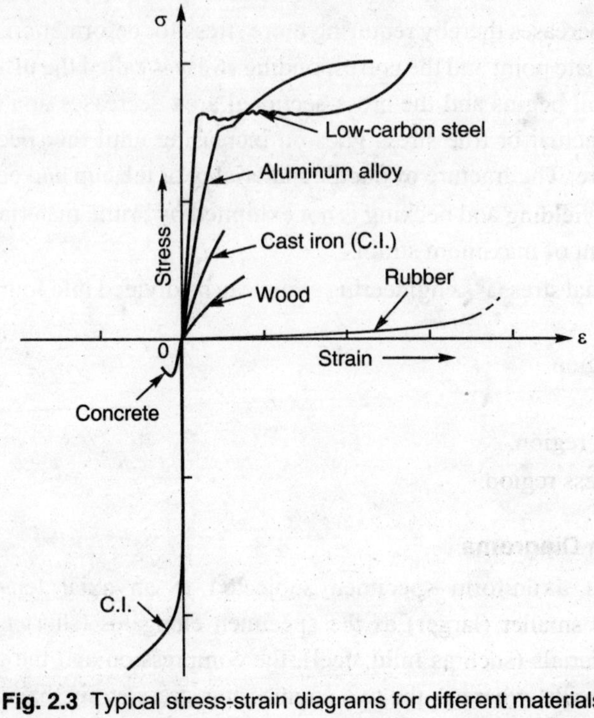

Fig. 2.3 Typical stress-strain diagrams for different materials.

2.3.3 Linear and Non-Linear Materials

An elastic material usually implies that stress is directly proportional to strain. Such materials are called ***linearly elastic*** or Hookean. A material responding in a nonlinear manner and yet, when unloaded, returning back along the loading path to its initial stress-free of deformation is also an elastic material. Such materials are called ***nonlinear elastic.*** The difference between the two types of elastic materials is shown in Fig. 2.4(a) and (b). If in stressing a material its elastic limit is exceeded, on unloading it usually responds approximately in a linearly elastic manner, as shown in Fig. 2.4(c). The area enclosed by the loop corresponds to dissipated energy released through heat.

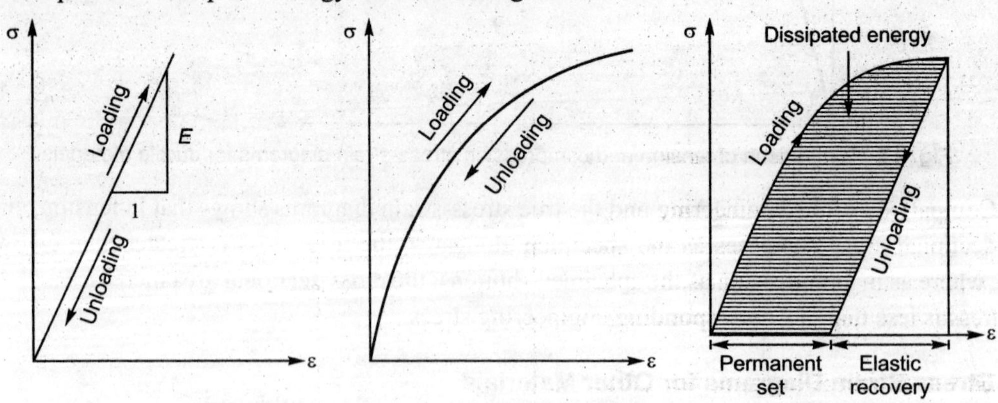

(a) Linear elastic material (b) Non-linear elastic material (c) Plastic material

Fig. 2.4 Stress-strain diagrams.

2.3.4 Idealized Stress-Strain Diagrams

The idealized stress-strain diagrams are shown in Fig. 2.5.

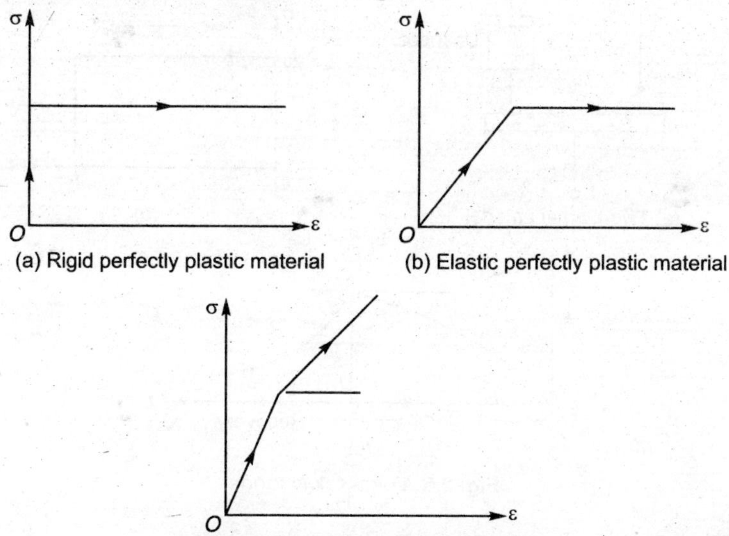

(a) Rigid perfectly plastic material (b) Elastic perfectly plastic material

(c) Elastic linearly hardening material

Fig. 2.5 Idealized stress-strain diagrams.

1. **Rigid perfectly plastic material :** The σ-ε diagram shown in Fig. 2.5 (a) is applicable to problems in which the elastic strains can be neglected as compared to plastic strains. Perfectly (ideally) plastic behaviour means that a large amount of unbounded deformation can take place at a constant stress.
2. **Elastic perfectly plastic material :** The σ-ε diagram shown in Fig. 2.5 (b) is particularly useful if both elastic and plastic strains have to be included. This situation frequently arises in analysis.
3. **Elastic linearly hardening material :** The σ-ε diagram shown in Fig. 2.5 (c) provides a reasonable approximation for many materials and is more accurate than the previous two models over a wider range of strain.

2.3.5 Viscoelastic Materials

Some of the materials like asphalt pavements, solid propellants in rocket motors, high-polymer plastics and concrete, as well as machine elements at elevated temperatures, gradually deform under stress, and such deformations are usually not fully recoverable. This is called time-dependent flow or creep phenomenon.

For viscous materials stress depends on strain as well as rate of strain. The simplest relation can be stated as :

$$\sigma = E\varepsilon + \eta\dot{\varepsilon} \qquad \ldots(2.1)$$

where η = coefficient of viscosity, E = Young's modules of elasticity, ε = strain, $\dot{\varepsilon}$ – Strain rate

There are two models for viscoelastic materials.

1. Voigt-Kelvin Solid

This model shown in Fig. 2.6 is obtained by placing a spring and a dashpot in parallel. Here $\varepsilon_s = \varepsilon_d = \varepsilon$, where the subscripts s and d designate the spring and dashpot respectively. Total stress, $\sigma = \sigma_s + \sigma_d$. Eq. (2.1).

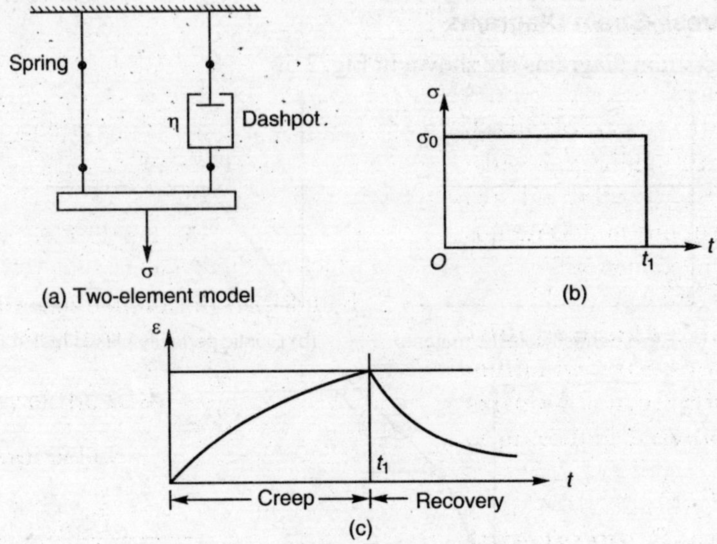

Fig. 2.6 Voigt-Kelvin model.

$$\dot{\varepsilon} + \left(\frac{E}{\eta}\right)\varepsilon = \frac{\sigma}{\eta} \quad \text{...(2.2)}$$

As time increases, the strain asymptotically approaches the maximum strain associated with the elastic spring until, finally all the applied stress is carried by the spring and the dashpot becomes inactive. If the stress is removed at an earlier time, an asymptotic recovery of the strain takes place. The Voigt-Kelvin model exhibits a delayed elastic response and is termed an anelastic material. An elastic sponge filled with a viscous fluid can be linked to such a behaviour.

2. Maxwell Solid

This model as shown in Fig. 2.7 consists of spring and deshpot in series.

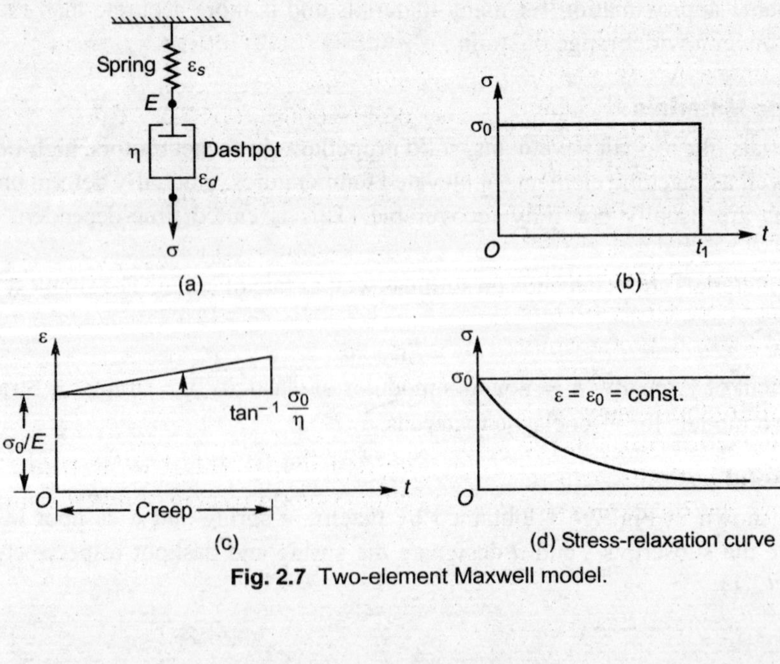

Fig. 2.7 Two-element Maxwell model.

Here, $\sigma_s = \sigma_d = \sigma$ and $\varepsilon = \varepsilon_s + \sigma_d$. Therefore,

$$\dot{\varepsilon} = \dot{\varepsilon}_s + \dot{\varepsilon}_d = \frac{\dot{\sigma}}{E} + \frac{\sigma}{\eta} \quad \text{or} \quad \dot{\sigma} + \left(\frac{E}{\eta}\right)\sigma = E\dot{\varepsilon} \qquad \ldots(2.3)$$

2.4 FACTOR OF SAFETY

It is the ratio of the maximum permissible stress to which a member can be subjected to the allowable or working stress. For ductile materials,

Factor of safety, $$\text{FOS} = \frac{\text{Yield strength}}{\text{Allowable stress}} = \frac{S_{yt}}{\sigma_a} \qquad \ldots(2.4a)$$

and for brittle materials,

$$\text{FOS} = \frac{\text{Ultimate strength}}{\text{Allowable stress}} = \frac{S_{ut}}{\sigma_a} \qquad \ldots(2.4b)$$

2.5 CLASSIFICATION OF ENGINEERING MATERIALS

The engineering materials may be classified as shown in Fig. 2.8

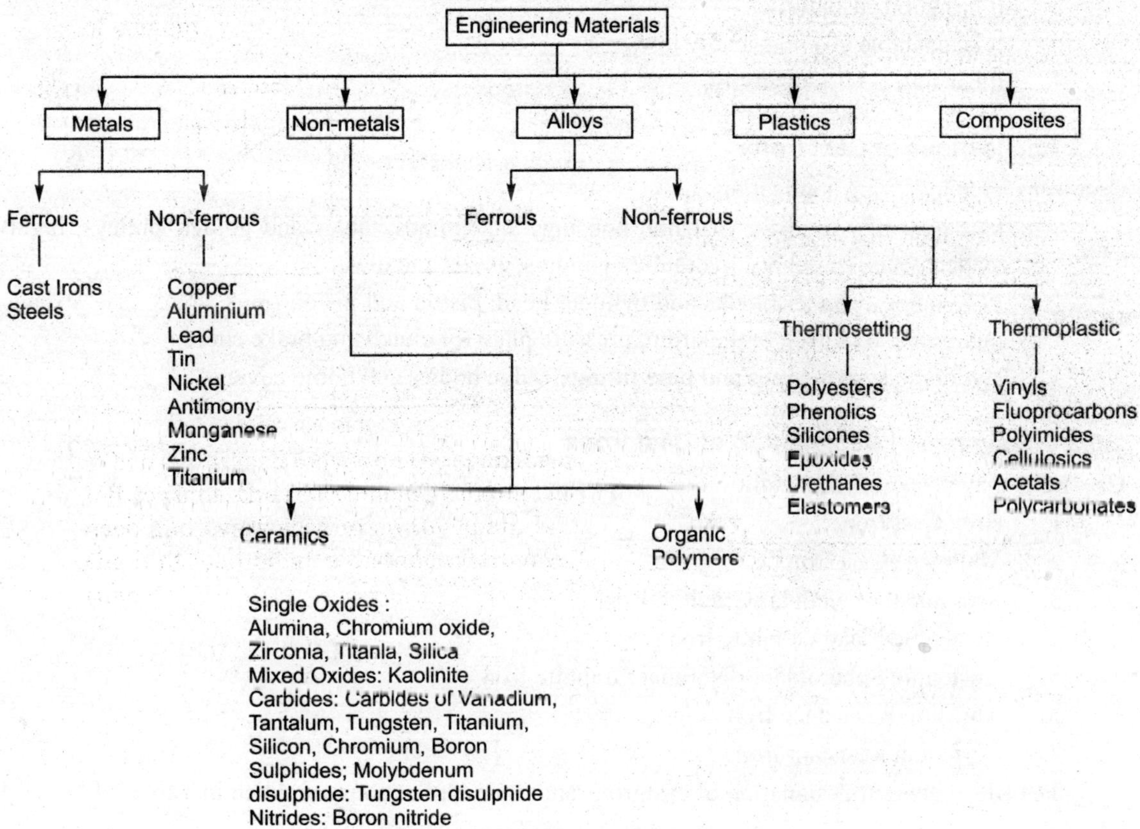

Fig. 2.8 Classification of engineerimg materials.

Ferrous metals are these which have free valence electron in the outermost shell. Examples of ferrous metals are : cast iron and steels.

Non-ferrous metals are those which contain a metal other than iron as their main constituent.

Non-metal are those which have no free valence electrons in the outermost shell, *e.g.*, carbon,boron, etc.

Alloys are a combination of two or more metals mixed together to obtain desired properties.

2.6 CAST IRONS

Cast iron is an alloy of iron, carbon and silicon. The percentage of carbon lies between 1.8 to 3 percent approximately. The carbon is present either in the form of graphite or iron carbide.

2.6.1 Advantages of Cast Irons

Cast irons offer the following advantages from designs pont of view :

1. It is very hard and brittle, therefore, it has good wear properties.
2. It is very strong in compression and weak in tension.
3. It has good vibration absorption proprieties.
4. It is easily machinable.
5. It has good castability.
6. It is available at a reasonably low cost.
7. It has low notch sensitivity.

2.6.2 Applications of Cast Irons

Cast irons have following applications :

1. It is generally used for machine housings and frames, tables and guides, pulleys, highly loaded gears, sprockets, hydraulic cylinders, anvils and dies.
2. I.C. engine cylinder blocks and cylinder head, piston and piston rings, crank case, exhaust manifolds, flywheel, brake drum, pressure plate for clutch and brake shoes.
3. Hydraulic system pipes and pipe fittings, valve bodies and dome covers.

2.6.3 BIS system of Designation of Cast Irons

Cast irons may be classified as follows :

1. Grey Cast Iron
2. Malleable Cast Iron
3. Spheroidal or Nodular Graphite Iron
4. Austenitic Flake Graphite Iron
5. Austenitic Spheroidal or Nodular Graphite Iron
6. Abrasion Resistance Iron
7. Corrosion Resistant Iron

The BIS system of designation of cast irons and their applications are given in Table 2.1.

Table 2.1 : BIS designation of cast irons and their applications.

Cast Irons	*Designation*	*Applications*
1. Grey	FG followed by the tensile strength in MPa, *e.g.*, FG 350 indicates Grey cast iron of tensile strength 350 N/mm^2	Machine tool bodies, automobile parts, pipe fittings, agricultural implements.
2. Malleable	Black heart, white heart and Perlitic malleable cast irons are designated respectively by BM, WM and PM followed by the tensile strength in MPa.	Gear housing, brake pedals, automotive and agricultural parts.
3. Spheroidal or Nodular Graphite	SG followed by the tensile strength in MPa and oblique percentage elongation, *e.g.*, SG 600/2 indicates a spheroidal and nodular graphite iron having tensile strength of 600 MPa and elongation of 2%	It is used for crankshafts, pistons, guides and control levers of heavy machines,wrenches and dies.
4. Austenitic Flake Graphite	AFG followed by the name and percentage of important elements of chemical composition.	Pressure covers for turbine-generator sets, housings for switch ger, insulator flanges, bushings, piston rings, pump components, valve castings, machine tools.
5. Austenitic Spheroidal or Nodular Graphite	ASG followed by the name and percentage of important elements of chemical composition.	-do-
6. Abrasion Resistant	ABR followed by the minimum tensile strength in MPa and the name and percentage of important elements of chemical composition.	Stone crushers, cement plants mixtures, rolling mills.
7. Corrosion Resistant	CSC followed by the minimum tensile strength in MPa and the name and percentage of important elements of chemical composition.	Industrial furnaces, acid plants, sea going ships.

2.6.4 Effect of Alloying Elements on Cast Iron

The effect of alloying elements on cast iron is given in Table 2.2.

Table 2.2 : Effect of alloying elements on cast iron.

Alloying Element	*Percentage*	*Effects*
Nickel	0.25-5%	Refines grain structure, increases strength and toughness, increases resistance to corrosion. Acts as a graphitizer.
Chromium	10-30%	Refines grain structure, increases strength, hardness and resistance to corrosion, increases wear resistance and heat resistance. Tends to present graphitization.
Copper	Upto 1%	Increases fluidity, imparts corrosion resistance and improves toughness and hardness. Improves machinability slightly.
Molybdenum	0.25-1.25%	Increases the strength and toughness. Improves high-temperature strength.
Vanadium	0.10-0.50%	Increases strength and resistance to fatigue stresses. Reduces graphitization.
Boron	Upto 0.05%	Increases surface hardness.

2.6.5 Effect of Impurities on Cast Iron

The impurities generally present in cast iron are : Sulphur, Manganese, Phosphorus and Silicon. Their effects on cast iron are given in Table 2.3.

Table 2.3 : Effect of impurities on cast iron.

Impurity	*Percentage*	*Effects*
Sulphur	Upto 0.1%	Lowers fluidity during pouring, decreases strength and makes the metal more brittle.
Manganeses	Upto 0.75%	Tends to whiten and harden cast iron.
Phosphorus	Upto 1.0%	Increases flowability of grey cast iron, increases brittleness.
Silicon	Upto 2.5%	Makes cast iron soft and easily machinable, decreases blow holes in castings and helps in producing sound castings.

2.7 STEELS

Carbon steels are alloys of iron and carbon only. It contains carbon upto a maximum of 1.8%. It is obtained by oxidizing the impurities in molten pig iron and then adding the required amount of carbon. Carbon exists in steel in the combined form as iron carbide.

2.7.1 Classification of Steels

Steels for machine applications may be classified as follows :

1. Carbon steel-used for wire for ropes, cold rolled sheets and strip, plates for general engineering use.
2. Alloy steel-used for bearings.
3. Stainless steel-used for spring washers, helical springs, valves for I.C. engines, cold forged rivets, bars and flats, plate, sheet and strips.
4. Special steel-used for pressure vessels, spring strips, volute, laminated and helical springs for railways and automobiles, naval vessels.

   Low carbon or mild steel $: C = 0.05 - 0.30\%$

   Medium carbon steel $: C = 0.30 - 0.60\%$

   High carbon steel $: C = 0.60 - 1.5\%$

2.7.2 Designation of Steel

Steel contains varying percentage of carbon and other alloying elements. To distinguish one type of steel from another, BIS has specified a peocedure for the nomenclature of steels. This is called giving designation to steel. The following procedures have been laid down by BIS for the designation of steels :

1. On the Basis of Mechanical Properties

Carbon and low alloy steels where the main criterion in the selection and inspection is the tensile or yield stress are designated in the following order :

(i) Symbol "Fe" or "FeE" depending on whether the steel has been specified on the basis of minimum tensile strength or yield strength in N/mm^2.

(ii) Figure indicating the minimum tensile strength or yield strength in N/mm^2. If no minimum tensile or yield strength is guranteed the figure shall be 00.

(iii) Chemical symbols for elements the presence of which characterize the steel.

(iv) Symbol indicating special characteristics covering method of deoxidation, degree pf purity, etc.

(v) Symbol indicating applications, if necessary. For example, Fe 410 Cu K indicates, killed steel containing copper as alloying element with a minimum tensile strength of 410 N/mm^2.

2. On the Basis of Chemical Composition

(a) Unalloyed steels : They are designated in the following order :

(i) Figure indicating 100 times the average percentage of carbon content.

(ii) Letter "C".

(iii) Figure indicating 10 times the average percentage of manganese content and rounded off to the nearest integer following the rule for rounding off.

(iv) Symbol indicating special characteristics including guaranteed hardenability for which symbol "G" is used at the end of the designation.

Examples :

1. 55 C4 indicates a plain carbon steel with 0.55% carbon and 0.4% manganese. A steel with 0.35-0.45% carbon and 0.7-0.9% manganese is designated as 40C8.

(b) Unalloyed tool steels : They are designated in the following order :

(i) Figure indicating 100 times the average percentage of carbon,

(ii) Symbol "T" for tool steel.

(iii) Figure indicating 10 times the average percentage of manganese content.

Examples :

1. 50T4 indicates an unalloyed tool steel with 0.50% carbon and 0.4% manganese.
2. A tool steel with 0.35-0.45% carbon and 0.7-0.9% manganese is designated as 40T8.

(c) Unalloyed free cutting steels : They are designated in the following order :

(i) Figure indicating 100 times the average percentage of carbon.

(ii) Letter "C".

(iii) Figure indicating 10 times the average percentage of manganese.

(iv) Symbol "S", "Se", "Te" or "Pb" depending on the element present which makes the steel free cutting, followed by the figure indicating 100 times the percentage content of the element. For phosphorized steels the symbol P is included.

(v) Symbol indicating special characteristics.

(d) Low and medium alloy steels : They are designated in the following order :

(i) Figure indicating 100 times the average percentage of carbon

(ii) Chemical symbol for alloying elements each followed by the figure for its average percentage content multplied by 10C.

(iii) When the alloying element is less than one percent, it is written upto two decimal places underlined by a bar.

Examples :

1. Consider an alloy steel with the following composition :

Carbon = 0.12-0.18%

Silicon = 0.10-0.35%

Manganese = 0.40-0.60%

Chromium = 0.50-0.80%

Average percentage of carbon = 0.15, which is denoted by 15. Silicon and manganese are not important alloying elements and are deleted. The significant element chromium has average percentage 0.65%, which is less than 1%. Hence, it is denoted as Cr65. Therefore, complete designation of steel is 15Cr65.

2. A steel designated by 40Cr14 has average carbon 0.4% and chromium 0.14%.
3. A steel having chemical composition : C = 0.15-0.25%, Si = 0.10-0.50%, Mn = 0.30-0.50%, Ni = 1.5-2.5%, Cr = 16-20% has major average alloying elements Cr = 18% and Ni = 2%. C = 0.20%. Hence its designation is : 20Cr18Ni2.

(e) High alloy steels : They are designated in the following order :

(i) Letter "X" for high alloy steel.

(ii) Figure indicating 100 times the average percentage of carbon content.

(iii) Chemical symbol for alloying elements each followed by the figure for its average percentage content rounded off to the nearest integer.

(iv) Chemical symbol to indicate specially added element to attain the desired propoerties.

(v) Symbol indicating specific characteristics.

(f) Alloy tool steels : They are designated as follows :

(i) Letter "T" for low alloy and medium alloy.

(ii) Letters "XT" for high alloy and as given for low and medium and high alloy steels.

(g) Free cutting alloy steels : They are designated as follows :

As stated above plus a symbol of the additional element percent followed by the figure indicating 100 times its content.

2.7.3 Applications of steels

The applications of steels are given in Table 2.4.

Table 2.4 : Applications of steels.

Designation of steel	*Applications*	*Ultimate tensile strength S_{ut}, MPa*	*Yield strength S_{yt}, MPa*
	1. Plain carbon steels		
7C4	Automobile bodies and hoods.	320	–
10C4	Cam and camshaft, worm, gudeon pin, chain wheel and spindle.	340	–
30C8	Socket, tie rod, yoke, lever and rocker arm.	500	400

40C8	Transmisison shaft, crankshaft, spindle connecting rod, stud and bolt.	580	380
45C8	Transmission shaft, machine tool spindle, bolt and gear of large dimensions.	630	380
50C4	Transmission shaft, worm, gear and cylinder.	760	460
55C8	Gear, cam, sprocket, cylinder and key.	720	466
60C4	Machine tool spindle, hardened bolt and pinion.	750	–
65C6	Flat and coil spring.	750	–
	2. Free cutting steels		
10C8S10	Bushes, tappets, gudgoon pin, pawls and worms.	460	–
14C14S14	Pinions, motorcycle and bicycle components.	520	–
25C12S14	Bolts and studs.	560	–
40C10S18	Bolts, crankshaft and connecting rod.	600	–
40C15S12	High strength bolts, axles and shafts	640	–
	3. Alloy steels		
40Cr1	Gear, connecting rod and axle.	680	520
40Ni3	Turbine blades and high strengh bolt.	780	580
35Ni1CrCr60	Crankshaft, connecting rod, gear shaft, cam shaft and gears.	680	520
30Ni14Cr1	High strength gears.	1510	1270
40Ni2Cr1Mo15	Transmission shaft, axle, gear and bolts.	780	580
27Mn2	Welded parts, spindle and shaft.	580	430
37Mn2	Transmissin shaft and axle.	580	430
50Cr1V23	Leaf and coil springs.	1880	1760
	4. Stainless steels		
04Cr13	Hydraulic pumps, valves and steam applications.	440	–
05Cr17		440	–
15Cr25V20		490	–
02Cr18Ni11	Cold press work	440	–
04Cr18Ni10		490	–
10Cr17Ni7		590	–
04Cr17Ni12Mo2		490	
02Cr16Ni12Mo2		440	–
12Cr13	Springs and high tempeature applications	590	–
20Cr13		690	–
30Cr13		780	–
15Cr16Ni2		830	–

2.7.4 Effect of Alloying Elements on Steel

Table 2.5 : Effect of alloying elements on steel.

Alloying element	*Percentage*	*Effects*
Chromium	Upto 30%	Increases toughness and hardness, increases resistance to abrasion, wear corrosion and oxidation.
Nickel	Upto 50%	Increases strength and toughness without decreasing ductility,
Silicon	Upto 0.8%	Acts as deoxidizing agent, produces high magnetic permeability.
Molybdenum	0.20-0.70%	Increases hardness and improves machining properties.
Vanadium	0.2-0.5%	Increases toughness and strength, increases life of tools, springs and members subjected to high temperature.
Tungsten	0.4-0.22%	Widely used in tool steels, increases toughness and hardness.
Cobalt	0.15-0.30%	Increases hot hardness.
Copper	0.15-0.30%	Improves resistance to atmospheric corrosion.
Aluminium	1-5%	Used as a deoxidizer, restricts grain growth.
Sulphur	0.06-0.30%	Increases machinability.

2.7.5 Reasons for use of Alloy Steels in Machines Design

The alloy steels consists of iron plus carbon and varying percentages of the following alloying metals :

Tungsten (W), Molybdenum (Mo), Chromium (Cr). Vanadium (V), Cobalt (Co), Nickel (Ni), Silicon (Si) Copper (Cu), Aluminium (Al), and Sulphur (S).

These alloying metals are added to plain carbon steels to improve one or the other properties so that the part may function to the desired expectancy. The properties to be improved upon are the following :

1. Toughness and hardness.
2. Abrasion, wear, corrosion and oxidation resistance.
3. Tensile strength and toughness without decreasing ductility.
4. Machinability.
5. High temperature strength.
6. Electrical and magnetic properties.
7. Hot hardness.
8. Grain structure.

2.7.6 Special Alloy Steels

1. Stainless Steels

These steels have high resistance to corrosion against atmospheres of air and water vapour, sea water, solutions of acids or salts, etc. There are three types of stainless steels.

(a) **Austenitic stainless steels :** These are chrome-nickel steels having the following chemical composition :

C = 0.12-0.14%

Cr = 17-20%

Ni = 8-11%

The most common steel in this group is the 18-8 stainless steel containing Cr = 18% and Ni = 8%. These steels are used in food processing and handling machinery, kitchen equipment, dairy plants, textile machinery, surgical and dental instruments, and nuclear engineering.

(b) **Ferritic stainless steels :** These steels mainly contain chromium from 14-30%. A thin film of chromium oxide formed on the surface of steel protects it against corrosion.

(c) **Martensitic stainless steels :** It contains chromium from 12-14%. These steels are used for valves, screens, pump shafts, cutlery, bolts, nuts and various other parts in chemical and petroleum industries, steam turbine blades, jet engine compressor blades, carburetor parts, instrument parts, surgical instruments, ball bearings, bushings, valve seats, etc.

2. Killed Steel

Steel that has been deoxidized with a strong deoxidizing agent, such as silicon or aluminium, in order to eliminate a reaction between carbon and oxygen during solidification. This steel has fewer blow holes and are suitable for sound forgings and heavy rolled sections.

3. High Speed Steel (HSS)

This is a high carbon steel tool material to which the various alloying elements added are : Tungsten (W), Molybdenum (Mo), Chromium (Cr), Vanadium (V), and Cobalt (Co). These alloying elements improve the hardness, toughness and wear resistance properties of steel. They are capable of retaining their hardness upto 600°C and can be operated at much higher cutting speeds as compared to alloy steels.

The 18-4-1 HSS contains : W = 18%, Cr = 4% and V = 1%, Cobalt = 5-8% is added to increase red hardness. Carbon is about 0.75%. This material is designated as T-4. The BIS designation of this steel is : T75W18Co5Cr4V1Mo70.

The most commonly used grades of HSS are : M-1, M-2, M-7, M-10, T-1, and T-2. The composition of these high speed steel tool materials is given in Table 2.6.

Table 2.6 : High speed steels.

Designation	*Percentage of constituents*					
	C	*W*	*Mo*	*Cr*	*V*	*Co*
M-1	0.80	1.75	8.50	3.75	1.15	–
M-2	0.85	6.00	5.00	4.00	2.00	–
M-7	1.00	1.75	8.75	4.00	2.00	–
M-10	0.90	–	8.00	4.00	2.00	–
T-1	0.75	18.00	–	4.00	1.00	–
T-2	0.85	18.00	–	4.00	2.00	–
T-4	0.75	18.00	–	4.00	1.00	5

4. Spring Steel

The cold coiled springs are made from circular section steel wire patented and cold drawn, unalloyed. The hot coiled springs are made from Grades 1, 2, and 3 steel wire of following chemical composition :

50Cr4V2, 60Cr4V2, 55Si7, 60Si7, and 65Si7

The railway rolling stock springs are made from hot rolled steel of following grades :

98C6, 113C6, 55Si7, 60Si7, 50Cr4V2, and 60Cr4V2

5. Free Cuttings Steels

These are carbon steels and carbon-manganese steels with a small percentage of sulphur. Sulphur improves the machinability of these steels.

2.8 HEAT TREATMENT METHODS

Heat treatment may be defined as combination of operation involving heating above the recrystalisation temperature with subsequent cooling to obtain certain desirable properties. During this process the chemical composition does not change but the crystalline structure does change.

The objectives of heat treatment are :

1. To relieve the residual stresses introduced in the component.
2. To improve machinability by softening the surface layer of component.
3. To obtain the desired surface hardness keeping the core soft.
4. To improve electrical and magnetic properties.
5. To modify the grain structure and size.
6. To improve the corrosion and wear qualities of surface.

Tempering : It is the process of heating hardened steel to a temperature below 723°C, holding it there for the proper time, and then cooling slowly.

Annealing : It is the process to obtain softness, improve machinability, increase ductility and toughness, relieve internal stresses, reduce or eliminate structural inhomogeneity, refine grain size, and to prepare steel for subsequent heat treatment.

Normalising : It consists in heating the steel to a temperature from 40-50°C above 785°C, holding at this temperature for a short time, and subsequent cooling in air. It is used to eliminate coarse-grained structure obtained in rolling and forging to increase the strength of medium carbon steels and reduce internal stresses.

Hardening : It consists in heating steel to a temperature above the critical point, holding at this temperature, and then quenched in water, oil, or molten salt bath. Austenite converts directly into martensite. Hardening must be followed by tempering to reduce brittleness, relieve internal stresses and increase wear resistance.

Surface Hardening : It is a selective heat treatment method in which the surface layers of a metal are hardened to a certain depth whilst a relatively soft core is maintained.

Induction Hardening : It is a surface hardening process in which high frequency current is accomplished by the thermal effect of current induced in the part being heated. The part is placed in an alternating magnetic field set up by the high frequency current. It does not produce any scaling and deformation is considerably reduced. It is extensively used for crankshafts, camshafts, gears, and many other automobile and tractor parts.

Flame Hardening : It is carried out by heating by an oxyacetylene flame having a temperature of 3000-3200°C and subsequent quenching hardens the surface layer. It is used for large cast gears, worms, shafts, steel guideways, mill rolls, etc.

Case Carburising : It is the process of saturating the surface layer of steel with carbon to a concentration of 0.75-1.2% and subsequent quenching. It increasing fatigue strength of materials.

Nitriding : It is the process of saturating the surface of steel with nitrogen by holding it for a prolonged period at a temperature from 480-650°C in an atmosphere of ammonia. It increases the surface hardness, wear resistance, fatigue strength, and corrosion resistance.

Cyaniding : It is a process in which both carbon and nitrogen are added to the surface layer of steel by heating the steel in liquid or solid media to increase its hardness and wear resistance.

2.9 NON-FERROUS METALS AND ALLOYS

Non-ferrous metals are those which contain a metal other than iron as their chief constituent like copper, aluminium, brass, nichel, etc.

Non-ferrous metals and alloys offer the following advantages :

1. Low density.
2. Attractive colours.
3. Softness and facility for cold working.
4. Good formability.
5. Resistance to corrosion.
6. Ease of fabrication.
7. Excellent machinability.
8. High thermal conductivity.
9. Excellent bearing properties.

2.9.1 Copper and Copper Alloys

Copper is a valuable metal having red colour. Commercial copper is a tough, ductile and malleable metal. It has high electrical conductivity and is used extensively for wire, cable and other electrical and electronic apparatus which must conduct the current. It is also a good conductor of heat and highly resistance to corrosion by liquids. It is used in water heating apparatus, water pipes, vessels in brewery, chemical and petroleum plants, condenser tubes and solar water heaters.

The alloys of copper form an important group of materials with a wide variety of properties. Three groups of copper alloys are used in industry. These are : Brasses, bronzes and monel metal.

1. Brass

Brass is an alloy of copper and zinc. The usefull range is 10-30% Zn and 70-90% Cu. It is a good bearing material. By adding small quantities of other elements like aluminium, iron, manganese and tin, the strength of brass may be greatly increased. To improve the machinability of brass, 1-2% of lead is added. Small amounts of tin are added to brass to increase its hardness and resistance against corrosive action of se water.

2. Bronze

It is an alloy of copper and tin. The useful range is 5-25% Sn and 75-95% Cu. Tin bronzes are divided into two groups : Wrought bronzes containg upto 8% Sn, and Casting bronzes containing over 8% Sn. Tin bronze is comparatively hard, resists surface wear and can be cast into shape or rolled into wire, rods and sheets very easily. It is used in hydraulic fittings, pump linings, in making utensils, bearings, and bushes, etc.

The bronze alloys are : Phosphor bronze, silicon bronze, aluminium bronze, manganese bronze, and beryllium bronze.

(i) Phosphor Bronze : When bronze contains phosphorus, it is called phosphor bronze. Its composition may vary as given below :

$$Cu = 80\text{-}90\%; Sn = 10\text{-}20\%; P = 1.5\% \text{ (max)}$$

Phosphor bronze is used for bearings in which wearing qualities are desired. It is used for pump parts, linings, propellers, heavily loaded bearings, worm gears, gears, nuts for machine lead screws and springs.

(ii) Gun Metal : It contains Cu = 88%, Sn = 10% and Zn = 2%. It is used extensively for casting boiler fittings, bushes, bearings glands, etc.

(iii) Babbit Metal : An alloy of tin, lead, copper and antimony is called babbit meal. It is the most common bearing material. The composition of common babbit metals and their uses are given in Table 2.7.

Table 2.7 : Common Babbit Metals

Percentage Composition				*Uses*
Tin	*Antimony*	*Copper*	*Lead*	
93	3.5	3.5	–	Big end bearings, motor and aero engines
80	10.5	3.5	–	Main bearings, motor and aero engines.
80	11	3	6	Bearings for heavy loads and high speeds
60	10	1.5	28.5	Bearings for engines and electric machines, railways and tramways
40	10	1.5	48.5	Heavy pressure and bearing speed
20	15	1.5	63.5	Medium pressure and speed
5	15	–	80	Long bearings with medium load

(iv) Monel metal : Monel metal is the most important nickel-copper alloy, having high strength, toughness and excellent corrosion resistance. It contains Ni = 67% and Cu = 30% with small amounts of iron, manganese, silicon and carbon. It is a white, tough and ductile metal that can be easily machined, welded and heat treated. It is used in the form of rod, sheet, wire and welded tubing. It is widely used for structural and machine parts which must have a high resistance to corrosion and have high strength at elevated temperatures, as in the case of steam turbine blades, high temperature valves, impeller of centrifugal pump, and for springs subjected to temperatures above 200°C.

2.9.2 Aluminium and Aluminium Alloys

Aluminium is silvery white in colour, it is used where a light non-corrosive metal is necessary as in aircraft, missile and automobile components. Due to its good electrical conductivity, it is used for electrical wires and overhead cables. Its high resistance to corrosion and its non-toxicity makes it a useful metal for cooking utensils. It can be beaten into foil which is widely used for reflectors, mirrors and telescopes.

Aluminium alloying elements are, copper, silicon, manganese, magnesium, iron, zinc and nickel. Aluminium alloys can be classified as : cast and wrought. Casting alloys are used in foundry practice and wrought alloys are used to make great variety of articles by rolling forming, drawing, forging and press working.

Duralumin : It is a wrought aluminium alloy having the following composition :

Cu = 2.2 - 5.2%; Mg = upto 1.75%; Mn = upto 1%; Si = upto 1%; Fe = upto 1%

It has good strength and corrosion resistance. It is used in aircraft, manufacture of pistons vanes, screws, housings, motors pump impellers, rings, disks and other parts.

Hindalium : It is an alloy of aluminium and magnesium with a small percentage of chromium. It is mainly used for anodized utensil manufacture.

2.10 NON-METALLIC MATERIALS

The commonly used non-metallic materials are : Wood, glass, rubber, leather, carbon and plastic.

1. Wood : It is used where light weight parts subjected to moderate shock loading are required, e.g., in circuit breaker operating rods where non-metallic bearing material is desirable.

2. Glass : It is considered to be a weak and brittle material but glass with newer compositions may be used for design components. Glass parts can be moulded by heating and blowing. Glass is used for all parts of centrifugal pumps for acids, piping for mechanical process, pipe fittings, heat exchangers and linings for tanks and fittings.

3. Rubber : By definition, a rubber is a substance that has at least 200% elongation in a tensile test and is capable of returning rapidly and forcibly to its original dimensions when load is removed. There are five types of rubber: natural rubber, neoprene, silicon, polyurethane and fluoroelastomers. It is an excellent material for seats and disphragms, for water lubricated bearings, for parts subjected to vibrations as vibration damper, flexible couplings and bearings, and for tubes and hose pipes.

4. Leather : It is flexible and can withstand considerable wear. It is used for belts and as a packing or a washer.

5. Carbon : It gives low friction losses and low wear rates against metals. It is used in chemical handling equipment, pump rotors, vanes and gears. Clutch plates and rings of metal and cork have been replaced by carbon parts. Self lubricated carbon bearings are being used.

6. Plastics : Plastics are synthetic materials processed by heat and pressure. The advantages and disadvantages of plastics are given below :

(a) Advantages

(i) Low specific gravity resulting in light weight construction.

(ii) Good resistance to shocks and vibrations.

(iii) Low coefficient of friction, resistance to wear and self-lubricating property.

(iv) Resistance to corrosive atmosphere.

(v) Low cost.

(b) Limitations

(i) Low strength and rigidity, low heat resistance and low dimensional stability.

(ii) Low temperature creep behaviour.

(iii) High thermal expansion coefficient.

(iv) Embrittlement with age.

(c) Types of Polymers

(i) Chain or Addition Polymers : They consist of a Monomer and an Initiator. The monomer contains double bond that can be opened. The initiator is an organic or inorganic substance or a catalyst. Hydrocabons form staraight chain (alphatic hydrocarbons) or Benzene rings (aromatic hydrocarbons). Other polymer molecules may contain N, O, S, P, or Si, in the side chains while the movement Cl, F, or Br may replace H.

(ii) Step-reaction or condensation Polymers : Two dissimilar monomers are joined into short groups, followed by the polymerization and cross—linking of these groups; a by product of low molecular weight is often released.

(iii) Linear Polymers : Straight chains Covalent bonds exist only between molecules which are held to each other by secondary bonds that could be broken and reformed during deformation.

(iv) Cross-linked Polymers : Covalent bonds exist between molecules and no permanent deformation is possible.

(d) Types of Plastics

There are two types of plastics : Thermoplastics and Thermosetting plastics.

(i) Thermoplastics : They are linear straight chain polymers which can be softened repetedly without undergoing change in chemical composition, e.g., polyimides, cellulosics, acetals, polycarbonates, PVC, nylon, plexiglass, Teflon, etc.

(ii) Thermosetting plastics : They are cross : linked polymers which undergo a chemical change with a application of heat and pressure and cannot be resoftened, e.g., polysters, phenolics, silicones, epoxies, urethanes elastomers, araldite, catalin, duraplex, beckosol, etc.

(e) Applications of Plastics

Table 2.8 : Applications of platics

Name of Plastic	*Applications*
(a) Thermoplastics	
Polyethylene	Blow moulded bottles, automotive gas, drums, trash and garbage cans, household and automotive, extruded pipes.
Polypropylene	Blow moulded, bottles, closures, utomotive parts, applications, housewares and toys.
Polystyrene	Toys displays and housewares, protective packing materials, insulated drinking cups and floating devices.

ABS (Arcylonitrile Butadiene Styrene)	Automotive parts, telephones, luggage, pipe fitting and conduit.
Polyvinyl chloride (PVC)	Window frames, gutters, pipes, wire insulation and bottles, garden house, boots, tubing, packing and raincoats.
Polysters	Gears, rollers, bearings, housings, for pumps and appliances, implellers pulleys, switch parts, automotive components, electrical/electronics components.
Polyimides (Nylons)	Fibres, automotive components, gears, cams, pulleys, rollers, boat propellers.
Polyacetals	Gear, bearing, bushings, automobile door handles, plumbing valves.
Polycarbonates	Safety glazing, safety shields, non breakable windows, automobile tail light, poer tool housings.
Polyimides	Gears, bushings, seals, insulators, structural components.
PTFE (Polytetrafluoroethylene)	Nonstick coating on cookwares, non-lubricated bearings, chemical resistant pipe fittings, valves, pump parts, gskets, seals and packings.
PCTFE	Pipe and pump parts, valve diaphragms, coatings for corrosive industries.
(b) Thermosetting Plastics	
Epoxy	Tools and dies, adhesive and coating, laminates for aircraft, photoelastic models.
Melamine-formaldehyde	Table wares, electric insulation, automotive ignition parts, cutlery handles, jars and bowels.
Phenol-formaldehyde	Industrial electric parts, automotive electrical components.
Alkyd	Boats, tanks, bins, housing, truck, trailer and tractor components, ducts, shrounds, vaulting poles, athletic goods.
Urea-formaldehyde	Lighting appliances, switch plates, reflectors, housing closures, containers, handles, knobs, tableware, and piano keys.

2.11 COMPOSITE MATERIALS

Composites are cohesive structures made by physically combining two or more compatible materials, differing in composition and characteristics and some times in form. The fibre reinforced composites are only those heterogeneous materials which are prepared by association and bonding in a single structure of material possessing quite different properties and when these are complementary, give a composite material possessing additional and or superior properties to individual components either alone or mixed together. These are ideal materials for structural applications where high strength-to-weight and stiffness-to-weight ratios are required. The most important class of composite materials is fibre-reinforced plastics (FRP). The matric material is invariably a plastic and the reinforcing materials are : glass, carbon, boron, silica, ceramic, metals, and organic material like bamboo or jute fibres.

The composites as a class of engineering materials provide almost unlimited potential for higher strength, stiffness, and corrosion resistance over the pure material system of metals, ceramics and polymers.

2.12 WEIGHTED FACTOR METHOD OF MATERIAL SELECTION

The weighted point method is a systematic method for the selection of materials. It consists of the following five steps :

1. Prepare a list of desirable properties.
2. Assign values to the desired properties. The desirable properties could be yield strength, endurance limit, hardness, corrosion resistance, etc.
3. The desirable properties are divided into two groups : "Go-no-go" parameters and "discriminating" parameters. The "Go-no-go" parameters are the constraints. If a materials is not available or cannot be fabricated into given shape, it is totally rejected.
4. The discriminating parameters are the properties of material which can be given quantitative values. The weightage depends upon the importance of that particular property in the given application. The weightage factor may vary from 1 to 5 with 1 for the poorest and 5 for the best.
5. Assign a rating to each property of the candidate material ranging from 1 to 5, depending upon how closely it meets the requirements. These ratings are multiplied by the weighting factors for each property. These numbers are finally added and materials are arranged in descending order of their total points. The main drawback of this method is the skill and judgement required to assign the weightage.

Consider eight different types of steel withh the desirable properties listed in Table 2.9.

Table 2.9 : Weightage factors for comparative properties of cast steels.

S. No.	*Material*	*Abrasive wear resistance*	*Dampling capability*	*Corrosion resistance*	*Welda-bility*	*Machin-ability*	*Castabi-lity*	*Forgabi-lity*
Rating factor	→	3	2	3	2	4	1	5
1.	Low C	P	P	P	E	F	F	F
2.	Medium C	F	P	P	G	F	F	F
3.	High C	G	VP	VP	P	F	F	G
4.	Low alloy	F	P	P	G	F	F	G
5.	High alloy	P	VP	F	VP	VP	P	P
6.	Heat resistant	VP	P	G	F	F	VP	VP
7.	Corrosion resistant	G	G	F	F	F	F	G
8.	Wear resistant	E	P	F	F	VP	P	P

E = Excellent = 5 points, G = good = 4, F = fair = 3, P = poor = 2, VP = very poor = 1

Table 2.10 : Multiplication of weightage factor and rating factor.

S. No.	Material	Abrasive wear resistance	Damping capability	Corrosion reistance	Welda-bility	Machi-nability	Casta-bility	Forga-bility	Total
1.	Low C	6	4	6	10	12	3	15	56
2.	Medium C	9	4	6	8	12	3	15	57
3.	High C	12	2	3	4	12	3	20	56
4.	Low alloy	9	4	6	8	12	3	20	62
5.	High alloy	6	2	9	2	4	2	10	35
6.	Heat resistant	3	4	12	6	12	1	5	43
7.	Corrosion resistant	12	8	9	6	12	3	20	68
8.	Wear resistant	15	4	9	6	4	3	10	51

Descending order of materials : 7-4-2-3-1-8-6-5.

We find that material number 7 is found to be the best based on weighted point method.

SOLVED QUESTIONS FROM UNIVERSITY EXAMINATINS

■ **Example 2.1** Define mechanical property of an engineering material. State any six mechanical properties, give their definitions and one example of the material passing the property.

[MDU, Dec., 2001]

Solution : Refer to art. 2.2

■ **Example 2.2** State the alloying elements added to steel to get alloy steels and the effect they produce. Give at least one example of each. [MDU, Dec., 2007]

Solution : Refer to art. 2.7.4. For example of alloy steels, refer to Table 2.4.

■ **Example 2.3** Discuss the factors which are considered in the selection of a material for a machine component. [MDU, Dec., 2008]

Solution : Refer to art. 2.2

■ **Example 2.4** Explain the importance of material selection for machine elements.

[MDU, Dec., 2009]

Solution : Refer to art. 2.2

■ **Example 2.5** How do you classify materials for engineering use? [MDU, May, 2008]

Solution : Refer to art. 2.5.

■ **Example 2.6** What are the factors to be considered for the selection of materials for the design of machine elements? Discuss. [MDU, May, 2008]

Solution : Refer to art. 2.2

■ **Example 2.7** Explain the various mechanical properties of commonly used engineering materials. [MDU, May, 2009]

Solution : Refer to art. 2.2.

EXERCISES

1. Designate the following materials
   (a) Grey cast iron with an ultimate tensile strength of 350 MPa.
   (b) Plain carbon steel with a minimum tensile strength of 340 MPa.
   (c) Plain carbon steel with a minimum yield strength of 250 MPa.
   (d) Plain carbon steel with 0.45% carbon and 0.8% manganese.
   (e) Alloy steel with carbon = 0.40-0.50% and chromium = 0.90-1.20%.
2. What do you understand by the following designation of materials?
   (i) FG 400 (ii) 15C8 (iii) 45C15S15
   (iv) 45Cr1V23 (v) 23Mn2
3. Recommend suitable materials for the following components :
   (a) Valve spring (b) Gasket (c) Screw jack spindle
   (d) Helical springs (e) Screw jack nut
4. Explain the following terms :
   (a) Ductility (b) Toughness
   (c) Hardness (d) Fatigue
5. Discuss the effect of alloying elements on cast iron.
6. How does impurities affect the properties of cast iron?
7. How steels are designated?
8. Differentiate between the following steels :
   (a) Killed steel (b) Free-cutting steel (c) Stainless steel
9. Explain the following heat treatment processes briefly :
   (a) Normalising (b) Martempering (c) Case-carburizing
10. Suggest the materials for the following :
    (a) I.C. engine piston (b) Turbine blade
    (c) Carburettor body (d) Helical springs

□□□

3

Chapter

Manufacturing Considerations in Design

Inside This Chapter

3.1 Assembly Considerations in Design
3.2 Support and Retainment of Rotating Parts
3.3 Frames and Housings
3.4 Castings
3.5 Welding
3.6 Fits and Tolerances

3.1 ASSEMBLY CONSIDERATION IN DESIGN

Most of the machines are obtained by a combination of various machine components. Therefore, while fabricating and designing machine components, due consideration must be given to the assembly aspects of design. The following points should be considered from assembly point of view :

1. Provide for proper relationship of position and orientation with respect to other parts of the machine.
2. Provide for proper mating of surfaces.
3. Design for the assembly to less costlier
4. Design the assembly in such a manner so that the replacement of worn out components involve minimum of effort and cost.
5. The product should have a datum surface or point on which the assembly may be based.
6. Provide for easy access to each part to enable easy assembly.
7. The product should be so designed that assembly can be affected by placing the parts one over another.
8. Minimise side fastening of the assembly.
9. Design the assembly to consist of a number of sub-assemblies.

3.2 SUPPORT AND RETAINMENT OF ROTATING PARTS

Rotation is the motion most commonly found in machinery. Gears, belt pulleys, electric motors, automobile wheels, etc., all represent rotating machine parts. The rotating parts often carry heavy loads, and must be carefully designed to obtain trouble free operation for a long period of time.

The fundamental requirements of support and retainment of rotating machine parts can best be expressed in the form of two brief rules.

1. Every part of rotating assembly must be supported radially.
2. Every part of a rotating assembly must be retained or held axially, even if there are no apparent axial loads on the rotating assembly.

The support and retainment of rotating parts may be achieved with the help of following aids :

1. A rotating machine part may be attached to a rotating shaft by means of a key, splines, pins or a force-fit assembly.
2. Shoulder or a spacer (sleeve) may be provided on the shaft to avoid sliding of the rotating machine part on the shaft.
3. The shaft may be supported on a sliding or rolling bearings provided at appropriate locations. The type of the rolling bearing to be used may be selected with the help of Table 3.1

Table 3.1 : Selection of rolling bearings

Type of Load	*Machine Parts*	*Recommended Bearing Types*
Purely radial	Flat and V-belt pulleys, spur gears, chains, ropes	Radial ball bearings, sliding bearings, straight roller bearings, needle roller bearings
Axial and Radial	Helical gears, worm gears, Bevel gears	Deep groove ball bearings, angular contact ball bearings. Tapered roller bearings (axial load in one direction only). Thrust bearings (carry no radial loads).

4. Provide shoulders in the bearing to avoid side slipping of the bearing in the housing.
5. Step down the shaft at the bearing to avoid sliding of the shaft sideways.
6. Provide retaining rings and locknuts to retain the shaft in the bearings.
7. Provide removal caps on the bearings for easy access and assembly of the bearings.
8. Provide shims (spacers) between the housing and the caps of the bearings to compensate for manufacturing tolerances.
9. Provide seals where the shafts enter or leave a housing to hold the lubricant in and to keep the dirt out.
10. For overhung rotating parts, mount the bearing directly on the hub of the member being supported to get the bearing close to the overhanging member and at the same time have a bearing large enough to carry the large loads imposed on this bearing.
11. For mounting bevel gear and bevel pinion, use cartridge mounting of the bevel pinion and shims to adjust the position of the pinion so as to obtain the correct gear tooth clearance.
12. For worm gears use shims to adjust the gear position axially at assembly.

3.3 FRAMES AND HOUSINGS

A housing encloses and supports the machine parts. It may be either a casting or a fabricated construction obtained by welding. Therefore all the design considerations as listed in this chapter for castings and welding may be adhered to. In addition to these the following points may also be considered :

1. On account of the advantage of machining and assembly requirements, multiple piece housings should be preferred over one piece housing.
2. Provide sufficient and substantial flanges on the housing for holes to be drilled for the fastening bolts.
3. Provide bolt holes as close to the wall of the housing as possible to minimise bending of flanges.
4. Provide the bolts as close to the bearing bores as possible so that bearing loads will not deflect the housing.
5. For small housing, through bolts may be used for fastening of different pieces of the housing to obtain a very rigid construction.
6. Means should be provided to get the various pieces of the housings to line up properly when the machine is assembled. This can be achieved by piloting one part of the housing inside the other.
7. Provide opening in the housing to allow inspection, facilitate assembly or to permit adjustments to be made. Openings on cylindrical surfaces must be flat.
8. Provide base as part of the housing to support the housing on a floor or foundation surface. Flanges should be provided on the base to drill holes for the foundation bolt. Invariably for small housing, the base should be in one part. For large and heavy housings, the base can be in two or three pieces.
9. Design the base so as to allow the machine to stand by itself.

3.4 CASTINGS

A casting may be defined as an object formed directly from molten metal by solidification in a suitably shaped mould. The most important casting processes for iron and steel are : sand moulding, permanent moulding, shell moulding and centrifugal casting. For nonferrous metals the die casting process is most widely used.

3.4.1 Iron Castings

Cast iron is available in several types depending on the composition and casting technique. The most widely used of all these types is grey cast iron. Experience has shown that many casting problems are related to design. Therefore the designer must see how these problems may be minimised. The following are some recommendations to be followed while designing grey iron casting :

1. Work closedly with the foundrymen while designing castings and the views of the foundrymen should be incorporated. Any constraints of the available foundry facility should also be taken into account.
2. Design for casting soundness. Castings should be designed in such a way that the defects in castings are automatically eliminated.

3. Choose the most satisfactory form from the point of view of strength. As the compressive strength of iron castings is nearly four times the tensile strength, iron casting should therefore be loaded mainly in compression. Although this condition can not always be achieved, it is usually possible to design so that direct tensile and bending tensile stresses are kept to a minimum value.
4. Avoid cast iron where impact and shock loading are involved. For ordinary grey cast iron, the elongation at fracture is so small that for all practical purposes it can be taken as zero. Cast iron is therefore not suitable for any part subjected to impact loads if the part concerned must respond to loading by deformation and not by suddenly fracturing.
5. Avoid cast iron at temperatures above 300°C. Upto about 350°C the strength of cast iron is little affected by temperature. Above 400°C, however, the combined carbon decomposes into ferrite and graphite. At the same time a dangerous increase in volume takes place, and this sets up stresses in the casting which superimose themselves on existing stresses and so lead to the formation of cracks at the critical cross-sections. The strength also naturally decreases to a marked extent above 400°C.
6. Keep thickness as uniform as possible. The strength of grey cast iron varies with thickness. Apart from the carbon and silicon content, the graphite separation is also affected by the rate of cooling. The later, however, depends on the thickness of the casting. Therefore sections of different thicknesses in one and the same casting will differ in strength, owing to variations in the crystalline structure. Casting should therefore be designed for maximum uniformity of thickness throughout.
7. Avoid sharp angles and corners to avoid hot spots. The cold casting has a smaller volume than the molten one. The total change of volume is termed concentration. The undersired consequences of concentration are shrinkage cavity formation and the setting up of stresses. These cavities result from the fact that the outside of the casting and the thinner parts are the first to cool and solidify. The concentration effect causes the molten cast iron to be drawn in. If this is prevented or restricted, voids called shrinkage cavities will be formed. Therefore provide gradual changes of cross-section and uniform thickness throughout.
8. Provide for the simplest possible patterns and cores for moulding; provide for holes to knock out cores; try to design casting of a form so that no cores will be required.
9. Provide for reinforcing ribs and brackets in the casting to increase its resistance to distortion upon redistribution of internal stresses which may be due to nonuniform cooling of the parts of the casting having different sections.
10. Provide for the required draft (0.5° to 3°) on the walls of the castings to enable the pattern to be withdrawn from the mould. For internal surfaces the draft values should be higher than for external surfaces.
11. Provide for reinforced walls of the castings where holes are to be drilled.
12. Use fillets and give angular forms ample radius to decrease stress concentration.
13. Do not use bosses, lugs and pads unless absolutely necessary.
14. Bring the minimum number of sections together.
15. Provide for a common centreline for various sections of the casting.
16. The changes in section should take pace entirely on one side of the thinner section.

Methods for improving casting design are shown in Fig. 3.1

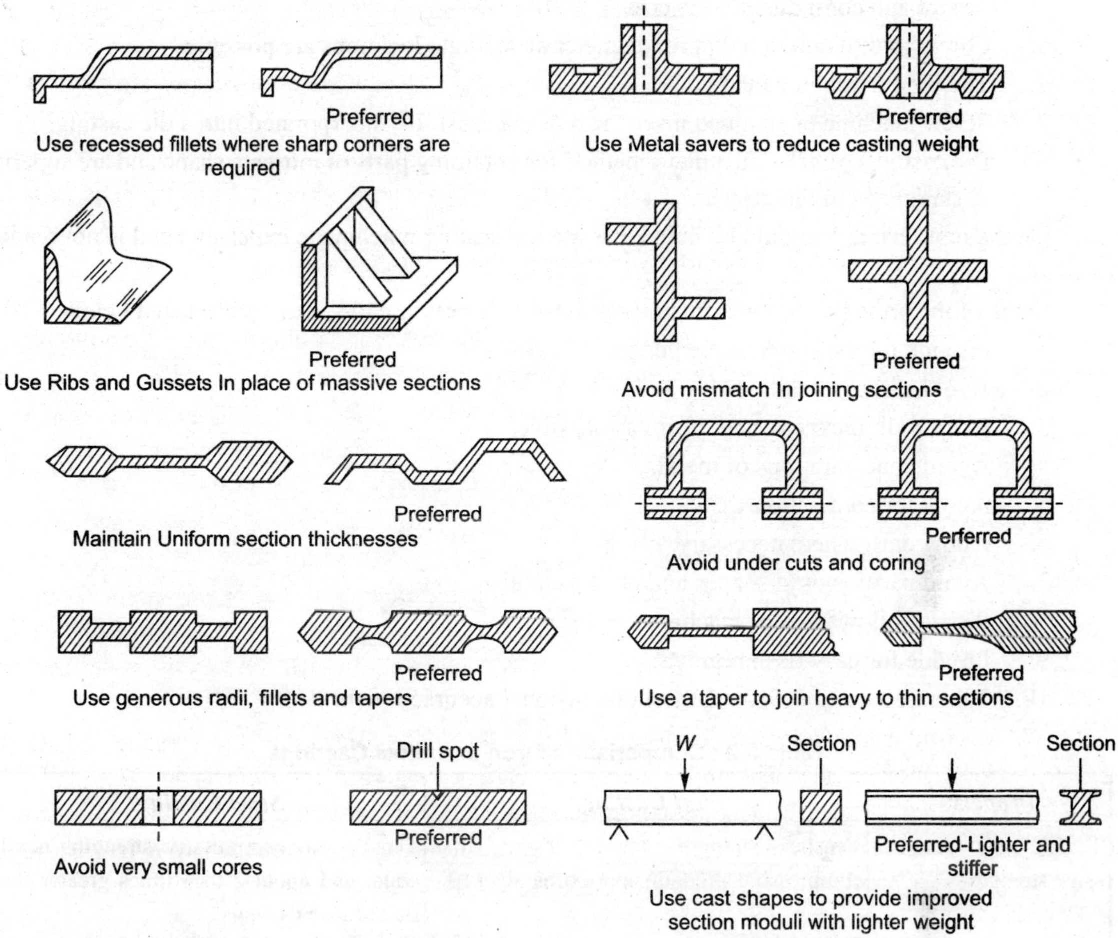

Fig. 3.1 Methods for improving casting design.

3.4.2 Steel Casting

The various design considerations as listed above for iron castings may also be considered for steel castings. However the steel castings and iron castings compare as shown in Table 3.2.

3.4.3 Die Casting

A die casting is an alloy part formed by injecting molten metal under high pressure into a steel mould called a die. The stationary die is called the cover die and the movable die is called the "ejector die". Zinc, aluminium, magnesium, etc. are the metals most widely used for die casting. Some of the advantages of the die casting as a production process are listed below ;

1. Dimensional tolerances can be held to remarkably close limits so that little or no machining is required.
2. Surfaces are smooth and of pleasing appearance and require a minimum of preparation for plating or other finishes.
3. Walls of various thicknesses can be held in the same casting.

4. Although the material costs are high than for cast iron, the thin walls serve to reduce the weight and consequently the cost.
5. Cored holes down to 0.75 mm diameter at accurate locations are possible.
6. The production rate is rapid.
7. Screw machine or stamped inserts can be successfully incorporated into a die casting.
8. Die castings offer a convenient method for obtaining parts of intricate shape and are superior to stampings in this respect.

On the other hand, it should be noted that the die casting machine is expensive and is not readily available.

Some of the principal factors affecting the design of pressure die castings are listed below.

1. Provide for the simplest die design.
2. Avoid abrupt transitions.
3. Keep wall thickness as uniform as possible.
4. Avoid concentrations of metal.
5. Allow for contraction.
6. Allow draft where necessary.
7. Avoid partly closed shapes and closed cavities.
8. Provide allowance for machined surfaces.
9. Provide for easy flash removal.
10. Take account of the standard of dimensional accuracy permitted.

Table 3.2 : Comparison of Iron and Steel Castings

Property	*Iron Castings*	*Steel Castings*
Tensile and compressive strength	Tensile strength much lower than compressive strength, approximately 4 : 1	Tensile and compressive strengths nearly equal and about 2 to 4 times greater than that of iron castings.
Bending strength	Half of steel casting	Twice that of cast iron. Gives smaller dimensions and lighter weight.
Torsional strength and	—	About 50% higher than that of cast iron.
Modulus of Elasticity	Has better damping capacity.	About twice that of cast iron. Better suited for structures requiring stiffness. Better fatigue resistance.
Elongation at Fracture	Very small. Not suitable for severely loaded parts.	About 12 to 20 percent. Can take sufficient elastic or permanent deformation and does not suddenly fracture.
Heat Resistance	Can not be used above 300°C due to the presence of free graphite	Can be used above 400°C.
Wear Resistance	Has good wear resistance characteristics	Gives better toughness and wear resistance when alloyed with manganese.
Contraction and Hot cracking.	Thin and large castings can be easily moulded.	If flows sluggishly than cast iron. Tends to form bubbles and has contraction twice of cast iron. So thicker castings and thin sections can not be moulded.

3.5 WELDING

Welding consists uniting two separate metal parts of like composition to form a whole, offering the maximum homogeneity. Welding results in great saving of material, low manufacturing costs and dependability of the medium. It gives the designer great latitude in planning and designing.

The main factors to be borne in mind while designing welded structures are summarized below :

1. Provide for a straight-line force pattern as far as possible.
2. Avoid laps, straps, and stiffening angles.
3. Use butt welds wherever possible.
4. Limit the number of welds used.
5. Make sure that ends to be welded together are of equal thickness.
6. Avoid placing welds in vulnerable cross-sections.

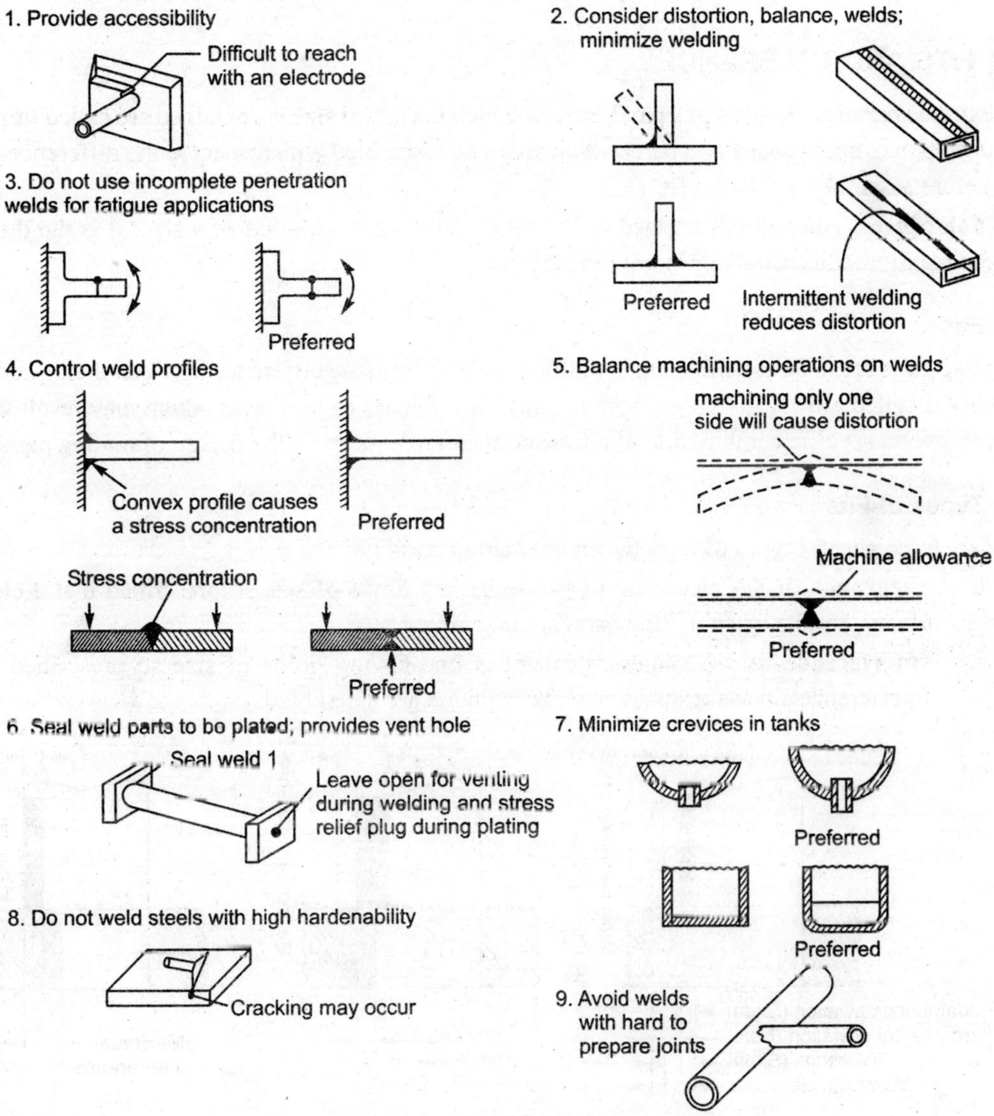

Fig. 3.2 Design hints for welding.

7. Facilitate assembly by means of shoulder, etc.
8. Avoid the use of welding fixtures as far as possible.
9. Provide for easy access to welds so that they are accessible for inspection.
10. Allow for the effects of thermal stresses.
11. For welds to be subjected to alternating loads, avoid running a weld at right angles to the direction of maximum principal stress owing to the low fatigue resistance offered by welds.
12. Distribute heavy loading over long welds in the longitudinal direction.
13. Avoid subjecting welds to bending loads.
14. Do not locate a weld at the point of maximum deformation.
15. Design ribs correctly to decrease undesirable intensification of notch effects.

Important design hints for welding are shown in Fig. 3.2.

3.6 FITS AND TOLERANCES

Two extreme permissible sizes of a part between which the actual size is contained are called **limits**. The relationship existing between two parts which are to be assembled with respect to the difference on their sizes before assembly is called a **fit.**

Tolerance : Tolerance is defined as the total permissible variation of a size. It is the difference between maximum limit and minimum limit of size.

3.6.1 Fits

When two parts are to be assembled the relation resulting from the difference between their sized before assembly is called a **fit.** The fit signifies the range of tightness or looseness which may result from the application of a specific combination of allowances and tolerances in the design of mating parts.

3.6.2 Types of Fits

There are three general types of fit between the mating parts :

1. **Clearance fit :** A clearance fit is one having limits of size so prescribed that a clearance always results when mating parts are assembled.
2. **Interference fit :** An interference fit is one having limits of size so prescribed that an interference always results when mating parts are assembled.

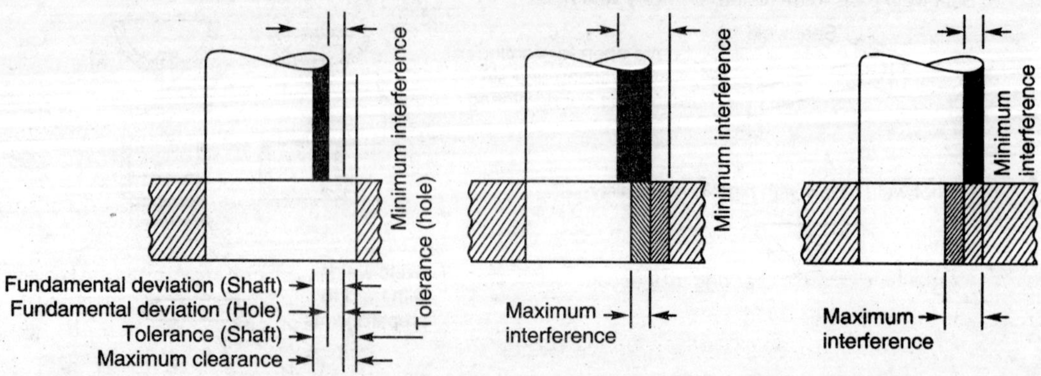

Fig. 3.3 Types of fits.

3. **Transition fit :** An transition fit is one having limits of size so prescribed that either a clearance or an interference may always result when mating parts are assembled.

The three types of fits are shown in Fig. 3.3. The disposition of tolerance zones for the three classes of fit are shown in Fig. 3.4.

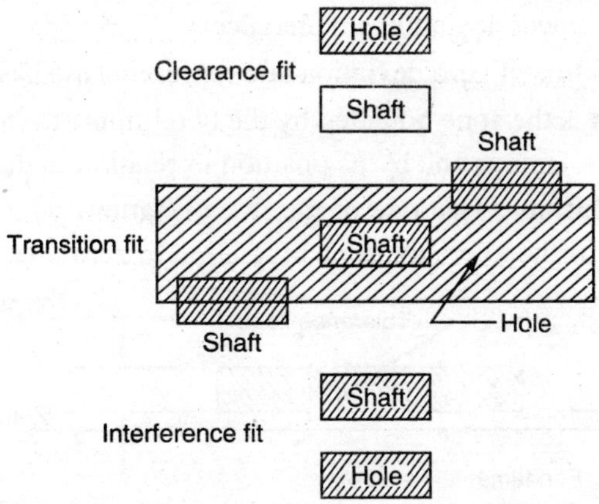

Fig. 3.4 Disposition of tolerance zones for the three classes of fit.

3.6.3 Terminology

The terminology used in fits and tolerances is shown in Fig. 3.5. The important terms are :

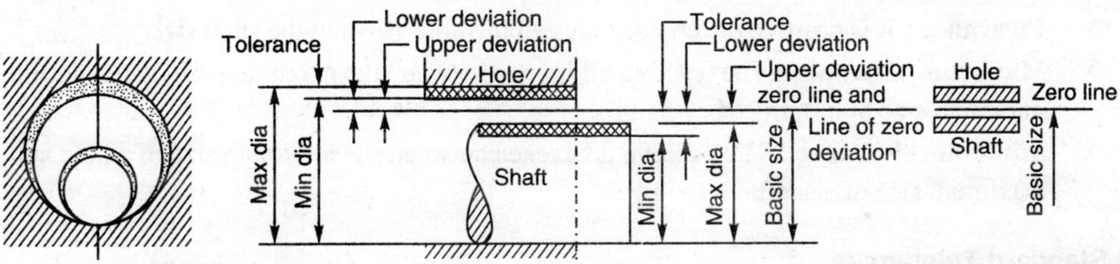

(a) Diagram illustrating basic size deviations and tolerances.

(b) Simplified schematic diagram of clearance fit.

Fig. 3.5

- **Basic size :** It is the exact theoretical size arrived at by design. It is also called nominal size.
- **Actual size :** The size of a part as may be found by measurement.
- **Maximum limit of size :** The greater of the two limits of size.
- **Minimum limit of size :** The smaller of the two limits of size.
- **Allowance :** It is an international difference between maximum material limits of mating parts. It is a minimum clearance or maximum interference between mating parts.
- **Deviation :** The algebraic difference between a size (actual, maximum, etc.) and the corresponding basic size.
- **Actual deviation :** The algebraic difference between the actual size and the corresponding basic size.
- **Upper deviation :** The algebraic difference between the maximum limit of size and the corresponding basic size.

  Upper deviation of hole = ES (Ecart Superior)

  Upper deviation of shaft = es

- **Lower deviation :** The algebraic difference between the minimum limit of size and the corresponding basic size.
  Lower deviation of hole = EI (Ecart Inferior)
  Lower deviation of shaft = ei
  Upper deviation = Lower deviation + Tolerance
- **Zone line :** It is the line of zero deviation and represents the basic size.
- **Tolerance zone :** It is the zone bounded by the two limits of size of the parts and defined by its magnitude, *i.e.,* tolerance and by its position in relation to the zero line.
- **Fundamental deviation :** That one of the two deviations which is conveniently chosen to define the position of the tolerance zone in relation to zero line, as shown in Fig. 3.6.

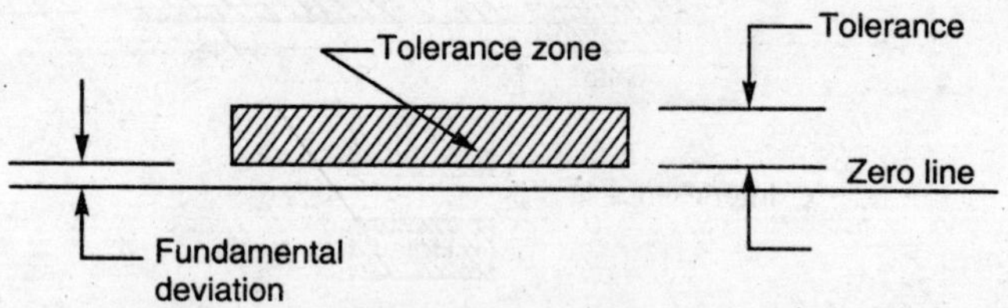

Fig. 3.6 Disposition of fundamental deviation and tolerance zone with respect to the zero line.

- **Basic shaft :** A shaft whose upper deviation is zero.
- **Basic hole :** A hole whose lower deviation of zero.
- **Clearance :** It is positive difference between the hole size and the shaft size.
- **Maximum clearance :** The positive difference between the maximum size of a hole and the minimum size of a shaft.
- **Minimum clearance :** The positive difference between the minimum size of a hole and the maximum size of a shaft.

3.6.4 Standard Tolerances

There are 18 standard grades of tolerances as specified by BIS with designations IT01, IT0 and IT 1 to IT16.

Standard tolerance unit, $i = 0.45\,D^{1/3} + 0.001\,D.$

where i = standard tolerance unit in microns

D = diameter in mm

The standard tolerances for the various grades are given in Table 3.3 and tolerance grades for various manufacturing processes in Table 3.4.

Table 3.3 : Standard tolerances

Grade	IT5	IT6	IT7	IT8	IT9	IT10	IT11	IT12	IT13	IT14	IT15	IT16
Value	7*i*	10*i*	16*i*	25*i*	40*i*	64*i*	100*i*	106*i*	200*i*	400*i*	640*i*	1000*i*

Table 3.4 : Tolerance grade in various manufacturing processes.

Tolerance grade	*Manufacturing process that can produce*
16	Sand casting : flame cutting
15	Stamping
14	Die casting or moulding; rubber moulding
13	Press work, tube rolling
12	Light press work; tube drawing
11	Drilling, rough turning, boring, precision tube drawing
10	Milling, slotting, planning, metal rolling, or extrusion.
9	Capstan or automatic; horizontal or vertical boring
8	Centre lathe turning and boring, reaming, capstan or automatic in good condition.
7	High quality turning, broaching, honing
6	Grinding or fine honing
5	Machine lapping diamond or fine boring, fine grinding.

3.6.5 Hole Basis and Shaft Basis for Fits

1. **Hole basis system :** In this system, the different clearances and interferences are obtained in associating various shafts with a single hole, whose lower deviation is zero.
2. **Shaft basis system :** In this system, the different clearances and interferences are obtained in associating various holes with a single shaft, whose upper deviation is zero.

3.6.6 Selection of Fits

Hole basis system is the most commonly used system because due to the fixed character of hole production tools, it is difficult to produce holes with odd sizes. Commonly used types of fits are given in Table 3.5. Shafts '*a*' to '*h*' produce clearance fit, '*j*' to '*n*' transition fit, and '*p*' onwards interference fit with hole.

Table 3.5 : Commonly used fits

Types of fit	*Class shaft*	*With holes*			*Remarks*
		H6	***H7***	***H8***	
Clearance	d		d8	d8	Loose running fit used for plummer block bearings, loose pullyes, etc.
	e	e7	e8	e8-e9	Easy running fit used for properly lubricated bearings. Finer grades are used for heavily loaded bearings of turbogenerators, electric motors, etc.
	f	f6	f7	f8	Normal running fit used for normal grease or oil lubricated bearings where temperature changes are not too much. This fit may be used for bearings for small electric motors, pumps, or bearings of gear box shaft, etc.
	g	g5	g6	g7	It is close running fit or sliding fit or spigot and location fit.

	h	h5	h6	h7-h8	It is precision sliding fit or fine spigot or location fit.
Transition	j	j5	j6	j6	It is very accurate location fit giving easy assembly and dismentling. It is used in case like coupling spigots and recesses.
	k	k5	k6	k7	It is light keying fit.
	m	m5	m6	m7	It is medium kelvin fit.
	n	n5	n6	n7	It is heavy keying fit for tight assembly of mating surfaces.
Interference	p	p5	p6		It is light press fit with easy dismantling for non-ferrous parts and is standard fit with easy dismantling for assembly of ferrous and non-ferrous parts.
	r	r5	r6		It is light drive fit or non-ferrous parts and medium drive fit for ferrous parts assembly.
	s	s5	s6	s7	It is heavy drive fit for ferrous parts giving permanent or semi-permanent assembly but it is standard press fit for non-ferrous parts. It is used for pressing collars on to shafts, valve seatings, etc.
	t	t5	t6	t7	It is force fit on ferrous parts for permanent assembly.

3.6.7 Dimensioning of Tolerances Rules

1. The upper deviation should be written above the lower deviation value irrespective of whether it is a shaft or a hole (Fig. 3.7(a)).
2. Both deviations are expressed to the same number of decimal places, except in the cases where the deviation in one direction is nil (Fig. 3.7(b)).
3. Tolerances should be applied either to individual dimensions or by a general note, assigning uniform or graded tolerances (Fig. 3.7(c)).

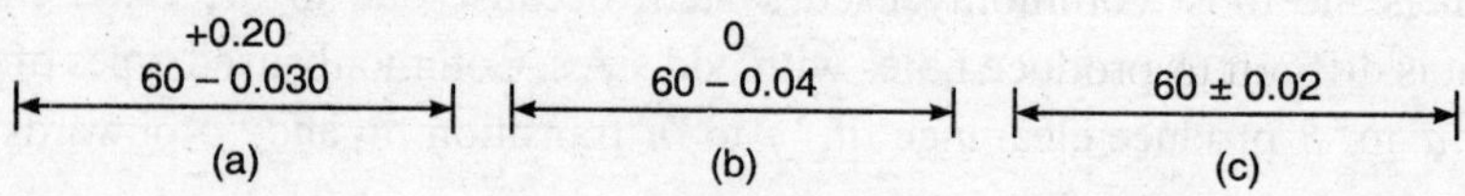

Fig. 3.7

4. The use of general tolerance not greatly simplifies the drawing and saves much labour in its preparation.

On the drawing, the limits on a dimension can be specified in two ways, *i.e.*, (i) unilateral, and (n) bilateral.

In unilateral tolerance system, the variation in size is permitted in one direction only from the design size. For example : 40 + 0.000 or 40 + 0.025
− 0.025 − 0.000

In the bilateral tolerance system, the variation in size is permitted in both directions from the design size. For example :

50 + 0.055
− 0.025.

SOLVED QUESTIONS FROM UNIVERSITY EXAMINATIONS

■ **Example 3.1** Discuss various types of fits and tolerances. [MDU, Dec., 2007]

Solution : Refer to art. 3.6.

■ **Example 3.2** Discuss with neat sketches the general principles in design of castings. [MDU, May, 2009]

Solution : Refer to Fig. 3.1.

■ **Example 3.3** Graphically represent the decrease, transition and indifference fits for hole and shaft base systems. What do you understand by allowance?

Solution : Refer to Fig. 3.4.

EXERCISES

1. What are the assembly considerations in design?
2. What are the considerations for the support and retainment of rotating parts?
3. What are the considerations for the design of frames and housings?
4. List the various considerations for the design of following type of castings?
   (a) Iron castings (b) Steel castings (c) Die castings.
5. Explain with sketches the methods for improving casting design.
6. Compare iron and steel castings.
7. What are the main factors to be borne in mind while designing welded structures?
8. Explain with sketches the design hints for welding.
9. What are fits and tolerances?
10. Explain with neat sketches the types of fits.
11. Explain the following with net sketch :
    Basic size, actual size, allowance, deviation, and tolerance.
12. Differentiate between basic shaft and basic hole.
13. Explain the hole basis and shaft basis systems for fits.
14. What are the rules for dimensioning of tolerances?

4

Chapter

Strength Considerations in Design

Inside This Chapter

4.1 Introduction
4.2 Review of Stresses in Machine Members
4.3 Simple and Principal Strains
4.4 Relationship between Elastic Constants
4.5 Working Stress
4.6 Factor of Safety
4.7 Modes of Failure
4.8 Theories of Failure
4.9 Introduction
4.10 Stress-Concentration
4.11 Completely Reversed and Fluctuating Stresses
4.12 Fatigue Failures
4.13 Fatigue Considerations in Design
4.14 Endurance Limit and Wohler's Curve
4.15 Fatigue Stress-Concentration Factor
4.16 Notch Sensitivity
4.17 Factors Afecting Endurance Limit
4.18 Design for Finite (or Limited) and Infinite (or Unlimited) Life
4.19 Working Stresses
4.20 Combined Fluctuating Stresses

Part-I : Static Loading

4.1 INTRODUCTION

Some of the basic definitions are :

- **Static Stress :** A stress which is applied to a structure very gradually, increasing from zero to their final value and then either do not change their magnitude, direction or point of application with time at all or change them very slightly, so that acceleration occuring in this case can be neglected.
- **Dynamic Stress :** A stress which changes with time. It is generally either of the reversible or alternating type.
- **Strength :** The strength of a material may be defined as the maximum resistance which a material can offer to the externally applied forces expressed per unit area.

- **Significant Stress :** This is the stress imposed on a member by service loads, assembly conditions and fabrication. This stress can be static, dynamic or a combination of the two.
- **Significant Strength :** It is the maximum stress which the material in its fabricated form can sustain when subjected to significant stress, *i.e.,* to service load, assembly conditions and fabrication strength, as defined here, take into account hardness of the material, size of the member, surface condition, type of service loading and life expectancy, etc.
- **Elastic Limit :** The least stress that will cause permanent set.
- **Yield Stress :** The lowest stress at which strain increases without increase in stress. If there is a decrease in the stress after yielding, a distinction is made between upper and lower yield points. The upper yield point is the stress at which the stress-strain diagram first becomes horizontal and the lower yield a point is somewhat lower and almost constant stress under which the metal continues to deform.
- **Ultimate Strength :** It is the maximum stress, based upon the original area of cross-section, to which the material can be subjected to in a simple tensile test.
- **Fatigue :** The reduced tendency of the material to offer resistance to applied stresses is called fatigue.
- **Fatigue Limit :** The number of cycles of stress that can be sustained by a material point to failure of a specified nature for a stated stress condition.
- **Endurance Limit :** The maximum stress below which a material can presumaly endure or withstand an infinite number of cycle of stress.

4.2 REVIEW OF STRESSES IN MACHINE MEMBERS

A machine component may be subject to the following stresses :

(a) Simple Stresses

1. Tension/Compression

Let P = load applied in tension or compression, N

A = area of cross-section, mm^2

Then tensile/compressive stress,

$$\sigma_t \text{ or } \sigma_c = \frac{P}{A} \text{ N/mm}^2 \text{ or MPa} \qquad \ldots(4.1)$$

2. Shear

Let F = shear force acting on the member, N

A = area under shear, mm^2

The shear stress,

$$\tau = \frac{F}{A} \text{ MPa} \qquad \ldots(4.2)$$

3. Bending

Let M = bending moment applied to the machine member, N-m.

z = section modulus of the cross-section, m^3

Using bending formula,

$$\frac{M}{I} = \frac{\sigma}{y} = \frac{E}{R}$$

where I = area moment of inertia about the bending axis, m^4

$= \frac{\pi d^4}{64}$ for a circular shaft of diameter d

y = distance of the farthermost fibres from the neutral axis, m

E = Young's modulus of elasticity of the material of machine member, N/m^2

R = radius of bent member, m

z = section modulus is bending $= \frac{I}{y} = \frac{\pi d^3}{32}$ for a circular shaft of diameter d, m^3

Then bending stress, $$\sigma_b = \frac{M}{z} = \frac{32M}{\pi d^3} \text{ N/m}^2 \qquad \text{...(4.3)}$$

4. Torsion

Let T = torque applied to the member, N-m

z_p = polar section modulus, m^3

Using the torsion formula,

$$\frac{T}{J} = \frac{\tau}{r} = \frac{G\theta}{L}$$

where J = polar moment of **inertia**, m^4

$= \frac{\pi d^4}{32}$ for a circular shaft of diameter, d

τ = shear stress, N/m^2

r = distance of the fibres from the centre of shaft, m

G = shear modulus of rigidity, N/m^2

θ = angle of twist, rad

L = length of the shaft, m

z_p = section modulus in torsion

$= \frac{J}{r} = \frac{\pi d^3}{16}$ for a circular shaft of diameter d, m^3

Then torsional shear stress, $$\tau = \frac{T}{z_p} = \frac{16T}{\pi d^3} \text{ N/m}^2 \qquad \text{...(4.4)}$$

(b) Eccentric Axial Loading

Consider a member subjected to an eccentric axial load P at an eccentricity e, as shown in Fig. 4.1(a). According to the principle of statics, the eccentric load P can be replaced by a parallel force P passing through the centroid along with a couple $P \times e$ as shown in Fig. 4.1(b) and (c). In Fig. 4.1(b), the force P causes a simple tensile stress of magnitude $\frac{P}{A}$. In Fig.4.1(c), the couple causes bending stress of magnitude $\frac{P \times e}{z}$. The resultant stresses at the cross-section, Fig. 4.1(d), are obtained by the principle of superposition of stresses. They are given by :

$$\sigma = \frac{P}{A} \pm \frac{P \times e}{z} \qquad \text{...(4.5)}$$

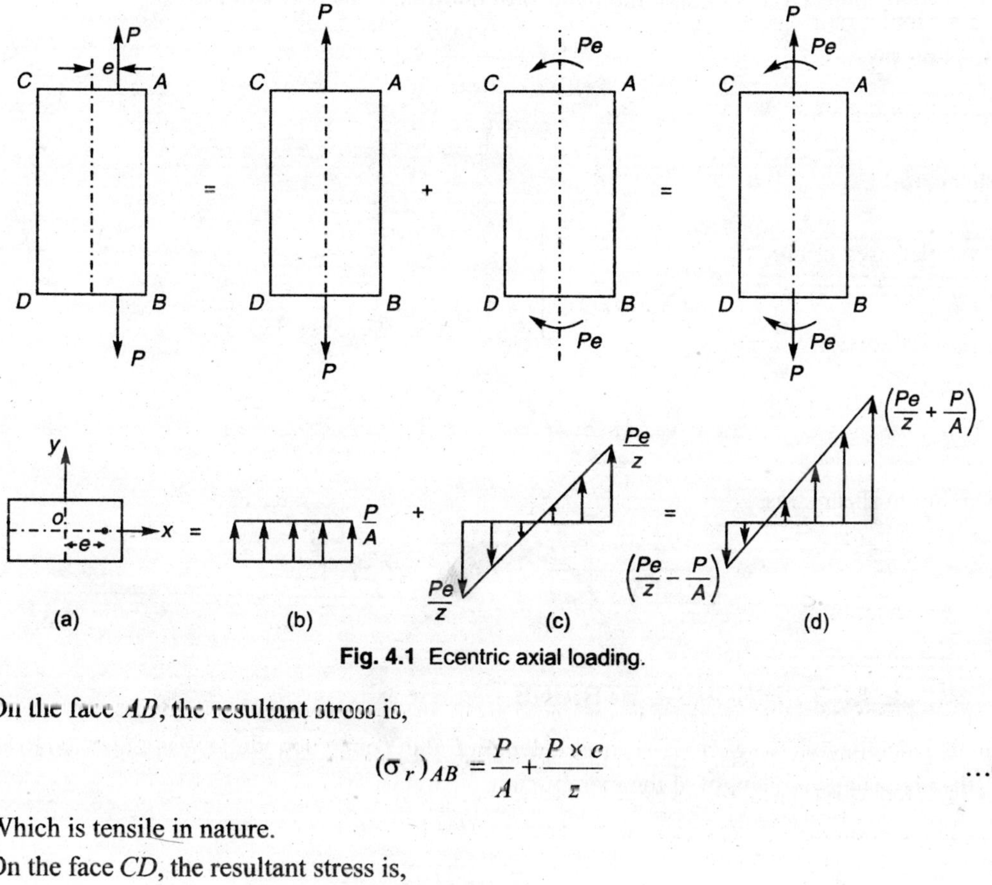

Fig. 4.1 Ecentric axial loading.

On the face AD, the resultant stress is,

$$(\sigma_r)_{AB} = \frac{P}{A} + \frac{P \times e}{z} \qquad \text{...(4.6a)}$$

Which is tensile in nature.

On the face CD, the resultant stress is,

$$(\sigma_r)_{CD} = \frac{P}{A} - \frac{P \times e}{z} \qquad \text{...(4.6b)}$$

Which may be tensile or compressive depending upon the magnitude of bending stress.

(c) Principal Stresses

However complex the state of stress may be at a point in a three-dimensional system, there are always three mutually perpendicular planes in a body on which the normal stresses are either maximum, or minimum. These planes are called the principal planes and the normal stresses acting on them are called the principal stresses

In a two-dimensional system, there are two principal stresses. If σ_x, σ_y and τ_{xy} are the stresses, then the principal stresses are given by :

$$\sigma_{1,2} = \left(\frac{1}{2}\right)(\sigma_x + \sigma_y) \pm \left(\frac{1}{2}\right)[(\sigma_x - \sigma_y)^2 + 4\tau_{xy}^2]^{\frac{1}{2}} \quad \text{...(4.7)}$$

Maximum shear stress,

$$\tau_{max} = \pm\left(\frac{1}{2}\right)[(\sigma_x - \sigma_y)^2 + 4\tau_{xy}^2]^{\frac{1}{2}} \quad \text{...(4.8)}$$

For a shaft subjected to bending moment M and torque T, the stresses are :

Bending stress, $$\sigma_x = \frac{32M}{\pi d^3}$$

$$\sigma_y = 0$$

Shear stress, $$\tau_{xy} = \frac{16T}{\pi d^3}$$

Principal stresses are,

$$\sigma_{1,2} = \left(\frac{1}{2}\right)(\sigma_x + \sigma_y) \pm \left(\frac{1}{2}\right)[(\sigma_x - \sigma_y)^2 + 4\tau_{xy}^2]^{\frac{1}{2}}$$

$$= \frac{16}{\pi d^3}[M \pm (M^2 + T^2)^{\frac{1}{2}}] \quad \text{...(4.9)}$$

Maximum shear stress,

$$\tau_{max} = \left[\frac{16}{\pi d^3}\right][\pm (M^2 + T^2)^{\frac{1}{2}}] \quad \text{...(4.10)}$$

4.3 SIMPLE AND PRINCIPAL STRAINS

If a machine member of area of cross-section A, length L and Young's modulus E is subjected to an axial load P, then the change in length of the member is,

$$\delta L = \frac{PL}{AE}$$

Longitudinal strain, $$\varepsilon = \frac{\text{Change in length}}{\text{Original length}} = \frac{\delta L}{L}$$

Poisson's ratio, $$\nu = \frac{\text{Lateral strain}}{\text{Longitudinal strain}} \quad \text{...(4.11)}$$

Since the lateral strain is opposite in sense to the longitudinal strain, therefore, Poisson's ratio is always negative.

If σ_1, σ_2 and σ_3 are the principal stresses at a point in a body, then the principal strains are :

$$\varepsilon_1 = \left(\frac{1}{E}\right)[\sigma_1 - \nu(\sigma_2 + \sigma_3)] \quad ...(4.12)$$

$$\varepsilon_2 = \left(\frac{1}{E}\right)[\sigma_2 - \nu(\sigma_3 + \sigma_1)]$$

$$\varepsilon_3 = \left(\frac{1}{E}\right)[\sigma_3 - \nu(\sigma_1 + \sigma_2)]$$

4.4 RELATIONSHIP BETWEEN ELASTIC CONSTANTS

Let G = shear modulus, K = bulk modulus, E = elastic modulus, ν = Poisson's ratio

Then

$$G = \frac{E}{2(1+\nu)} \quad ...(4.13)$$

$$K = \frac{E}{3(1-2\nu)} \quad ...(4.14)$$

4.5 WORKING STRESS

While designing machine components, it is desirable to keep the induced stress below the ultimate strength or yield strength of the material. This stress is known as the working (or design, or safe, or allowable) stress.

4.6 FACTOR OF SAFETY

The term factor of safety or design factor are commonly used synonymously to indicate the ratio of estimated strength to computed stress.It is defined as the ratio of maximum strength to the working stress. In case of ductile materials. It is based upon the yield strength of the material. In the case of brittle materials, it is based upon the ultimate strength of the material. Thus

Factor of safety,

$$n = \frac{\text{Ultimate or Yield strength of the material}}{\text{Design stress}}$$

The factor of safety depends upon the various factors as listed below :

1. Variation in material properties.
2. Effect of size in stating material strength properties.
3. Type of loading.
4. Effect of machining and forming processes.
5. Effect of heat treatment upon the physical properties of the material.
6. Effect of wear upon the functions and life of a machine member.
7. Effect of time and environment in which the device is expected to operate.

8. Specific requirements for life and reliability.
9. Overall concern for human safety.

For steady loads, the factor of safety "n" may be calculated from the following formula.

$$n = a \times b \times c \qquad \text{...(4.15)}$$

Where Factor, 'a' = elastic ratio

$$= \frac{\text{Ultimate strength}}{\text{Yield strength}}, \text{ for static loading}$$

$$= \frac{\text{Ultimate strength}}{\text{Endurance limit}}, \text{ for dynamic loading}$$

Factor, 'b' = shock or service factor

$$= \frac{\text{Stress produced by a shock or suddenly applied load}}{\text{Stress produced by a gradual load}}$$

$$= \left[1 + \left(1 + \frac{2h}{\delta_{st}}\right)^{\frac{1}{2}}\right] \text{ for linear impact}$$

$$= \left[1 + \left\{1 + \frac{2h}{r\theta_{st}}\right\}^{\frac{1}{2}}\right] \text{ for angular impact}$$

where h = height fallen by the load

δ_{st}, θ_{st} = static linear or angular deflection respectively

For sudden loading, $h = 0$ and $b = 2$. Average values of b are :

$b = 1.25 - 1.5$ for light shocks

$= 2 - 3$ for heavy shocks

$= 5$ for extreme cases

Factor, 'c' = real margin of safety to account for unreliable material and to account for unforseen causes of failure. Average values of 'c' are :

$c = 1.5 - 2$ for ductile materials of uniform structure

$= 2.5 - 3.0$ for brittle materials

$= 3 - 4$ for timber

Design stress, $$\sigma_d = \frac{S_{ut}}{n} \qquad \text{...(4.16)}$$

$$\text{Margin of safety} = n - 1 \qquad \text{...(4.17)}$$

$$\text{Factor of utilization} = \frac{1}{n} \qquad \text{...(4.18)}$$

Suggested values for factor of safety are given in Table 4.1.

Table 4.1 : Suggested values for factor of safety

Working conditions	Factor of safety
1. Exceptionally reliable materials subjected to well defined loads under controlled conditions.	1.25–1.5
2 Well known materials subjected to known loads under reasonably constant environmental conditions.	1.5–2
3. Average materials subjected to known loads under ordinary environmental conditions.	2–2.25
4. Less tried or brittle materials under average conditions of environment and load.	2.5–3
5. Untried materials under average conditions of environment and load.	3–4
6. Better known materials under uncertain environment and load.	3–4
7. Repeated loads.	1–6
8. Impact loads.	3–6
9. Brittle materials.	Twice the value in 1 to 6

4.7 MODES OF FAILURE

A static load is defined as a force which is gradually applied to a machine component and which remains unaltered over a span of time. The engineering materials can be broadly classified into two groups : *ductile* and *brittle* materials. A material is considered ductile if the tensile strain before fracture is relatively high as compared to a brittle material. As a thumb rule if the tensile strain is more than 5% then the material is considered ductile otherwise brittle. For example structural steel, copper and aluminium, etc. are ductile materials whereas cast iron in brittle in nature.

The ductile materials are distinguished by their clearly defined yield point in the simple tensile test stress—strain diagram where as brittle materials have no clearly defined yield point. For materials having no clearly defined yield point the offset method is used to determine the yield point stress. In this method, a straight line parallel to the linear portion of the stress-strain diagram is drawn at 0.2% of maximum strain to intersect the stress-strain curve. Then a line parallel to the strain-axis is drawn at this point to locate the yield strength of the material on the stress axis.

By failure of a material we do not mean total fracture but whenever the stress exceeds beyond a pre-specified stress level, like the yield strength or ultimate strength, then the material is said to have failed. The only failures of concern to the designer are those in which the design can be assumed to be faulty. The general procedure in this case is to check the balance between the stress and the strength. In the broadest sense, failure of a machine part means any action resulting from service conditions, causing it to cease to function satisfactorily.

There are five types of structural damage which machine parts may experience as listed below :

1. Excessive elastic deformation
2. Elastic buckling.
3. Plastic deformation.

4. Creep, and
5. Fracture.

There are three common type of fractures observed in materials.

(i) Grain boundary fractures

(ii) Cleavage fractures across the grains, and

(iii) Shear fracture across the grains. In iron, the grain boundary fractures are rare at ordinary temperatures, shear fractures are associated with ductile behaviour, and brittle behaviour is characterized by the cleavage mode fracture. Cleavage fractures follow the cube faces of ferrite crystals whereas in the shear fractures, translation always occurs in the direction having the minimum inter-atomic distance and usually along the planes having the greatest atomic density. Consequently, either of the two modes of fracture can be identified by a microscopic examination. The shear mode of fracture is associated with a grey and silky appearance of the fracture surface whereas the cleavage mode causes the surface to be bright and granular.

The factors influencing the type of failure for metals are given in Table 4.2

Table 4.2 : Factors conductive for fracture.

Factors conductive to shear fracture	*Factors conductive to cleavage fracture*
1. High shear stresses	1. High tensile stresses.
2. Low tensile stresses.	2. Low shear stresses.
3. High temperature.	3. Low temperatures
4. Slow rate of loading	4. High rate of loading
	5. High stress concentration
	6. Repeated loads.

4.8 THEORIES OF FAILURE

When a material is subjected to a single type of stress, *i.e,.* axial, bending or torsional stress, then it is a simple matter to anticipate the likely occurrence of failure. However, when the material is subjected to a complex system of stresses then it is difficult to predict the failure of the material straight away. In order to predict the failure of the material under combined stresses, the following theories of failure have been postulated.

(a) For brittle materials :
  (i) Maximum normal stress theory
  (ii) Mohr's theory

(b) For ductile materials :
  (i) Maximum shear stress theory.
  (ii) Maximum distortion energy theory.

1. Maximum Normal Stress Theory

According to this theory failure occurs in a multi-axial system when either a principal tensile stress reaches the uniaxial tensile ultimate strength S_{ut}, or a principal compressive stress reaches the uniaxial ultimate compressive strength, S_{uc}. According to this theory, the strength of the material depends upon only one of the principal stresses (*i.e.,* the maximum tensile or compressive stress) and is entirely independent of the other two principal stresses.

If $\sigma_1 > \sigma_2 > \sigma_3$ are the three principal stresses at a pint, then according to this theory, we have

$$\sigma_1 = S_{ut} \text{ or } \sigma_1 = S_{uc} \quad \text{...(4.19)}$$

Whichever is applicable.

This theory has been represented for the two-dimensional stress system ($\sigma_3 = 0$) in Fig. 4.2. The maximum normal stress theory is represented by the square *ABCD* if it is assumed that $S_{ut} = S_{uc}$.

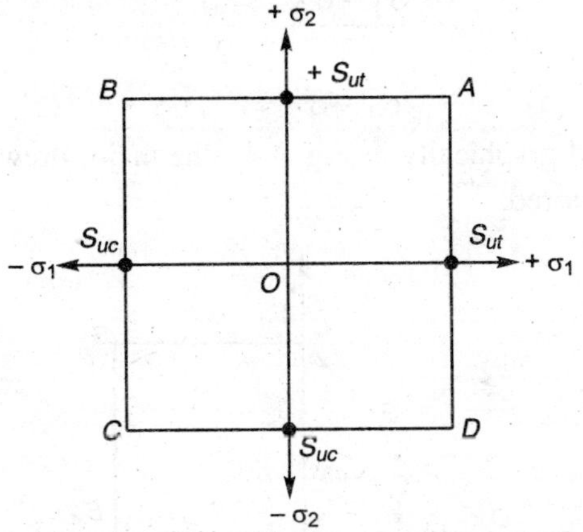

Fig. 4.2 Maximum normal stress theory.

2. Mohr's Theory

This theory suggests that principal Mohr's circles be drawn representing every available test condition and that the envelope of these circles be taken as the envelope of any and all principal Mohr's circles representing stress states on the verge of failure.

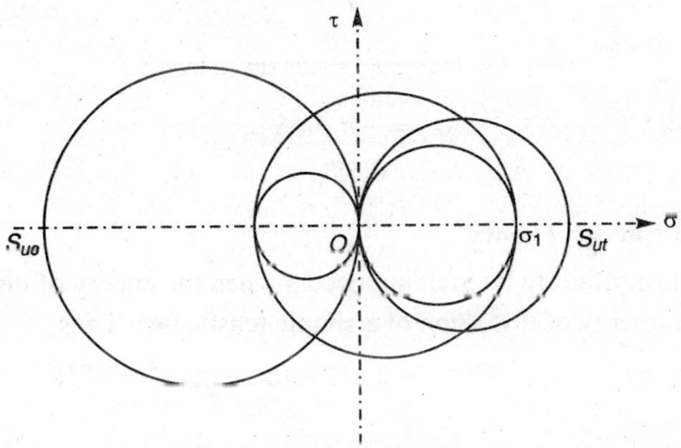

Fig. 4.3 Mohr's theory.

The Mohr's theory is illustrated in Fig. 4.3. Circles of S_{ut} are drawn as shown. Any other stress state for which the Mohr's circle (which envelops circles of diameter σ_1 and σ_2) is tangent to the line of tendency of the S_{ut} and S_{uc} circles, represents a condition of failure. The following condition holds good for the first quadrant.

$$\frac{\sigma_1}{S_{ut}} + \frac{\sigma_2}{S_{uc}} = 1 \qquad \text{...(4.20)}$$

3. Maximum Shear Stress Theory

This theory states that a material subjected to any combination of stresses shall fail (by yielding or fracturing) whenever the maximum shear stress exceeds the shear strength (yield or ultimate) of the material. The shear strength is usually assumed to be determined from the standard uniaxial tension test.

If $\sigma_1 > \sigma_2 > \sigma_3$ and $\sigma_3 = 0$, then according to this theory, we have

$$\frac{\sigma_1 - \sigma_2}{2} = \frac{S_{yt}}{2} \qquad \text{...(4.21)}$$

or

$$\sigma_1 - \sigma_2 = S_{yt}$$

This theory is represented graphically in Fig. 4.4. The major drawback of this theory is that the intermediate stress is totally ignored.

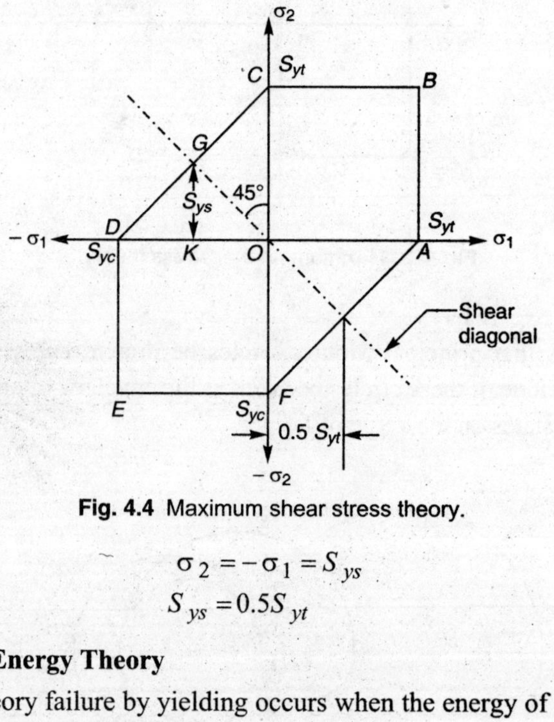

Fig. 4.4 Maximum shear stress theory.

For the torsion case,

$$\sigma_2 = -\sigma_1 = S_{ys}$$

$$S_{ys} = 0.5 S_{yt} \qquad \text{...(4.22)}$$

4. Maximum Distortion Energy Theory

According to this theory failure by yielding occurs when the energy of distortion in a multi-axial stress system exceeds the energy of distortion of a simple tensile test. Thus

$$(\sigma_1 - \sigma_2)^2 + (\sigma_2 - \sigma_3)^2 + (\sigma_3 - \sigma_1)^2 = 2S_{yt}^2 \qquad \text{...(4.23)}$$

$\sigma_3 = 0$, then

If

$$\sigma_1^2 + \sigma_2^2 - \sigma_1\sigma_2 = S_{yt}^2 \qquad \text{...(4.24)}$$

This theory is represented graphically in Fig. 4.5

For the torsion case,

$$\sigma_2 = -\sigma_1 = \tau \qquad \text{...(4.25)}$$

$$S_{ys} = 0.577 S_{yt} \qquad \text{...(4.26)}$$

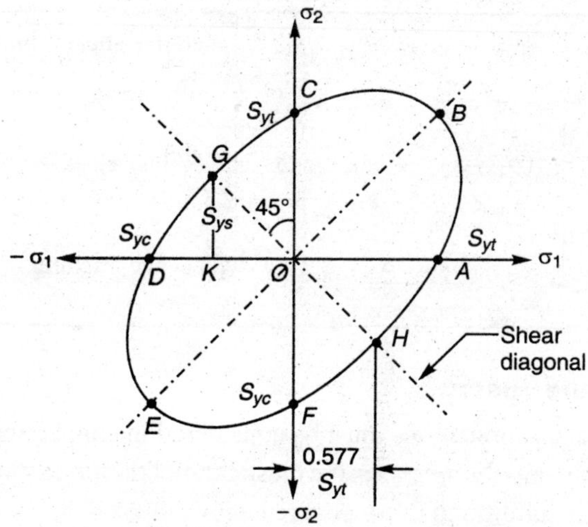

Fig. 4.5 Maximum distortion energy theory.

Other failure of failure are : maximum strain theory and maximum strain energy theory.

5. Maximum Strain Theory

According to this theory failures shall occur when the maximum normal strain under the three principal strains exceeds the normal strain in a uniaxial tensile test. Thus

$$\left(\frac{1}{E}\right)[\sigma_1 - \nu(\sigma_2 + \sigma_3)] = \frac{S_{yt}}{E}$$

or

$$\sigma_1 - \nu(\sigma_2 + \sigma_3)] = S_{yt} \qquad \ldots(4.27)$$

6. Maximum Strain Energy Theory

According to this theory failure shall occur when the strain energy under the three principal stresses shall exceed the strain energy under a uniaxial tensile test. Thus

$$\left[\frac{1}{2E}\right][\sigma_1^2 + \sigma_2^2 + \sigma_3^2 - 2\nu(\sigma_1\sigma_2 + \sigma_2\sigma_3 + \sigma_3\sigma_1)] = \frac{S_{yt}^2}{2E}$$

or

$$\sigma_1^2 + \sigma_2^2 + \sigma_3^2 - 2\nu(\sigma_1\sigma_2 + \sigma_2\sigma_3 + \sigma_3\sigma_1) = S_{yt}^2 \qquad \ldots(4.28)$$

A relative comparison of theories of failure is given in Table 4.3.

Table 4.3 : Comparison of theories of failure.

Theory of failure	*Governing equation*	*Mode of failure*	*Applications*
1. Maximum principal stress theory	$\sigma_1 - S_{ut}$ or S_{uc}	Cleavage fracture	Brittle materials
2. Maximum principal strain theory	$\sigma_1 - \nu(\sigma_2 + \sigma_3) = S_{yt}$	Cleavage fracture	Brittle materials
3. Maximum shear stress theory	$\sigma_1 - \sigma_2 = S_{yt}$	Yielding, shear failure	Ductile materials

4. Maximum strain energy theory	$\sigma_1^2 + \sigma_2^2 + \sigma_3^2 - 2\nu(\sigma_1\sigma_2 + \sigma_2\sigma_3 + \sigma_3\sigma_1) = S_{yt}^2$	Yielding, shear failure	Ductile materials
5. Maximum shear strain energy theory	$(\sigma_1 - \sigma_2)^2 + (\sigma_2 - \sigma_3)^2 + (\sigma_3 - \sigma_1)^2 = 2S_{yt}^2$	Yielding, shear failure	Ductile materials
6. Mohr's theory	$\frac{\sigma_1}{S_{ut}} + \frac{\sigma_2}{S_{uc}} = 1$	Cleavage fracture	Brittle materials

4.8.1 Selection of a Failure Theory

In engineering practice, many implications must be considered in connection with the selection and use of a failure theory. Situations involving excessive elastic deflection, bucking, creep, impact and fatigue all require considerations in addition to those covered above in art. 4.8.

For the predication of failure by yielding of ductile materials, distortion energy theory is generally considered to give the most accurate results. For the common case of biaxial stresses, the Mohr's theory is preferred. The maximum shear stress theory is simple to apply for both biaxial and triaxial stresses, but is more conservative.

For predicting the fracture of brittle materials, Mohr's theory can be used for biaxial stress states. This theory gives quite accurate but conservative results. The maximum normal stress theory gives results which err somewhat on the non-conservative side for biaxial tension/compression.

■ **Example 4.1** An offset link subjected to a force of 30 kN is shown in Fig. 4.6. It is made of grey cast iron FG 350 and the factor of safety is 2.5. Determine the dimensions of the cross-section of the link.

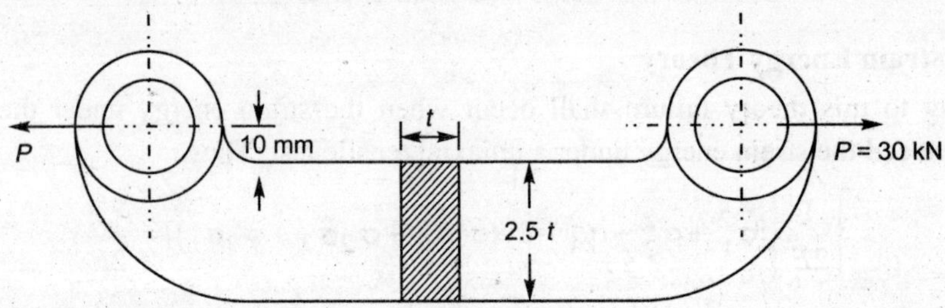

Fig. 4.6

Solution : Working stress for the link,

$$\sigma_w = \frac{\text{Tensile strength}}{\text{Factor of safety}} = \frac{S_{ut}}{n} = \frac{350}{2.5} = 140 \text{ MPa}$$

The cross-section is subjected to direct tensile stress and bending stresses. The top fibres are highly stresses.

Area of cross-section, $A = t \times 2.5t = 2.5t^2 \text{ mm}^2$

Moment of inertia about N-A, $I = t \times \frac{(2.5t)^3}{12} = 1.3t^4 \text{ mm}^4$

Direct stress, $\sigma_d = \frac{P}{A} = \frac{30{,}000}{2.5t^2} = \frac{12{,}000}{t^2} \text{ N/mm}^2$

Eccentricity of load, $e = (10 + 1.25t)$ mm

Bending moment,

$$M = Pe = 30{,}000 \times (10 \times 1.25t) = 30{,}000 + 37500t \text{ N-mm}$$

Bending tensile stress,

$$\sigma_t = \frac{M \times 1.25t}{I} = \frac{30{,}000\,(10 + 1.25t) \times 1.25t}{1.3t^4}$$

$$= \frac{28846(10 + 1.25t)}{t^3} \text{ N/mm}^2$$

Maximum tensile stress at the top fibres

$$= \sigma_d + \sigma_t = \frac{12{,}000}{t^2} + \frac{28846(10 + 1.25t)}{t^3} = 140$$

$$140t^3 = 12{,}000t + 288460 + 36057$$

or $\quad t^3 - 343.26t - 2060.43 = 0$

We solve the cubic equation by hit and trial method.

t, mm	20	21
$f(t)$	-925.63	-7.89

We adopt $t = 21$ mm

■ **Example 4.2** A cylindrical shaft, made of steel of yield strength 760 MPa, is subjected to a static load consisting of bending moment of 15 kN-m and a torsional moment of 25 kN-m. Determine suitable diameter of shaft using maximum shear stress theory and maximum strain energy theory, assuming a factor of safety 2.5 (Take $E = 200$ GPa and Poisson's ratio $= 0.25$).

Solution : Given : $S_{yt} = 760$ MPa, $M = 15$ kN-m, $T = 25$ kN-m, factor of safety, $n = 2.5$,

$$E = 200 \text{ GPa}, \; v = 0.25$$

Design stress, $\sigma_d = \dfrac{S_{yt}}{n} = \dfrac{760}{2.5} = 340$ MPa

Let d = diameter of shaft, mm

Bending stress, $\sigma_b = \dfrac{32M}{\pi d^3} = \dfrac{32 \times 15 \times 10^6}{\pi \times d^3}$ N/mm^2

Shear stress, $\tau = \dfrac{16T}{\pi d^3} = \dfrac{16 \times 25 \times 10^6}{\pi \times d^3}$ N/mm^2

Principal stresses, $\sigma_{1.2} = \dfrac{\sigma_b}{2} \pm \left(\dfrac{1}{2}\right)[\sigma_b^2 + 4\tau^2]^{\frac{1}{2}}$

$$= \left[\frac{16 \times 10^6}{\pi \times d^3}\right][15 \pm \{15^2 + 25^2\}^{1/2}]$$

$$= \left[\frac{16 \times 10^6}{\pi \times d^3}\right][44.155, -14.155]$$

(i) Maximum shear stress theory :

$$\tau_{max} = \frac{\sigma_1 - \sigma_2}{2}$$

$$= \left[\frac{16 \times 10^6}{\pi \times d^3}\right][29.155]$$

$$= S_{ys} = \frac{\sigma_d}{2} = \frac{304}{2} = 152$$

$$d^3 = 0.97687 \times 10^6$$

$$d = 99.2 \approx 100 \text{ mm}$$

(ii) Maximum strain energy theory :

$$\sigma_1^2 + \sigma_2^2 - 2\nu\sigma_1\sigma_2 = \sigma_d^2$$

$$\left[\frac{16 \times 10^6}{\pi \times d^3}\right][(44.155)^2 + (-14.155)^2 + 2 \times 0.25 \times 44.155 \times 14.155] = 304^2$$

$$\left[\frac{16 \times 10^6}{\pi \times d^3}\right][2462.53] = 304^2$$

$$d = 94 \text{ mm}$$

Adopt d 100 mm

■ **Example 4.3** A shaft is required to transmit 15 kW at 200 rpm. It is supported on two bearings 600 mm apart and two gears are keyed to it. A 18 teeth. 14.5° involute, 5 mm module pinion is located at 120 mm to the right of right bearing and delivers power to a gear directly below the shaft. A 72 teeth, 14.5° involute, 5 mm module gear is located at 150 mm to the right of the left bearing and receives power from the gear, directly above it. Determine the shaft diameter if permissible normal stress is 80 MPa.

Solution : Given : $P = 15$ kW, $n_1 = 200$ rpm, $l = 600$ mm, $z_2 = 18$, $m_2 = 5$ mm, $\alpha = 14.5°$, $l_2 = 120$ mm, $z_1 = 72$, $m_1 = 5$ mm, $\alpha_1 = 14.5°$, $l_1 = 150$ mm, $\sigma_d = 80$ MPa

The location of Gears on the shaft is shown in Fig. 4.7 (a).

Gear 1 :

Pitch diameter, $d_1 = m_1 z_1 = 5 \times 72 = 360$ mm

Pitch line speed,
$$v_1 = \frac{\pi d_1 n_1}{60 \times 10^3}$$

$$= \frac{\pi \times 360 \times 200}{60 \times 10^3} = 3.77 \text{ m/s}$$

Tangential force, $F_{t1} = \frac{10^3\, P}{v_1} = \frac{10^3 \times 15}{3.77} = 3978.87$ N (Outwards)

Radial force, $F_{r1} = F_{t1} \tan \alpha_1 = 3987.87 \times \tan 14.5° = 1029$ N (Downwards)

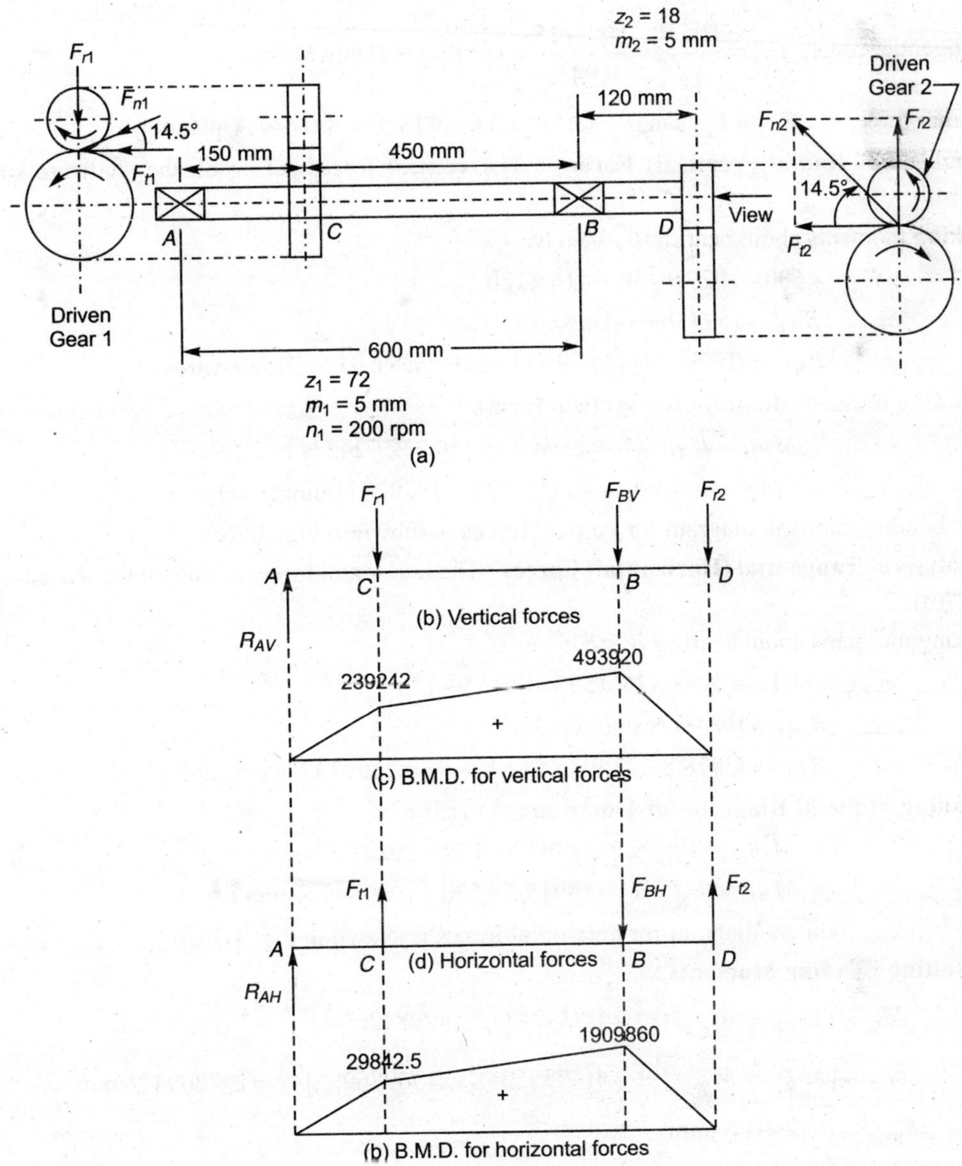

Fig. 4.7

Gear 2 :

Pitch diameter, $d_2 = m_2 z_2 = 5 \times 18 = 90$ mm

Pitch line speed,
$$v_2 = \frac{\pi d_2 n_2}{60 \times 10^3}$$
$$= \frac{\pi \times 90 \times 200}{60 \times 10^3} = 0.9425 \text{ m/s}$$

Tangential force, $F_{t2} = \dfrac{10^3 P}{v_2} = \dfrac{10^3 \times 15}{0.9425} = 15915.5$ N (Outwards)

Radial force, $F_{r2} = F_{t2} \tan \alpha_2 = 15915.5 \times \tan 14.5° = 4116$ N (Upwards)

Analysis of Radial (Vertical) Forces : The vertical forces acting on the shaft are shown in Fig. 4.7(b).

Taking moments about bearing B, we have

$$R_{AV} \times 600 = 1029 \times 450 + 416 \times 120$$

$$R_{AV} = 1594.95 \text{ N (Upwards)}$$

$$R_{BV} = (1029 - 4116) - 1594.95 = -4681.95 \text{ N (Downwards)}$$

Bending moment diagram for vertical forces :

$$M_{CV} = R_{AV} \times l_1 = 1594.95 \times 150 = 239242 \text{ N-mm (+ve)}$$

$$M_B = F_{r2} \times l_2 = 4116 \times 120 = 493920 \text{ N-mm (+ve)}$$

The bending moment diagram for vertical forces is shown in Fig. 4.7(c).

Analysis of Tangential (Horizontal) Forces : The horizontal forces acting on the shaft are shown in Fig. 4.7(d).

Taking moments about bearing B, we have

$$R_{AH} \times 600 = -3978.87 \times 450 + 15915.5 \times 120$$

$$R_{AH} = 198.95 \text{ N (Outwards)}$$

$$R_{BH} = (3978.87 + 15915.5) + 198.95 = 20093.32 \text{ N (inwards)}$$

Bending Moment Diagram for Horizontal Forces :

$$M_{CH} = R_{AH} \times l_1 = 198.95 \times 150 = 29842.5 \text{ N-mm +ve)}$$

$$M_{BH} = F_{t2} \times l_2 = 15915.5 \times 120 = 1909860 \text{ N-mm (+ve)}$$

The bending moment diagram for horizontal forces is shown in Fig. 4.7(e)

Resulting Bending Moments :

$$M_C = [M_{CV}^2 + M_{CH}^2]^{1/2} = [(239243)^2 + (29842.5)^2]^{1/2} = 241096 \text{ N-mm}$$

$$M_B = [M_{BV}^2 + M_{BH}^2]^{1/2} = [(493920)^2 + (1909860)^2]^{1/2} = 1972694 \text{ N-mm}$$

$$M_{max} = 1972694 \text{ N-mm}$$

Torque in the Shaft :

$$T = \frac{10^3 \times P}{2\pi n_1} = \frac{10^3 \times 15}{2\pi \times 200} = 11.937 \text{ N-m}$$

Equivalent bending moment,

$$M_e = \left(\frac{1}{2}\right)[M + \{M^2 + T^2\}^{1/2}]$$

$$= \left(\frac{1}{2}\right)[1972694 + \{(1972694)^2 + (11937)^2\}^{1/2}]$$

$$= 107212 \text{ N-mm}$$

$$= \left(\frac{\pi}{32}\right) d^3 \sigma_d$$

$$= \left(\frac{\pi}{32}\right) \times d^3 \times 80$$

$$d^3 = 254473.5$$

$$d = 63 \text{ mm}$$

Part II : Fluctuating Loading

4.9 INTRODUCTION

The loads in machine members are rarely static. The loads generally vary with time, *i.e.,* fluctuating, alternating or variable loads. Many of the machine components such as axles, shafts, crankshafts, connecting rods, springs, gear teeth, etc. are subjected to fluctuating loads. This problem is further complicated by the presence of stress raisers like notches, grooves, holes and fillet radii, etc. in the machine parts. We shall discuss all these aspects for the design of machine components.

4.10 STRESS-CONCENTRATION

The elementary formulae used in design are based on the assumption that the cross-section of the machine member is constant without any discontinuities. Such conditions are however hardly ever attained. Throughout the stressed region of actual machine members, there are always present some discontinuities in the cross-section in the form of holes, notches grooves, fillet radius, keyway, etc. The stress level at these discontinuities is higher than that of the stress calculated by the normal stress formulae.

Stress-concentration may be defined as the localization of maximum stress at the location of these discontinuities. The cause of this highly localized stress is termed as stress raiser. Stress-concentration is measured by the stress concentration factor, defined by :

$$K_t = \frac{\text{Maximum stress}}{\text{Nominal stress}} = \frac{\sigma_{max}}{\sigma_{nom}} \qquad \ldots(4.29)$$

where subscript 't' stands for 'theoretical'.

K_t is called the geometric or theoretical stress-concentration factor which is based on the usual assumptions in theory of elasticity, *i.e.*, the material is homogeneous, isotropic and linearly elastic. The nominal stress is based on the knowledge from strength of materials. The nominal stress may be either based upon the gross or net area of cross-section.For axial loading the theoretical stress-concentration factor is denoted by K_{ta}, for bending by K_{tb} and for torsion by K_{ts}.

Consider a long thin flat plate of finite width w and thickness t with a central circular hole of diameter d subjected to uniaxial tensile force P, as shown in Fig. 4.8.

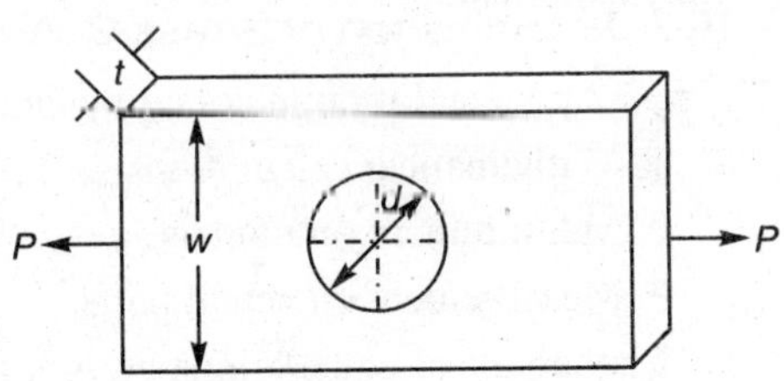

Fig. 4.8 Thin plate with a circular hole.

Gross area of cross-section of plate, $A_g = wt$

Net area of plate, $A_n = (w - d)\,t$

Gross nominal tensile stress in plate, $\sigma_g = \dfrac{P}{A_g}$

Net nominal tensile stress in plate, $\sigma_n = \dfrac{P}{A_n}$

Stress-concentration factor based on gross area,

$$K_{tg} = \frac{\sigma_{max}}{\sigma_g} \qquad \text{...(4.30)}$$

Stress-concentration factor based on net area,

$$K_{tn} = \frac{\sigma_{max}}{\sigma_n} \qquad \text{...(4.31)}$$

$$\frac{K_{tn}}{K_{tg}} = \frac{\sigma_g}{\sigma_n} = \frac{A_n}{A_g}$$

$$= \frac{w-d}{w}$$

$$= 1 - \frac{d}{w}$$

$$= 1 - \lambda \qquad \text{...(4.32)}$$

where $\lambda = \dfrac{d}{w}$ is a fraction by which area is decreased.

4.10.1 Causes of Stress-Concentration

The principal cause of stress-concentration is the presence of some sort of a discontinuity. The principal sources of such discontinuities to produce disturbances in a uniform stress field and most commonly encountered are :

1. Geometric discontinuities as represented by holes, notches, keyways, grooves, threads and fillet radii, etc.
2. Load discontinuities, such as those due to concentrated and discontinuously distributed surface loads.
3. Material discontinuities, such as those due to inclusions or reinforcements by materials whose elastic properties differ from the surrounding matrix.

4.10.2 Determination of Stress-Concentration Factors

The stress-concentration factors may be determined by the following methods :

1. Mathematical calculations.
2. Measurement on models.
3. Measurement on actual parts.

Mathematical calculations are generally limited to relatively simple two-dimensional problems. For stress raisers not solvable by mathematical analysis, stress distribution pattern may be evaluated from experimental measurements on models. Measurements of local stresses on actual parts may be made for which it is difficult to simulate by models. The experimental methods generally used are : photoelasticity, strain gauges, brittle coatings and analogous methods.

4.10.3 Mitigation of Stress-Concentration

The following methods may be used to minimize the stress-concentration :

1. Removal of undesirable material, (Fig. 4.9(a))
2. Added holes, (Fig. 4.9(b))
3. Drilled holes, (Fig. 4.9(c))
4. Using large fillet radius, (Fig. 4.9(d))
5. Combining grooves and fillet radii, (Fig. 4.9(e))
6. Using undercut shoulders, (Fig. 4.9(f))
7. Using larger diameter for threaded length, (Fig. 4.9(g))

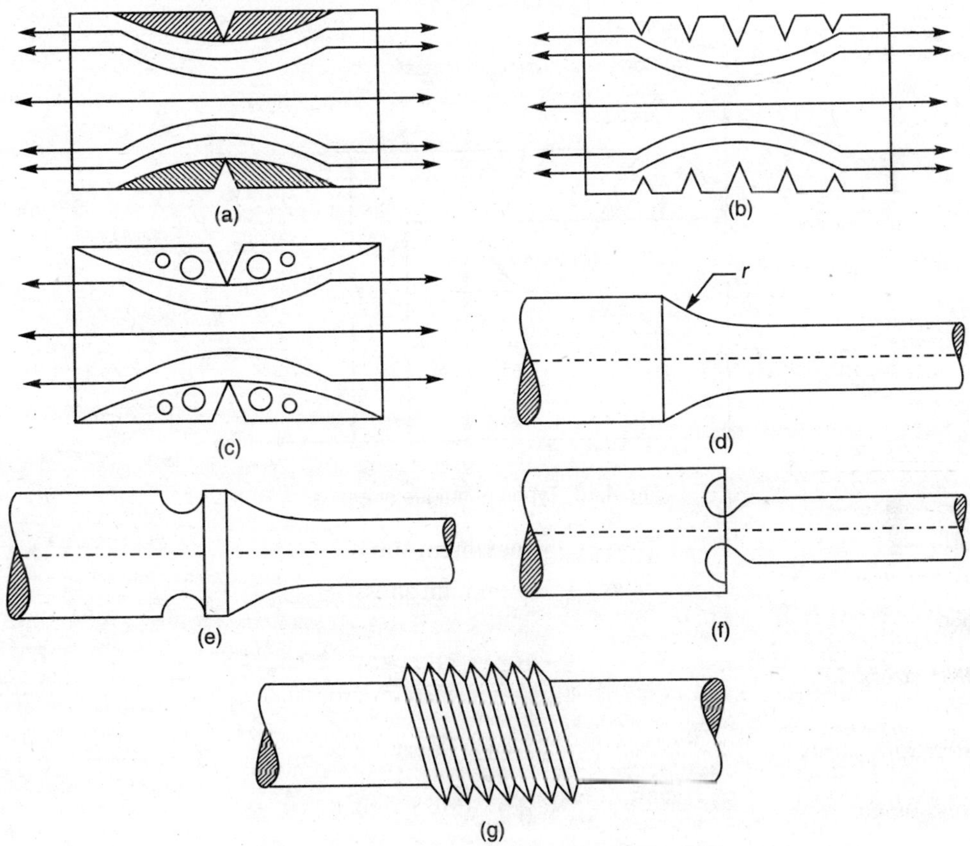

Fig. 4.9 Method of reducing stress-concentration.

4.11 COMPLETELY REVERSED AND FLUCTUATING STRESSES

There are two types of problems in fatigue design :

(i) Components subjected to completely reversed stresses, and

(ii) Components subjected to fluctuating stresses.

The mean stress is zero in case of completely reversed stresses. The stress distribution consists of tensile stresses for the first half cycle and compressive stresses for the remaining half cycle, passing through zero.

The stresses induced in machine components due to forces which are not static but vary in magnitude with respect to time are called fluctuating stresses. These stresses are the combination of static and alternating stresses. The fluctuating stress in which the mean or static stress is zero is called completely reversed stress, as shown in Fig. 4.10(a). The fluctuating stress is shown in Fig. 4.10(b).

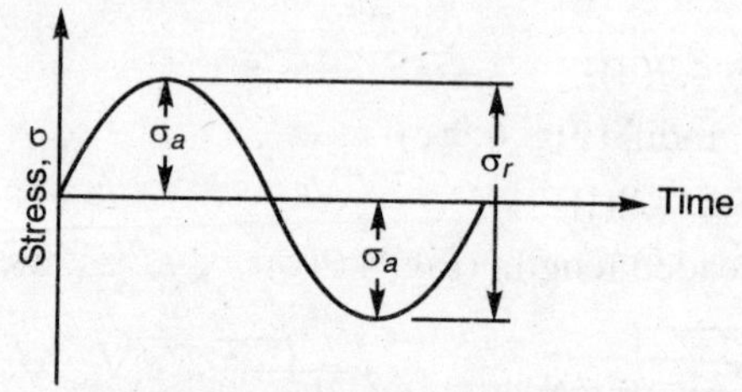

(a) Completely reversed stress

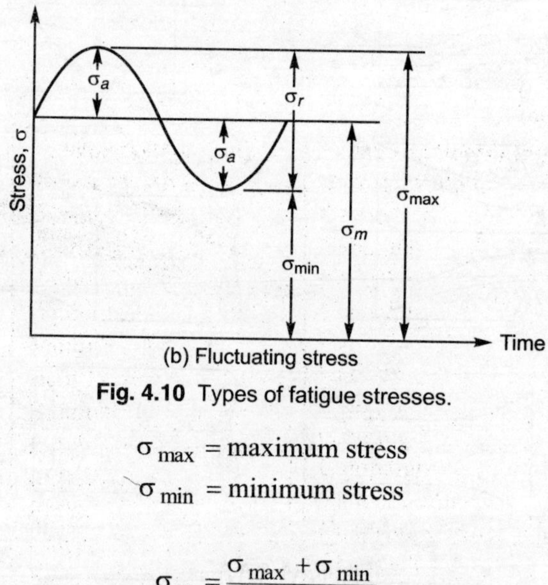

(b) Fluctuating stress

Fig. 4.10 Types of fatigue stresses.

Let σ_{max} = maximum stress

σ_{min} = minimum stress

Then

Mean stress,
$$\sigma_m = \frac{\sigma_{max} + \sigma_{min}}{2} \quad ...(4.33)$$

Alternating stress,
$$\sigma_a = \frac{\sigma_{max} - \sigma_{min}}{2} \quad ...(4.34)$$

Stress range,
$$\sigma_r = \sigma_{max} - \sigma_{min} \quad ...(4.35)$$

4.12 FATIGUE FAILURES

Fatigue failures are progressive in nature which are caused due to repeatedly imposing stresses generally below the yield point of the material. Fatigue failures occur without perceptible yielding or necking. They originate at a localized point and progress gradually through a portion of the member. Ultimate fatigue failures take place because the reduced cross-section can no longer withstand the load. It has been found that the magnitude of the stress causing fatigue failure decreases as the number of stress cycles increases. This phenomenon of decreased resistance of materials to fluctuating stresses is called fatigue. The differences between static and fatigue failures are listed in Table 4.4

Table 4.4 : Static v's fatigue failures

Static failures	*Fatigue failures*
1. The load is gradually applied and there is sufficient time for elongation of fibres.	The fatigue failure begins with a crack at some point in the material.
2. In ductile materials, there is considerable plastic flow prior to fracture.	The crack spreads due to fluctuating stresses.
3. The fracture is silky fibrous structure due to the stretching of crystals at the fractured surface.	Fatigue failure may be either due to region indicating slow growth of crack with a fine fibrous appearance or region of sudden fracture with a coarse granular appearance.
4. Static failures give warning well in advance.	The fatigue failure is sudden and total.

4.13 FATIGUE CONSIDERATIONS IN DESIGN

Fatigue : The reduced tendency of a material to offer resistance to the applied stresses is called fatigue.

The resistance to fatigue failures can be improved by the following methods;

1. Eliminating stress raisers.
2. Designing to take stress raisers into account when they cannot be eliminated.
3. Avoiding surface tears such as those caused by punching, shearing, etc.
4. Controlling heat treatment to prevent decarburization, oxidation, etc.
5. Improving fabricating and fastening details in the design.
6. Protecting the part from corrosion, erosion, etc.
7. Guarding against the formation of surface defects in service.
8. Minimize tensile mean stresses during assembly.

4.14 ENDURANCE LIMIT AND WOHLER'S CURVE

Endurance Limit (S_e) : It may be defined as the maximum value of completely reversed bending stress that the standard specimen of a material can sustain for an infinite number of cycles without fatigue failure.

Fatigue Strength (S_f) : It may be defined as the limiting stress below which a material will withstand a specified number of cycles of stress without fracture.

Fatigue Life (N) : It is the number of cycles of stress that can be sustained by a material prior to failure of a specified nature for a stated condition.

The endurance limit of a material is determined by applying to several specimens repeated bending loads which cause completely reversed stresses of known values, and recording the number of stress reversals each specimen did endure before it failed. Each specimen is subjected to a lower stress than was the preceding one, and it breaks after a greater number of cycles. Finally, a stress is reached that does not cause failure, regardless of how many reversed cycles are applied. This test is performed on R.R. Moore rotating beam fatigue testing machine. The number of stress cycles to determine the endurance limit are very large as compared to the stress. Therefore, the experimental results are plotted on a log-log graph sheet with the stress as ordinates and the number of cycles as abscissas, as shown in Fig. 4.11. This plot is called the *S-N* or Wohler's curve.

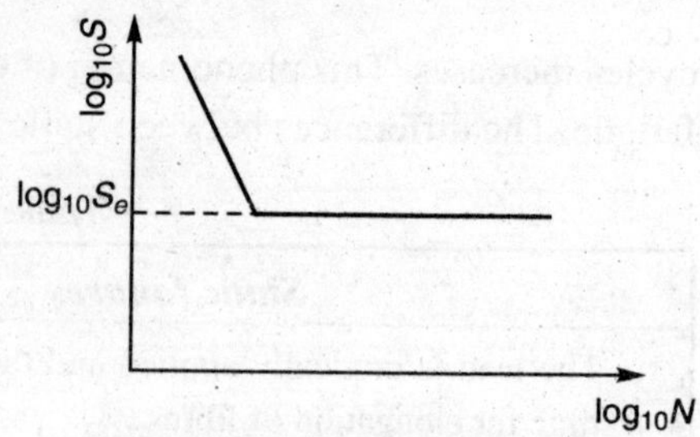

Fig. 4.11 S-N (or Wohler's) curve.

The uncorrected endurance limit in bending or axial loading is represented by S'_e and by S'_{es} in torsional shear. The endurance limit of some of the materials as a ratio of the ultimate tensile strength, S_{ut}, is given in Table 4.5

Table 4.5 : Ratios of S'_e / S_{ut}

Material	S'_e / S_{ut}
1. Wrought iron at 10^6 cycles	0.4
2. Cast steel at 10^6 cycles	0.5
3. Cast iron at 10^6 cycles	0.4 – 0.5
4. Wrought aluminium alloys at 5×10^8 cycles	0.4
5. Copper alloys at 10^8 cycles	0.25 – 0.50

4.15 FATIGUE STRESS-CONCENTRATION FACTOR

Experimental results confirm that reductions in fatigue strength caused by stress raisers are less than the predicted reduction based on theoretical stress-concentration factor, K_t. To deal with the situation, it is customary to use the so called fatigue stress-concentration factor, K_f, defined as :

$$K_f = \frac{\text{Fatigue limit of unnotched specimen (axial or bending)}}{\text{Fatigue limit of notched specimen (axial or bending)}} \quad \text{...(4.36)}$$

4.16 NOTCH SENSITIVITY

The effect of a notch on the fatigue strength of a part varies considerably with materials and notch geometry. It has been observed that it is usually less than the effect that would be predicted by using the value of theoretical stress-concentration factor. This general phenomenon is called 'notch sensitivity'. Notch sensitivity may be considered as a measure of the degree to which the theoretical effect of stress-concentration is obtained. It is defined as :

$$q = \frac{K_f - 1}{K_t - 1} \quad \text{...(4.37)}$$

or

$$K_f = 1 + q(K_t - 1) \quad \text{...(4.38)}$$

The values of q range from 0 to 1. For a completely notch sensitive material, $q = 1$ and experiences the full effect of K_t. For a completely non-notch sensitive material, $q = 0$ and experiences no effect of the stress raiser.

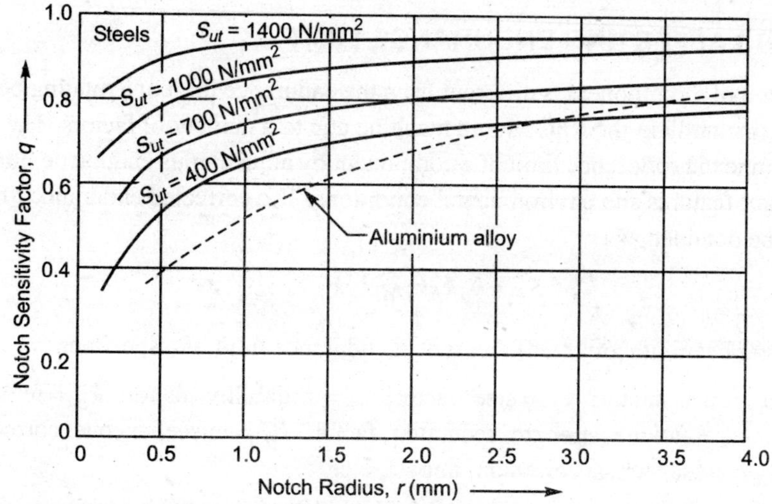

Fig. 4.12 Notch sensitivity charts for reversed bending and reversed axial stresses.

Similarly for torsional loading, we can define that,

$$q = \frac{K_{fs} - 1}{K_{ts} - 1} \quad \text{...(4.39)}$$

where $K_{fs} = \dfrac{\text{Fatigue limit of un-notched specimen (torsion)}}{\text{Fatigue limit of notched specimen (torsion)}}$ ...(4.40)

K_{ts} = Stress-concentration factor (shear stress)

Thus

$$K_{fs} = 1 + q(K_{ts} - 1) \quad \text{...(4.41)}$$

The values of notch sensitivity factor can be obtained from Fig. 4.12 for axial and reversed bending, and for reversed torsion from Fig. 4.13.

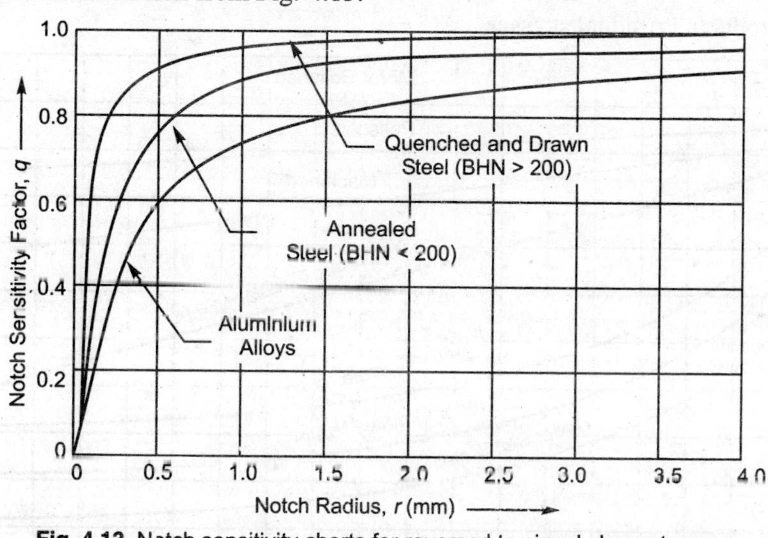

Fig. 4.13 Notch sensitivity charts for reversed torsional shear stresses.

4.17 FACTORS AFFECTING ENDURANCE LIMIT

The endurance limit of a component is different from the endurance limit of a rotating beam specimen as determined experimentally in the R.R. Moore machine due to a number of factors. The derating factors are used to determine the endurance limit of a component by adjusting its enduracne limit to account for loading effects, part features and environmental conditions. The corrected endurance limit of a machine component may be obtained as :

$$S_e = k_a k_b k_c k_d k_e k_f k_g S'_e \quad ...(4.42)$$

where S'_e = rotating beam (uncorrected) theoretical endurance limit of a specimen

k_a = surface finish factor; k_b = size factor; k_c = reliability factor; k_d = temperature factor; k_e = load factor; k_f = fatigue strength reduction factor; k_g = miscellaneous correction factor for press-fitted collar, residual stress, corrosion, impact, etc.

For steel,

$$S_{ut} = 3.5 \text{ BHN MPa} \quad ...(4.43)$$

$$S'_e = 0.5\, S_{ut} \quad ...(4.44)$$

For fluctuating torsional shear stresses, endurance limit is given by;

$$S'_{es} = 0.5\, S'_e, \text{ based on maximum shear stres theory} \quad ...(4.45a)$$

$$= 0.577\, S'_e, \text{ based on distortion energy theory} \quad ...(4.45b)$$

1. Surface Finish Factor, k_a

The surface finish of a part may affect its endurance limit three ways :

(i) by introducing stress-concentration resulting from surface roughness,

(ii) by altering the physical properties of the surface layer of the material, and

(iii) by introducing residual stresses.

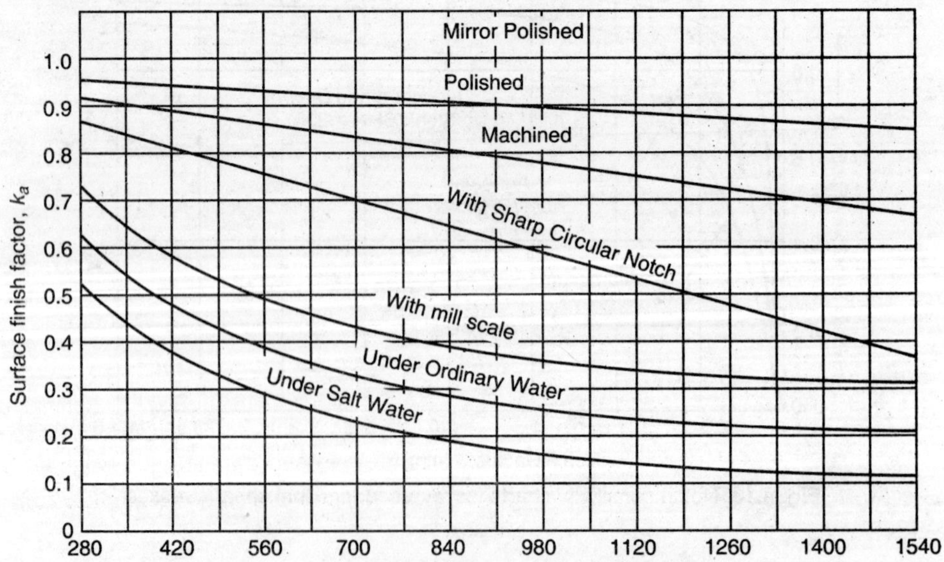

Fig. 4.14 Surface finish factor for various surface conditions.

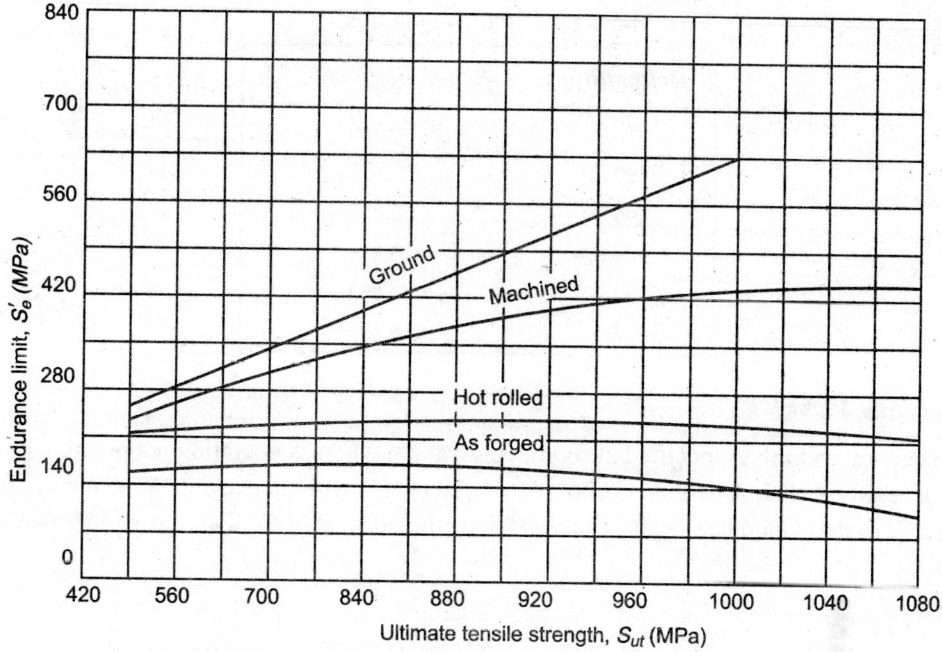

Fig. 4.15 Endurance limit of steel corresponding to ultimate tensile strength.

The surface finish factor is defined as the ratio between the endurance limit obtained with arbitrary surface finish and that obtained with the standard polished finish test specimen. The surface finish factor may be obtained from Figs. 4.14 or 4.15.

2. Size Factor, k_b

The endurance limit under bending and torsional loads tends to decrease as size increases. The size factor may be assumed as given in Table 4.6.

Table 4.6 : Size factor, k_b

Type of loading	*Size factor, k_b*
1. Reversed bending and torsional loads :	
(i) $d \leq 7.5$ mm	1.00
(ii) $7.5 < d \leq 50$ mm	0.85
(iii) $d > 50$ mm	0.75
2. Reversed axial loads for $d < 50$ mm	1.0

3. Reliability Factor, k_c

The rotary beam test data generally reflect a survival rate of 50%. To ensure that more than 50% of the parts will survive, the stress amplitude that the part experiences must be lower than the tabulated endurance limit. The reliability factor is used to achieve this reduction. The reliability factors are given in Table 4.7.

Table 4.7 : Reliability factor, k_c

Reliability, %	*Reliability factor, k_c*
50.00	1.000
90.00	0.897
95.00	0.868
99.00	0.814
99.90	0.753
99.99	0.702

4. Temperature Factor, k_d

The endurance limit is specified at room temperature, whereas operating temperature of machine parts are generally higher. A decline in endurance limit occurs with increasing temperature. To account for reduction in the endurance limit of steels for temperature above 71°C, the temperature factor is defined as :

$$k_d = \frac{344}{273 + T} \qquad \text{...(4.46)}$$

where T = temperature, °C

5. Load Factor, k_e

The load factor is the ratio of the endurance limit under the given loading conditions to the enduracne limit under reversed bending. The load factor may be obtained from Table 4.8.

Table 4.8 : Load factor, k_e

Type of loading	*Load Factor, k_e*
1. Reversed or rotating bending	1.0
2. Reversed axial loading with no bending	0.9
3. Reversed axial loading with indeterminate bending	0.6-0.85
4. Reversed torsion :	
(i) Ductile materials	0.577
(ii) Brittle materials	0.80

6. Fatigue Strength Reduction Factor, k_f

$$k_f = \frac{1}{K_f} \qquad \text{...(4.47)}$$

7. Miscellaneous Correction Factors, k_g

In addition to the above stated factors, there are more factors like impact factor, press-fitted collar factor, residual stresses, corrosion factor, etc., which have to be taken into account depending upon the conditions prevailing in a given problem.

4.18 DESIGN FOR FINITE (OR LIMITED) AND INFINITE (OR UNLIMITED) LIFE

1. Completely reversed stresses

For completely reversed stresses, $\sigma_m = 0$, $\tau_m = 0$ and only σ_a and τ_a are present.

The design problems for completely reversed stresses are divided into two groups :

(i) Design for infinite life, and

(ii) Design for finite life.

(i) Design for Infinite Life : When a component is to be designed for infinite life (number of cycles $\geq 10^6$), the endurance limit becomes the criterion for failure. The stresses induced in such components should be lower than the endurance limit in order to withstand the infinite number of cycles. Such components are designed with the help of the following equations :

Induced stress in bending, $$\sigma_a = \frac{S_e}{n} \qquad ...(4.48)$$

Induced stress in torsion, $$\tau_a = \frac{S_{es}}{n} \qquad ...(4.49)$$

S_e, S_{es} = corrected endurance strengths in bending and torsion respectively.

σ_a, τ_a = amplitude (or alternating) normal and shear stresses respectively.

n = factor of safety

(ii) Design for Finite Life : When a component is to be designed for finite life (number of cycles $< 10^6$), the $\log_{10} S_f - \log_{10} N$ curve for steels, as shown in Fig. 4.16, can be used. It consists of a straight line AB, drawn from $\log_{10}(0.9 S_{ut})$ at $\log_{10}(10^3) = 3$ cycles to $\log_{10}(S_e)$ at $\log_{10}(10^6) = 6$ cycles. The design procedure for problems is as follows :

(a) Locate point A with coordinates $[3, \log_{10}(0.9\, S_{ut})]$.

(b) Locate point B with coordinated $[6, \log_{10}(S_e)]$.

(c) Join AB, which is used as criterion of failure for finite life problems.

(d) For the desired life N cycles of the machine component, draw a vertical line passing through $\log_{10}(N)$ on the abscissa, intersecting line AB at C.

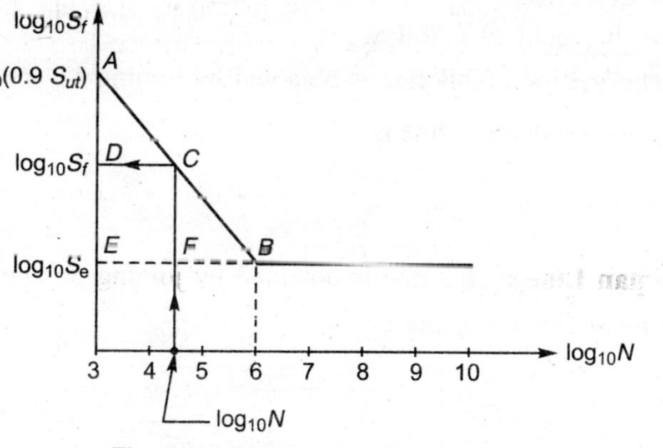

Fig. 4.16 S-N curve for finite file.

(e) Draw a line CD parallel to the abscissa. The ordinate at point D gives the endurance strength corresponding to N cycles. This value of fatigue strength S_f is used for the design calculations.

For finite life,

$$S_f = \frac{10^b}{N^m} \quad ...(4.50)$$

For $10^3 \leq N \leq 10^6$, the values of b and m are given by :

$$b = \log\left[\frac{(0.9\, S_{ut})^2}{S_e}\right] \quad ...(4.51)$$

$$m = \left(\frac{1}{3}\right)\log\left[\frac{0.9S_{ut}}{S_e}\right] \quad ...(4.52)$$

For $N \geq 10^6; S_f = S_e$ ...(4.53)

2. Fluctuating Stresses

When a component is subjected to fluctuating stresses, the stresses are resolved into two parts; mean stress, σ_m and alternating stress amplitude, σ_a. A $\sigma_m - \sigma_a$ diagram can be plotted to predict the failure under fluctuating stresses. In this diagram, the mean stress is plotted on the abscissa, with tensile stresses to the right of origin and compressive stresses to the left of origin. The alternating stress amplitude is plotted on the ordinate.

(i) **Ductile Materials under Axial or Bending Loads :** The $\sigma_m - \sigma_a$ diagram for axial and bending loading for ductile materials having tensile stresses is shown in Fig. 4.17. If the stress is purely static, $\sigma_a = 0$ and a mean stress of S_{yt} will initiate yielding and a mean stress of S_{ut} will cause fracture. If the stress is completely reversed, $\sigma_m = 0$ and the criterion of failure is endurance limit S_e, which is plotted on the ordinate. When the component is subjected to both types of stresses σ_m and σ_a, the failure criterion can be obtained as follows :

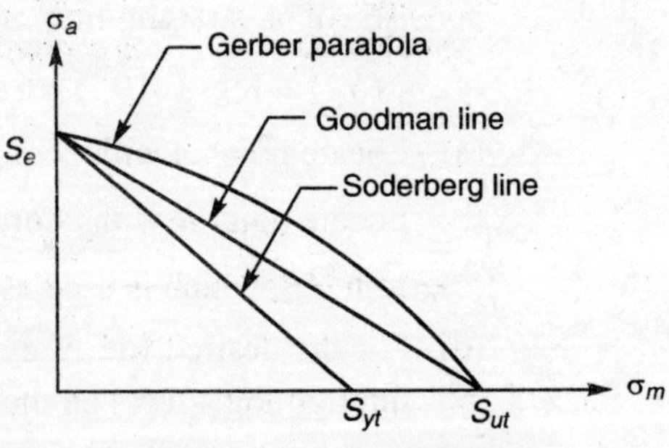

Fig. 4.17 $\sigma_m - \sigma_a$ diagrams.

(a) **Soderberg Line :** This line is obtained by joining S_e and S_{yt} by a straight line. The equation of Soderberg line is :

$$\frac{\sigma_m}{S_{yt}} + \frac{\sigma_a}{S_e} = 1 \quad ...(4.54)$$

(b) **Goodman Line :** This line is obtained by joining S_e and S_{ut} by a straight line. The equation of Goodman line is :

$$\frac{\sigma_m}{S_{ut}} + \frac{\sigma_a}{S_e} = 1 \quad ...(4.55)$$

(c) Gerber Parabola : This is obtained by joining S_e and S_{ut} by a parabolic curve. The equation of Gerber parabola is :

$$\left(\frac{\sigma_m}{S_{ut}}\right)^2 + \frac{\sigma_a}{S_e} = 1 \qquad \text{...(4.56)}$$

Soderberg line is used as a very conservative approximation and has the feature that its use automatically predicts the possibility of failure by general yielding. Its use is recommended sparingly. Goodman line is used where a generalized curve must be used because of lack of data. Gerber parabola is used when some small eccentricities of axial loads exist.

(d) Modified Goodman Diagram for Axial or Bending Loads : To ensure that a machine component does not fail by yielding as well as fatigue, a modified Goodman diagram is used by combining the Soderberg and Goodman lines.

In Fig. 4.18 Goodman line *AB* is the criterion for fatigue fracture, and line *CD* the criterion of static yielding. The equation of line *CD* is :

$$\sigma_m + \sigma_a = S_{yt}$$

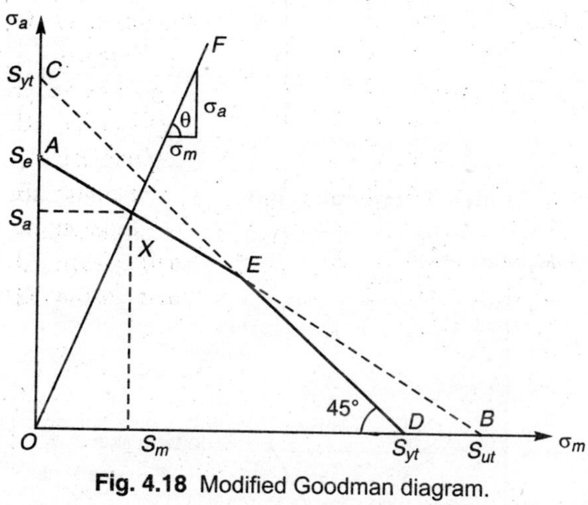

Fig. 4.18 Modified Goodman diagram.

All points below *AED* correspond to fluctuating stresses which should cause neither fatigue fracture nor plastic deformation. Area *OAED* represents the region of safety for components subjected to fluctuating stresses. Line *CD* is inclined at 45° to the abscissa. While solving a problem, a line *OF* with slope $\tan\theta = \frac{\sigma_a}{\sigma_m}$ is constructed.

Since $$\frac{\sigma_a}{\sigma_m} = \frac{\left(\frac{P_a}{A}\right)}{\left(\frac{P_m}{A}\right)} = \frac{P_a}{P_m}$$

Therefore $$\tan\theta = \frac{P_a}{P_m}$$

Similarly $$\tan\theta = \frac{M_a}{M_m}$$

The point of intersection of line OF and AE is X, which indicates the dividing line between the safe region and the region of failure. The coordinates of point $X(S_m, S_a)$ represent the limiting values of stresses which are used to calculate the dimensions of the component. The design stresses are :

$$\sigma_{dm} = \frac{S_m}{n} \text{ and } \sigma_{da} = \frac{S_a}{n}$$

where n = factor of safety.

The complete diagram for both tensile and compressive stresses is shown in Fig. 4.19.

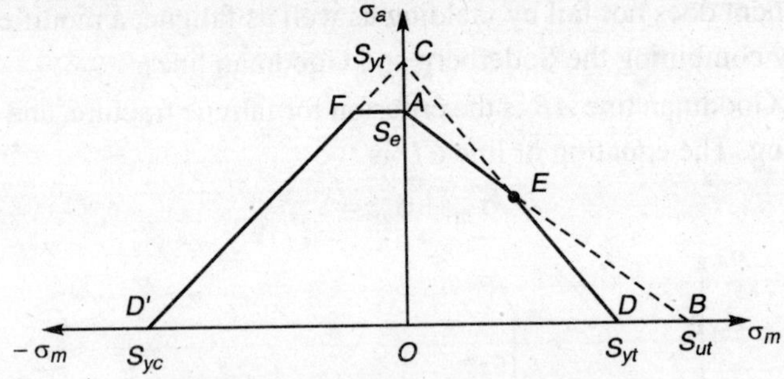

Fig. 4.19 Complete Goodman diagram.

(ii) Ductile Materials Under Torsional Loads : The various failure criterion are shown in Fig. 4.20. AED is the modified Goodman line for torsional loads.

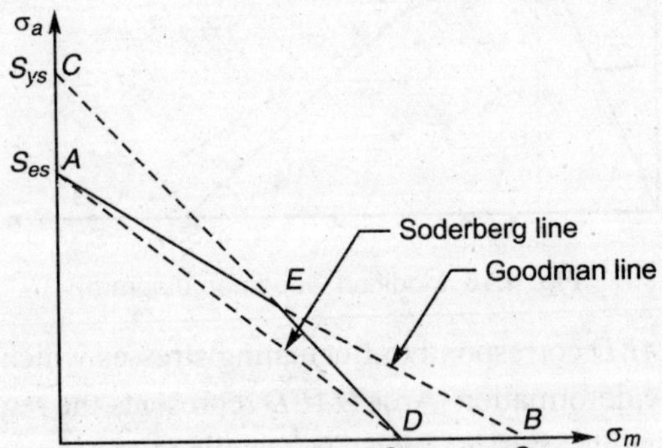

Fig. 4.20 $\tau_m - \tau_a$ diagram for torsional loads.

(a) Soderberg Line : This line is obtained by joining S_{es} and S_{ys} by a straight line. The equation Soderberg line is :

$$\frac{\tau_m}{S_{ys}} + \frac{\tau_a}{S_{es}} = 1 \qquad ...(4.57)$$

(b) Goodman Line : This line is obtained by joining S_{es} and S_{us} by a straight line. The equation of Gordman line as :

$$\frac{\tau_m}{S_{us}} + \frac{\tau_a}{S_{es}} = 1 \qquad \text{...(4.58)}$$

(c) Gerber Parabola : This is obtained by joining S_{es} and S_{us} by a parabolic curve. The equation of Gerber parabola as :

$$\left(\frac{\tau_m}{S_{us}}\right)^2 + \frac{\tau_a}{S_{es}} = 1 \qquad \text{...(4.59)}$$

(d) Modified Goodman Diagram for Torsional Loads : The modified Goodman diagram of fluctuating torsional shear stresses is shown in Fig. 4.21. In this diagram, the torsional yield strength S_{ys} is plotted on both the axes and a line *AB* is constructed to join these two points. This line is inclined at 45° to the abscissa. A line *CD* is drawn through S_{es} on the ordinate and parallel to the abscissa to intersect line *AB* at *D*. The area *OCDB* represents the region of safe design.

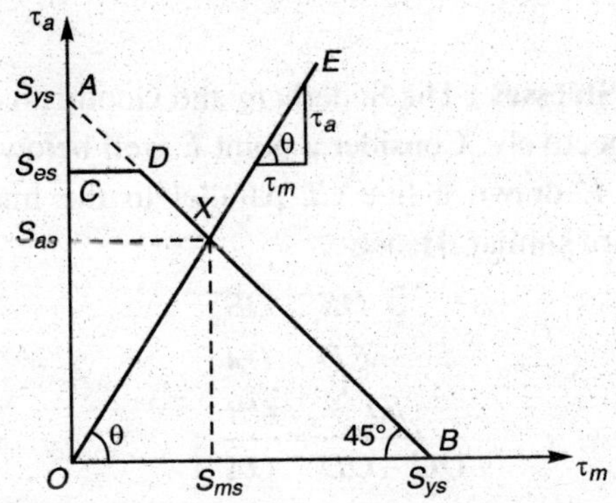

Fig. 4.21 Modified Goodman diagram for torsional loads.

To solve a problem, a line *OE* with slope $\tan\theta = \frac{\tau_a}{\tau_m} = \frac{T_a}{T_m}$ is constructed.

The point of intersection of line *OE* and *DB* is *X*, which indicates the dividing line between the safe region and the region of failure. The coordinates of point $X(S_{ms}, S_{as})$ represent the limiting values of stresses which are used to calculate the dimensions of the component. The design stresses are :

$$\tau_{da} = \frac{S_{as}}{n} \text{ and } \tau_{dm} = \frac{S_{ms}}{n}$$

(iii) $\sigma_m - \sigma_a$ Diagram for Brittle Materials : The $\sigma_m - \sigma_a$ diagram for brittle materials for axial loading is shown in Fig. 4.22. In this case no yielding criteria is involved and a

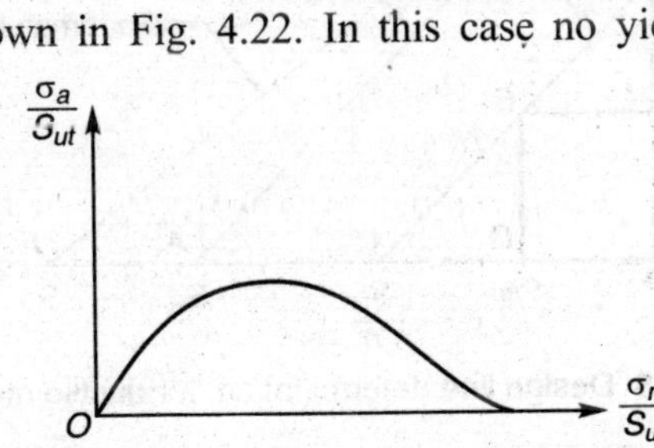

Fig. 4.22 $\sigma_m - \sigma_a$ diagram for brittle materials.

compressive mean stress permits great increase in alternating stress. The diagram for bending loads would be similar. For torsional loads $S_{us} = 0.8\, S_{ut}$ and $S_{es} = 0.8\, S_e$. The Goodman line for torsional loads would be;

$$\frac{\tau_m}{S_{us}} + \frac{\tau_a}{S_{es}} = 1 \quad ...(4.60)$$

4.19 WORKING STRESSES

When a machine component is subjected to fluctuating stresses then the failure of the material shall take place with stress values lying along a line of failure, *i.e.*, Soderberg line or Grodman line. The fatigue stress-concentration factor K_f is applied only to the alternating component of stress in case of ductile materials. However, in case of brittle materials, the fatigue stress-concentation factor is applied to both mean as well as alternating components of stress.

(i) Ductile Materials

(a) Axial or Bending Stresses : The Soderberg and Goodman diagrams are shown in Fig. 4.23 by *AB* and *FB* respectively. Consider a point *E* well below the line *AB* to have a factor of safety *n*. Through *E* drawn a line *CL* parallel to the line *AB*. For the Soderberg line, Δ*sCDE* and *AOB* are similar. Hence

$$\frac{DE}{CD} = \frac{OB}{OA}$$

$$\frac{JO}{OC - OD} = \frac{OB}{OA}$$

$$\frac{K_f \sigma_a}{\frac{S_{yt}}{n} - \sigma_m} = \frac{S_e}{S_{yt}}$$

or,

$$\frac{1}{n} = \frac{\sigma_m}{S_{yt}} + \frac{K_f \sigma_a}{S_e} \quad ...(4.61)$$

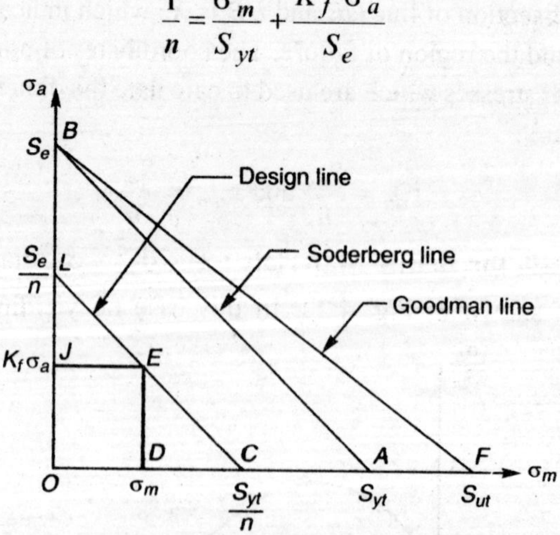

Fig. 4.23 Design line determination, for ductile materials.

Similarly for the Goodman line,

$$\frac{1}{n}=\frac{\sigma_m}{S_{ut}}+\frac{K_f\,\sigma_a}{S_e} \qquad \text{...(4.62)}$$

(b) Torsional Stresses :

Soderberg line :

$$\frac{1}{n}=\frac{\tau_m}{S_{ys}}+\frac{K_{fs}\,\tau_a}{S_{es}} \qquad \text{...(4.63)}$$

Goodman line;

$$\frac{1}{n}=\frac{\tau_m}{S_{us}}+\frac{K_{fs}\,\tau_a}{S_{es}} \qquad \text{...(4.64)}$$

(ii) Brittle Materials

(a) Axial or Bending Streses :

Soderberg line :

$$K_t\left[\frac{\sigma_m}{S_{yt}}+\frac{\sigma_a}{S_e}\right]=\frac{1}{n} \qquad \text{...(4.65)}$$

Goodman line :

$$K_t\left[\frac{\sigma_m}{S_{ut}}+\frac{\sigma_a}{S_e}\right]=\frac{1}{n} \qquad \text{...(4.66)}$$

(b) Torsional Stresses :

Soderberg line :

$$K_{ts}\left[\frac{\tau_m}{S_{ys}}+\frac{\tau_a}{S_{es}}\right]=\frac{1}{n} \qquad \text{...(4.67)}$$

Goodman line

$$K_{ts}\left[\frac{\tau_m}{S_{us}}+\frac{\tau_a}{S_{es}}\right]=\frac{1}{n} \qquad \text{...(4.68)}$$

where $S_{us}=0.8\,S_{ut}$ and $S_{es}=0.8\,S_{us}$

(iii) Equivalent Stresses

(a) Axial or Bending Stresses :

(i) Soderberg line

Equivalent mean stress,

$$\sigma_{me}=\frac{S_{yt}}{n}=\sigma_m+K_t\left(\frac{S_{yt}}{S_e}\right)\sigma_a \qquad \text{...(4.69)}$$

Equivalent alternating stress.

$$\sigma_{ae}=\left(\frac{S_e}{S_{yt}}\right)\sigma_m+K_f\sigma_a \qquad \text{...(4.70)}$$

(ii) Goodman line

Equivalent mean stress,

$$\sigma_{me}=\frac{S_{ut}}{n}=\sigma_m+K_f\left(\frac{S_{ut}}{S_e}\right)\sigma_a \qquad \text{...(4.71)}$$

Equivalent alternating stress,

$$\sigma_{ae} = \left(\frac{S_e}{S_{ut}}\right)\sigma_m + K_f \sigma_a \qquad \text{...(4.72)}$$

(b) Torsional Stresses : Equivalent mean shear stress,

$$\tau_{me} = \frac{S_{ys}}{n} = \tau_m + K_{fs}\left(\frac{S_{ys}}{S_{es}}\right)\tau_a \qquad \text{...(4.73)}$$

Equivalent alternating shear stress,

$$\tau_{ae} = \left(\frac{S_{es}}{S_{ys}}\right)\tau_m + K_{fs}\tau_a \qquad \text{...(4.74)}$$

Caution : While calculating the corrected endurance limit, the fatigue reduction factor should be ignored as it is beaing accounted for while calculating working stresses.

4.20 COMBINED FLUCTUATING STRESSES

(a) Soderberg Equation

1. Axial and Bending Loading

Consider a bar subjected to the combined action of fluctuating axial and bending loads shown in Fig. 4.24. The combined stress in the bar is,

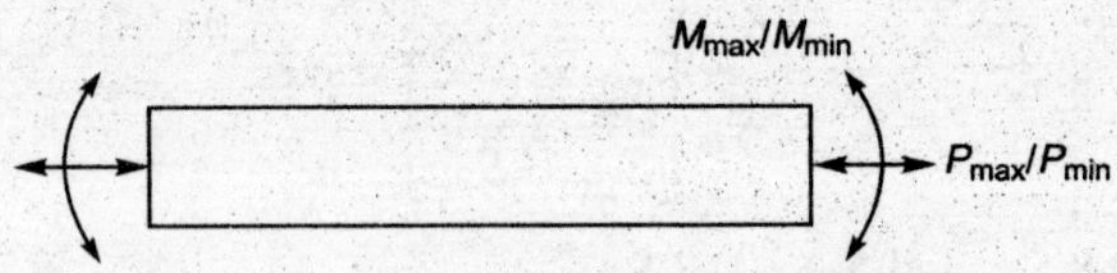

Fig. 4.24 Bar under combined fluctuating axial and bending loads.

$$\frac{P_{eq}}{A} + \frac{M_{eq}}{z} = \frac{S_{yt}}{n} \qquad \text{...(4.75)}$$

where $P_{eq} = P_m + \left(\frac{S_{yt}}{S_{ae}}\right)P_a$ = equivalent axial load. ...(4.76a)

$M_{eq} = M_m + \left(\frac{S_{yt}}{S_e}\right)M_a$ = equivalent bending moment. ...(4.76b)

$S_{ae} = 0.8\, S_e$ ...(4.77)

2. Two-dimensional Normal Stresses

Consider a body subjected to the combined action of fluctuating two dimensional normal stresses shown in Fig. 4.25.

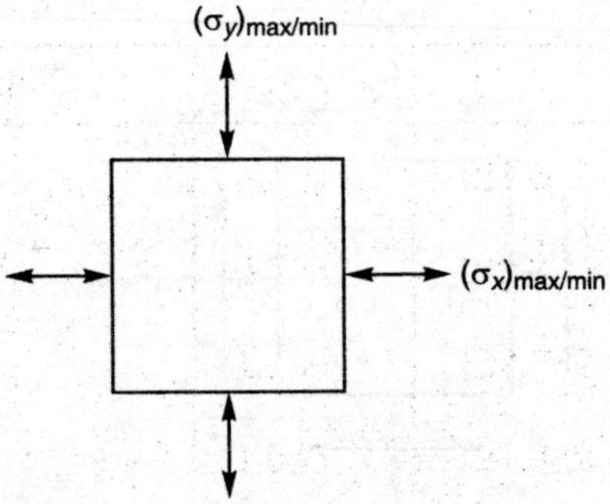

Fig. 4.25 Two-dimensional fluctuating normal stresses on an element.

$$\sigma_{xm} = \frac{(\sigma_x)_{max} + (\sigma_x)_{min}}{2}$$

$$\sigma_{xa} = \frac{(\sigma_x)_{max} - (\sigma_x)_{min}}{2}$$

$$\sigma_{ym} = \frac{(\sigma_y)_{max} + (\sigma_y)_{min}}{2}$$

$$\sigma_{ym} = \frac{(\sigma_y)_{max} - (\sigma_y)_{min}}{2}$$

The distortion energy theory for general three-dimensional state of stress is given by,

$$(\sigma_x - \sigma_y)^2 + (\sigma_y - \sigma_z)^2 (\sigma_z - \sigma_x)^2 + 6(\tau_{xy}^2 + \tau_{yz}^2 + \tau_{zx}^2) = 2\sigma^2$$

For $\sigma_z = \tau_{xy} = \tau_{yz} = \tau_{zx} = 0$, we get

$$\sigma^2 = \sigma_x^2 + \sigma_y^2 - \sigma_x \sigma_y$$

For two-dimensional fluctuating state of stress, we have

$$\sigma_m = [\sigma_{xm}^2 + \sigma_{ym}^2 - \sigma_{xm}\sigma_{ym}]^{1/2}$$

$$\sigma_a = [\sigma_{xa}^2 + \sigma_{ya}^2 - \sigma_{xa}\sigma_{ya}]^{1/2}$$

Soderberg equation using maximum distortion energy theory gives,

$$\sigma_m + \left(\frac{S_{yt}}{S_e}\right)\sigma_a = \frac{S_{yt}}{n} \quad \text{...(4.78)}$$

3. Two-dimensional Normal Stresses and Shear Stress (Fig. 4.26)

$$\tau_{xym} = \frac{(\tau_{xy})_{max} + (\tau_{xy})_{min}}{2}$$

$$\tau_{xya} = \frac{(\tau_{xy})_{max} - (\tau_{xy})_{min}}{2}$$

$$\sigma_m = [\sigma_{xm}^2 + \sigma_{ym}^2 - \sigma_{xm}\sigma_{ym} + 3\tau_{xym}^2)^{1/2}$$

$$\sigma_a = [\sigma_{xa}^2 + \sigma_{ya}^2 - \sigma_{xa}\sigma_{ya} + 3\tau_{xya}^2)^{1/2}$$

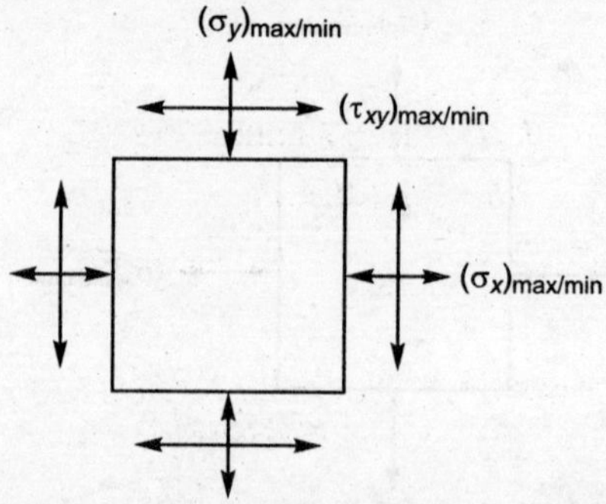

Fig. 4.26 Two-dimensional fluctuating complex stresses on an element.

Soderberg equation using maximum distortion energy theory gives,

$$\frac{\sigma_m}{S_{yt}} + \frac{\sigma_a}{S_e} = \frac{1}{n} \quad \text{...(4.79)}$$

If stress-concentration factors are given in bending and torsion, then for ductile materials, the stress-concentration factors are applied only to the alternating components of stress.

$$\sigma_m = [\sigma_{xm}^2 + \sigma_{ym}^2 - \sigma_{xm}\sigma_{ym} + 3\tau_{xym}^2]^{1/2} \quad \text{...(4.80a)}$$

$$\sigma_a = [(K_{tb}\sigma_{xa})^2 + (K_{tb}\sigma_{ya})^2 - K_{tb}^2\sigma_{xa}\sigma_{ya} + 3(K_{ts}\tau_{xya})^2]^{1/2} \quad \text{...(4.80b)}$$

■ **Example 4.4** A stepped thin flat plate with a circular hole is subjected to a uniaxial tensile force of 6 kN as shown in Fig. 4.27. The material of plate is *FG* 200 and a factor of safety of 2.5 is desired. Determine the thickness of plate.

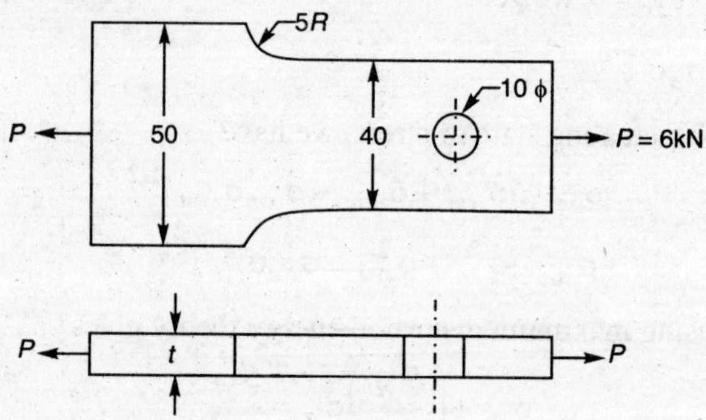

Fig. 4.27

Solution : Given : $P = 6$ kN, $r = 5$ mm, $d = 40$ mm, $D = 50$ mm, $d = 10$ mm, $w = 40$ mm,

$$S_{ut} = 200 \text{ MPa}, n = 2.5$$

Fillet :

$$\frac{r}{d} = \frac{5}{40} = 0.125 \text{ and } \frac{D}{d} = \frac{50}{40} = 1.25$$

For a stepped plate under tension, $K_t = 1.75$

Maximum stress at the fillet,

$$\sigma_{max} = \frac{K_t P}{dt} = \frac{1.75 \times 6000}{(40 \times t)} = \frac{262.5}{t} \text{ N/mm}^2$$

Hole : $\frac{d}{w} = \frac{10}{40} = 0.25$

For a finite width plate under tension with a hole, $K_t = 2.42$

Maximum stress at hole

$$\sigma_{max} = \frac{K_t P}{(w-d)\,t} = \frac{2.42 \times 6{,}000}{(40-10)\,t} = \frac{484}{t}$$

We observe that the stress at the hole is maximum, which shall be the basis for design to calculate the plate thickness.

Design stress, $\sigma_d = \frac{S_{ut}}{n} = \frac{200}{2.5} = 80$ MPa

$$\frac{484}{t} = 80$$

$$t = 6.05 \approx 6 \text{ mm}$$

■ **Example 4.5** A non-rotating stepped axle supports a transverse load of 3 kN at the centre as shown in Fig. 4.28. The tensile yield strength of the axle material is 300 MPa and a factor of safety of 3 is desired. Determine the diameter of the axle.

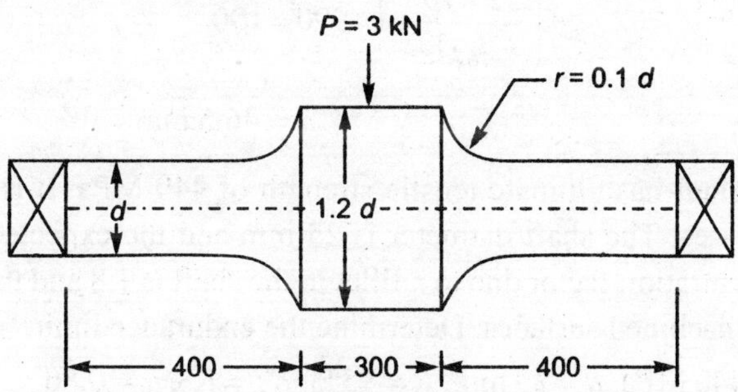

Fig. 4.28

Solution : Given : $\frac{r}{d} = 0.1, \frac{D}{d} = 1.2, P = 3$ kN, $S_{yt} = 300$ MPa, $n = 3$

Design stress, $\sigma_d = \frac{S_{yt}}{n} = \frac{300}{3} = 100$ MPa

For a circular memeber with a fillet radius under bending, $K_t = 1.65$

Bearing reactions due to symmetry of loading $= 1.5$ kN

Bending moment at the fillet, $M_f = 1.5 \times 400 = 600$ kN. mm

Bending moment at the centre,

$$M_c = \frac{PL}{4} = \frac{3 \times 1100}{4} = 825 \text{ kN mm}$$

Maximum bending stress at the fillet,

$$\sigma_{max} = \frac{K_t \times 32 M_f}{\pi d^3}$$

$$= \frac{1.65 \times 32 \times 600 \times 10^3}{\pi \times d^3}$$

$$= \left(\frac{32 \times 10^3}{\pi d^3} \right) \times 990 \text{ MPa}$$

Maximum bending stress at the centre,

$$\sigma_{max} = \frac{32 M_c}{\pi d^3}$$

$$= \frac{32 \times 825 \times 10^3}{\pi \times d^3}$$

$$= \left(\frac{32 \times 10^3}{\pi d^3} \right) \times 825 \text{ MPa}$$

We observe that the stress is maximum at the fillet radius.

$$\sigma_{max} = \sigma_d$$

$$\left(\frac{32 \times 10^3}{\pi d^3} \right) \times 990 = 100$$

$$d = 46.5 \text{ mm}$$

■ **Example 4.6** A steel shaft has ultimate tensile strength of 440 MPa. It is subjected to completely reversed torsional stresses. The shaft diameter is 25 mm and the expected reliability is 95%. The theoretical stress concentration factor due to a fillet in the shaft is 1.8 and the notch sensitivity factor is 0.86. The shaft has machined surfaces. Determine the endurance limit.

Solution : Given : $S_{ut} = 440$ MPa, $d = 25$ mm, $R = 95\%$, $K_t = 1.8$, $q = 0.86$

$$S'_e = 0.5\, S_{ut} = 0.5 \times 440 = 220 \text{ MPa}$$

For machined surface, surface finish factor for $S_{ut} = 440$ MPa,

$$k_a = 0.82$$

For $d = 25$ mm, size factor, $k_b = 0.85$

For $R = 95\%$, reliability factor, $k_c = 0.868$

Temperature factor, $k_d = 1.0$

For torsion, load factor, $k_e = 0.577$

$$K_f = 1 + q(K_t - 1)$$

$$= 1 + 0.86(1.8 - 1) = 1.688$$

Fatigue strength reduction factor,

$$k_f = \frac{1}{K_f} = \frac{1}{1.8} = 0.592$$

Endurance limit, $S_e = k_a k_b k_c k_d k_f S'_e$

$$= 0.82 \times 0.85 \times 0.868 \times 1.0 \times 0.577 \times 0.592 \times 220$$

$$= 45.46 \text{ MPa}$$

■ **Example 4.7** A steel having ultimate tensile strength of 440 MPa is hot-rolled and normalized. It is 60 mm wide and has a central circular hole of 12 mm diameter as shown in Fig. 4.29. It is subjected to a completely reversed axial load of 20 kN. The notch sensitivity factor is 0.8 and the expected reliability is 90%. The factor of safety is 2. Determine the thickness of plate.

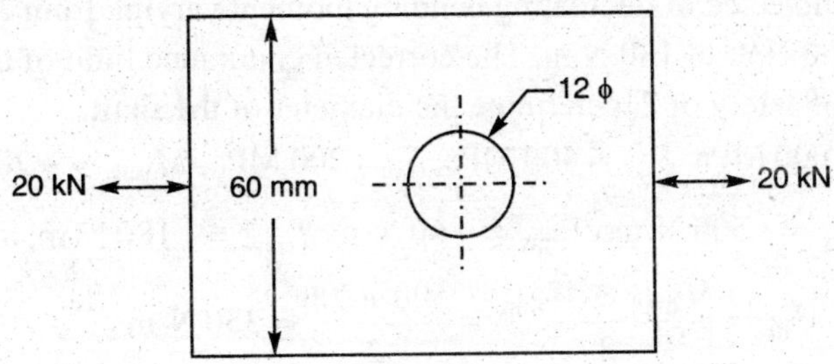

Fig. 4.29

Solution : Given : $S_{ut} = 440$ MPa, $w = 60$ mm, $d = 12$ mm, $P = 20$ kN, $q = 0.8$, $R = 90\%$, $n = 2$

$$S'_e = 0.5\, S_{ut} = 0.5 \times 440 = 220 \text{ MPa}$$

For hot rolled and noramlized condition, surface finish factor for $S_{ut} = 440$ MPa,

$$k_a = 0.67$$

Assuming $t < 50$ mm, size factor, $k_b = 0.85$

For $R = 90\%$, reliability factor, $k_c = 0.897$

Temperature factor, $k_d = 1.0$

For axial loading, load factor, $k_e = 0.9$

For the circular hole, $\frac{d}{w} = \frac{12}{60} = 0.2$

$$K_t = 2.52$$

$$K_f = 1 + q(K_t - 1)$$

$$= 1 + 0.8\,(2.52 - 01) = 2.216$$

Fatigue strength reduction factor, $k_f = \frac{1}{K_f} = \frac{1}{2216} = 0.451$

Endurance limit, $S_e = k_a k_b k_c k_d k_e k_f S'_e$

$$= 0.67 \times 0.85 \times 0.897 \times 1.0 \times 0.9 \times 0.451 \times 220$$

$$= 45.64 \text{ MPa}$$

Design stress, $\sigma_d = \frac{S_e}{n} = \frac{45.64}{2} = 22.82$ MPa

Stress in plate at hole location,

$$\sigma_{max} = \frac{P}{(w-d)t}$$

$$= \frac{20{,}000}{(60-12)t}$$

$$= 22.82$$

$$t = 18.26 \simeq 20 \text{ mm}$$

Example 4.8 A shaft made of steel having ultimate tensile strength 600 MPa and tensile yield strength 400 MPa is subjected to fluctuating bending moment varying from 200 N.m to 500 N.m and torque varying from 60 N.m to 180 N.m. The corrected endurance limit of the shaft material is 200 MPa. Using a factor of safety of 2, determine the diameter of the shaft.

Solution : Given : $S_{ut} = 600$ MPa, $S_{yt} = 400$ MPa, $S_e = 200$ MPa, $M_{min} = +200$ N.m,

$M_{max} = +500$ N.m, $T_{min} = +60$ N.m, $T_{max} = +180$ N.m, $n = 2$

$$M_m = \frac{M_{max} + M_{min}}{2} = \frac{500+200}{2} = 350 \text{ N.m}$$

$$M_a = \frac{M_{max} - M_{min}}{2} = \frac{500-200}{2} = 150 \text{ mN.m}$$

$$T_m = \frac{T_{max} + T_{min}}{2} = \frac{180+60}{2} = 120 \text{ N.m}$$

$$T_a = \frac{T_{max} - T_{min}}{2} = \frac{180-60}{2} = 60 \text{ N.m}$$

$$\sigma_{xm} = \frac{32M_m}{\pi d^3} = \left[\frac{32\times10^3}{\pi d^3}\right] \times 350 \text{ MPa}$$

$$\sigma_{xa} = \frac{32M_a}{\pi d^3} = \left[\frac{32\times10^3}{\pi d^3}\right] \times 150 \text{ MPa}$$

$$\tau_{xym} = \frac{16T_m}{\pi d^3} = \left[\frac{32\times10^3}{\pi d^3}\right] \times 60 \text{ MPa}$$

$$\tau_{xya} = \frac{16T_a}{\pi d^3} = \left[\frac{32\times10^3}{\pi d^3}\right] \times 30 \text{ MPa}$$

$$\sigma_m = (\sigma_{xm}^2 + 3\tau_{xym}^2)^{1/2}$$

$$= \left[\frac{32\times10^3}{\pi d^3}\right] [350^2 + 3\times60^2]^{1/2}$$

$$= \left[\frac{32 \times 10^3}{\pi d^3}\right] \times 365.1\,\text{MPa}$$

$$\sigma_a = (\sigma_{xa}^2 + 3\tau_{xya}^2)^{1/2}$$

$$= \left[\frac{32 \times 10^3}{\pi d^3}\right] [150^2 + 3 \times 30^2]^{1/2}$$

$$= \left[\frac{32 \times 10^3}{\pi d^3}\right] \times 158.745\ \text{MPa}$$

$$\tan\theta = \frac{\sigma_a}{\sigma_m} = \frac{158.745}{365.1} = 0.4348$$

$$\theta = 23.5°$$

The modified Goodman diagram is shown in Fig. 4.30

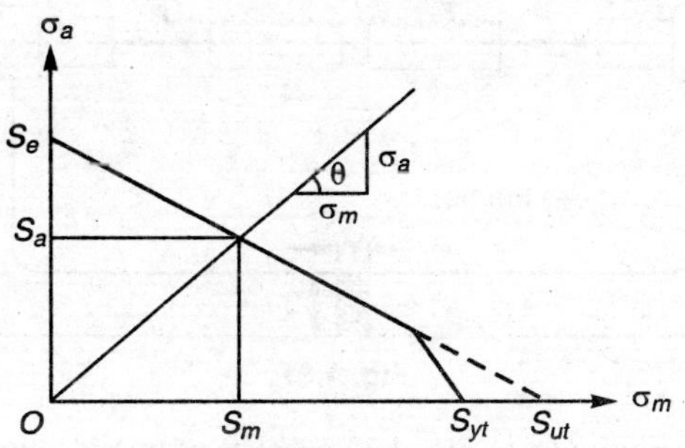

Fig. 4.30

$$\frac{S_m}{S_{ut}} + \frac{S_a}{S_e} = 1$$

$$\frac{S_m}{600} + \frac{S_a}{200} = 1$$

Also

$$\frac{S_a}{S_m} = 0.4348$$

$$\frac{S_m}{600} + \frac{0.4348 \times S_m}{200} = 1$$

$$S_m = 260.37\ \text{MPa}$$

$$S_a = 113.21\ \text{MPa}$$

Hence

$$\sigma_a = \frac{S_a}{n}$$

$$\left[\frac{32 \times 10^3}{\pi d^3}\right] \times 158.745 = \frac{113.21}{2}$$

$$d = 30.57 \approx 31\ \text{mm}$$

SOLVED QUESTION FROM UNIVERSITY EXAMINATIONS

■ **Example 4.1** What is factor of safety? Explain the factor influenceing the choice of factor of safety.

[MDU, Dec., 2008]

Solution : Refer to art. 4.6.

EXERCISES

1. The frame of a hacksaw blade shown in Fig. 4.31 is of U-shape and rectangular in cross-section having thickness 3 times the width. The vertical leg of the frame is 150 mm. The initial tension in the blade is to be 300 N. The tensile yield strength of the frame is 380 MPa and a factor of safety 3 is desired. Determine the cross-section.

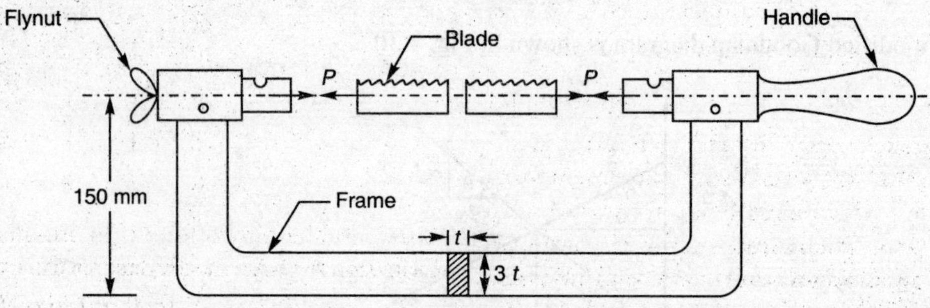

Fig. 4.31

2. The frame of a hydraulic press shown in Fig. 4.32 consists of two identical steel plates 150 mm wide. The maximum load on the plates is 25 kN at an eccentricity of 275 mm. The tensile yield strength of frame is 400 MPa. Calculate the thickness of plates if a factor of safety of 2.5 is desired.

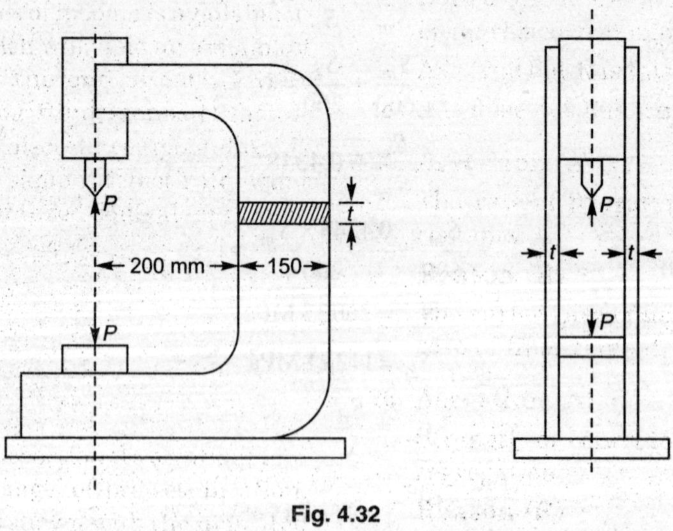

Fig. 4.32

3. A shaft is subjected to the combined action of bending and torsion. The maximum bending moment is 400 N.m and the maximum torque is 300 N.m. Calculate the diameter of the shaft based on maximum shear stress theory if the permissible shear stress in the shaft is 80 MPa.

4. A component machined from a plate made of steel is having ultimate tensile strength 650 MPa. The component has a step from 120 mm to 60 mm with a fillet radius of 5 mm. It is subjected to a completely reversed axial force of 60 kN. The surface of the component may be assumed to be machined and the expected reliability is 90%. The notch sensitivity is 0.85. If the factor of safety desired is 2.5, determine the thickness of the plate for unlimited life.

5. A forged steel bar of 63 mm salimeter is subjected to a reversed bending stress of 300 MPa. Calculate the life of the bar for 90% reliability if it is made of steel having ultimate tensile strength equal to 650 MPa.

6. A 30 mm diameter forged steel shaft having 600 ultimate tensile strength has a step to 20 mm diameter with a fillet radius of 5 mm. The notch sensitivity factor is 0.85. Calculate the endurance strength of the shaft if it is subjected to reversed bending moment.

7. A solid circular shaft is made of steel having $S_{ut} = 650$ MPa and $S_{yt} = 380$ MPa. It is subjected to an alternating torsional moment varying from −150 N.m to + 350 N.m. The shaft is ground keyway is 1.3 and the notch sensitivity is 0.90. Calculate the shaft diameter for unlimited life. A factor of safety of 2.5 is desired. Use maximum distortion energy theory of failure.

8. A solid circular shaft 20 mm in diameter is subjected to bending stress varying from 20 MPa to + MPa and torsional shear stress varying from −15 MPa to + 30 MPa. The material of the shaft has $S_{ut} = 540$ MPa and $S_{yt} = 360$ MPa. The corrected endurance strength of shaft material is 200 MPa. Calculate the factor of safety.

9. Determine the size of a piston rod made of hot rolled carbon steel ($S_{ut} = 560$ MPa, $S_{yt} = 350$ MPa subjected to axial load fluctuations from 20 kN is compression to 30 kN in tension. Minimum factor of safety desired is 1.5. Impact factor has been estimated to be 1.25. Assume suitable data not given.

10. A cantilever shown in Fig. 4.33 is of diameter 1.5 *d* for the first 100 mm and diameter *d* for the remaining 150 mm with a fillet radius of $r/d = 0.20$. It is subjected to an axial load *P* which varies from −200 N to + 500 N and also a transverse load *F* at its free end which varies from 100 N up and 150 N down. Determine the diameters for a minimum safety margin of 2. The shaft is made of 40C8 steel.

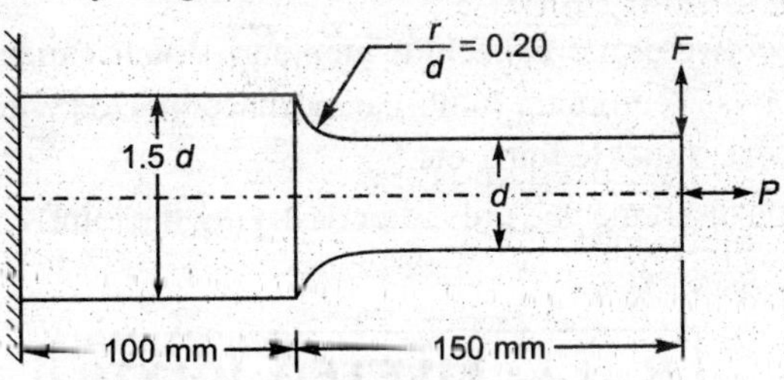

Fig. 4.33

5

Chapter

Mechanical Joints

Inside This Chapter

5.1 Introduction
5.2 Riveted Joints
5.3 Welded Joints
5.4 Cotter and Knuckle Joints
5.5 Spigot and Socket Cotter-Joint
5.6 Gib and Cotter Joint
5.7 Knuckle Joint

5.1 INTRODUCTION

Joints are required to connect two or more machine members. Joints can be classified as : (i) Temporary joints, and (ii) Permanent joints. Temporary joints can be disconnected without destroying the joint, *e.g.*, keys, bolted joints, cotter joint, knuckle joint, etc.

Permanent joints can be disconnected only after destrying the joint, *e.g.*, riveted and welded joints.

Part-I : RIVETED JOINTS

5.2 INTRODUCTION

A rivet is a short cylindrical bar with a head at one end, tail on the other end and shank or body in between. The rivets are used to make permanent joints between two plates. They are widely used in structural work, ship building, bridges, tanks and pressure vessels.

5.2.1 Types of Riveted Joints

There are two types of riveted joints :

(i) Lap joint, and (ii) Butt joint

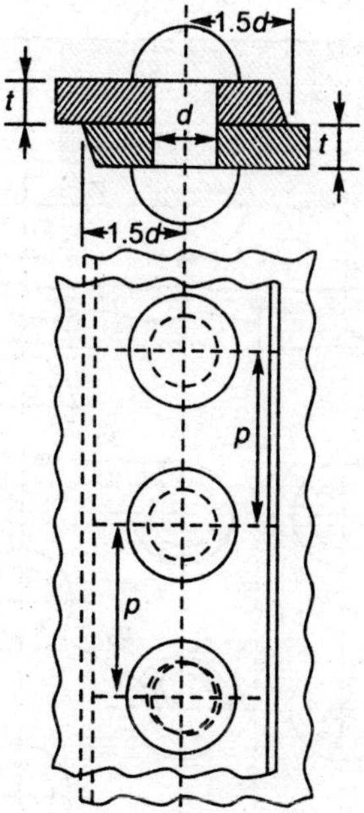

(a) Single riveted lap joint

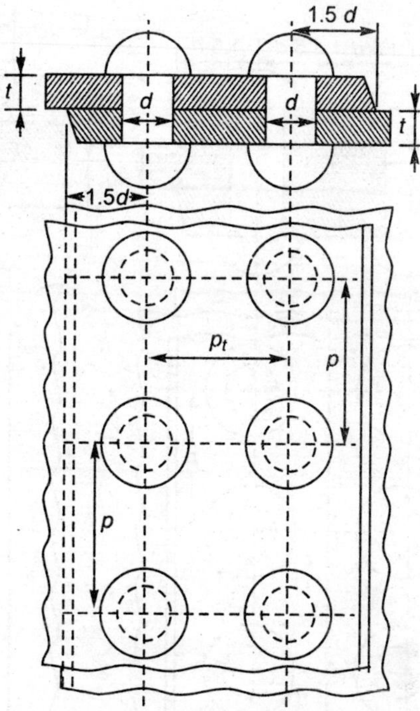

(b) Double riveted lap joint(chain riveting)

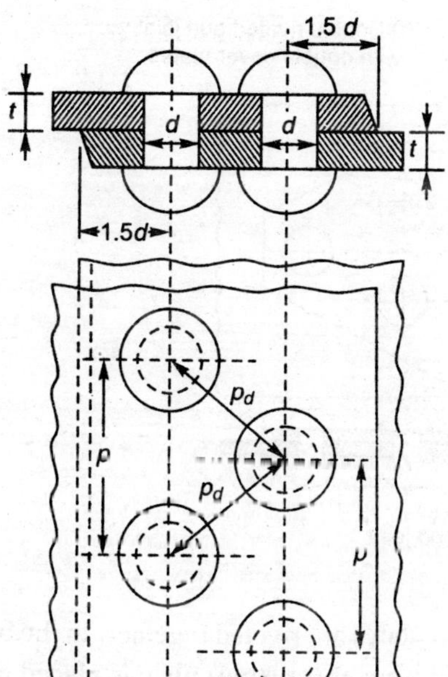

(c) Double riveted lap joint (zigzag riveting)

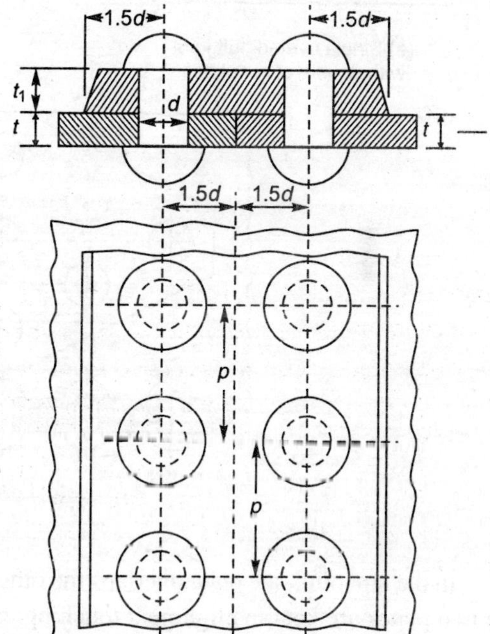

(d) Double riveted butt joint
with single cover plate

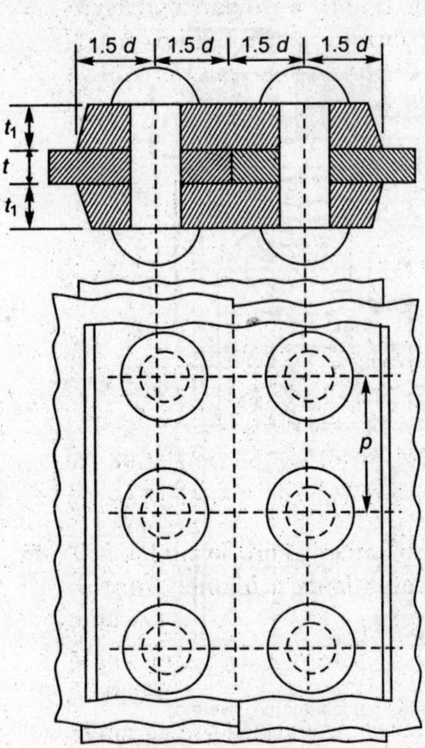

(e) Single riveted butt joint with double cover plates

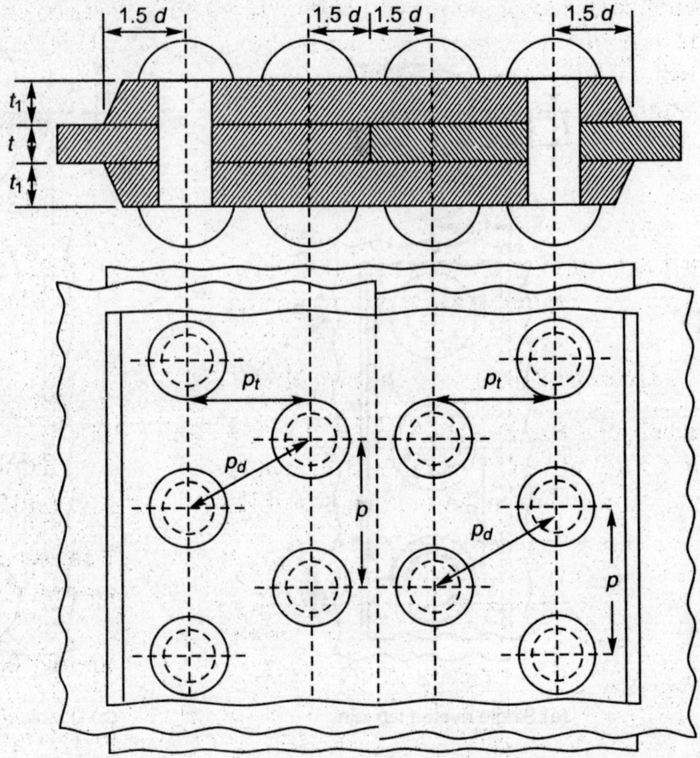

(f) Double riveted butt joint with double cover plates

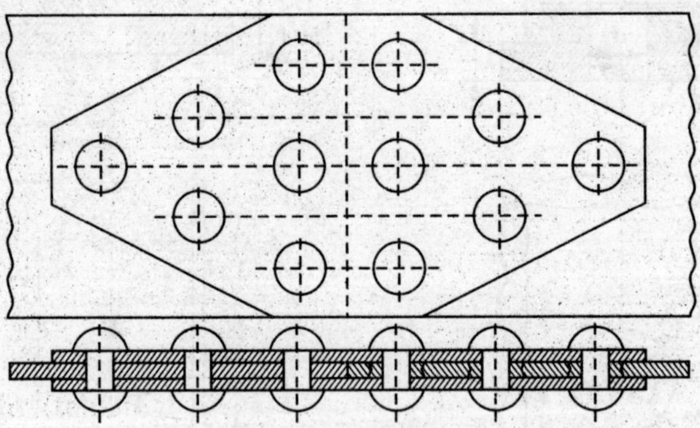

(g) Lozenge or diamond joint

Fig. 5.1

In the lap joint, one plate overlaps the other and the two plates are riveted together. In the butt joint the two plates are kept in alignment touching each other and a cover (or strap) plate is placed either on one side or on both sides of the main plates. The cover plate (s) is then riveted together with the main plate. Accordingly the butt joints may be :

(i) Single strap butt joint, and

(ii) Double strap butt joint.

In addition to the above, the riveted joints may be :

(i) Single riveted joint

(ii) Double riveted joint

(iii) Triple riveted joint, and so on.

In the double and higher riveted joints, the joint may be :

(i) Chain riveted, and

(ii) Zig-zag riveted.

In the chain riveted joint, the rivets in the various rows are opposite to each other. On the other hand, in the zig-zag riveted joint, the rivets in the adjacent rows are stagerred in such a way that every rivet is in the middle of the two rivets of the opposite row. Lap joints are used to transmit small loads, whereas butt joints are used to transmit heavy loads.

Various types of riveted joints are shown in Fig. 5.1.

5.2.2 Terminology

Pitch, p : It is the distance from the centre of one rivet to the centre of the next rivet measured parallel to the seam.

Back pitch, p_b: It is the perpendicular distance between the centre lines of the adjacent rows.

Diagonal pitch, p_d: It is the distance between the centres of rivets in the adjacent rows of zig-zag riveted joint.

Margin, m : It is the distance between the centre of rivet hole to the nearest edge of the plate.

5.2.3 Modes of Failure of a Riveted Joint

A riveted joint may fail in the following ways :

1. **Tearing of the plate at an edge :** A joint may fail due to tearing of teh plate at an edge as shown in Fig. 5.2. This can be avoided by keeping the margin, $m = 1.5d_0$, where d_0 = diameter of rivet hole.

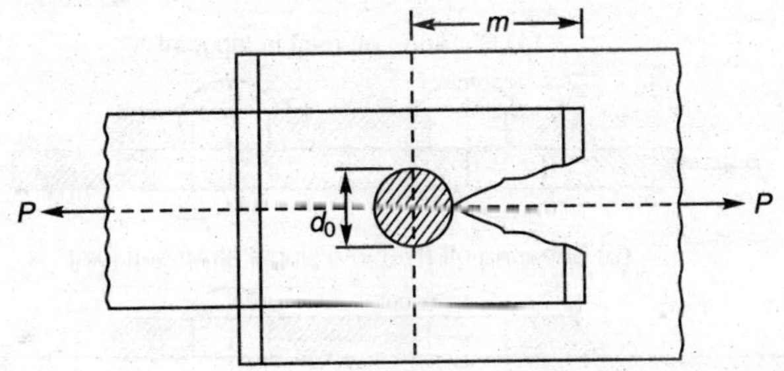

Fig. 5.2 Tearing of plate at an edge.

2. **Tearing of the plate across a row of rivets :** The main plate or cover plate may tear off across a row of rivets due to tensile stresses set up in the main plate, as shown in Fig. 5.3. The resistance offered by the plate against tearing is known as tearing resistance or tearing strength of the plate. It is determined across one pitch length only as every rivet is responsible for that much length.

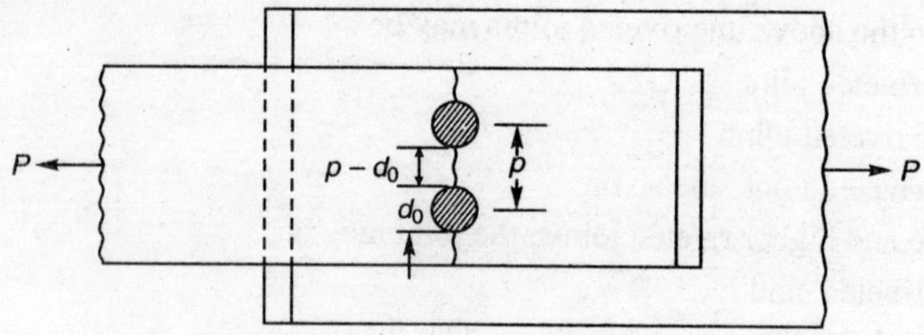

Fig. 5.3 Tearing of plate across the rows of rivets.

Let P = pitch of rivets

d_0 = diameter of rivet hole

t = thickness of plate

σ_t = allowable tensile stress for plate material

Tearing area per pitch length, $A_t = (p - d_0)\,t$

Tearing strength of plate per pitch length,

$$P_t = A_t \times \sigma_t = (p - d_0)\,t\sigma_t \qquad \ldots(5.1)$$

The joint shall fail when the applied load P is more than P_t. The rivet hole diameter is taken for all calculations because the rivet shall fill the rivet hole completely after the joint has been made.

3. **Shearing of the rivets :** The axis of the rivets is perpendicular to the load line. As a result of this the rivets can be either in single shear or double shear, as shown in Fig. 5.4. The rivets shall be in single shear in case of lap joint or butt joint with single cover strap. In case of butt joint with double cover straps, the rivets shall be in double shear.

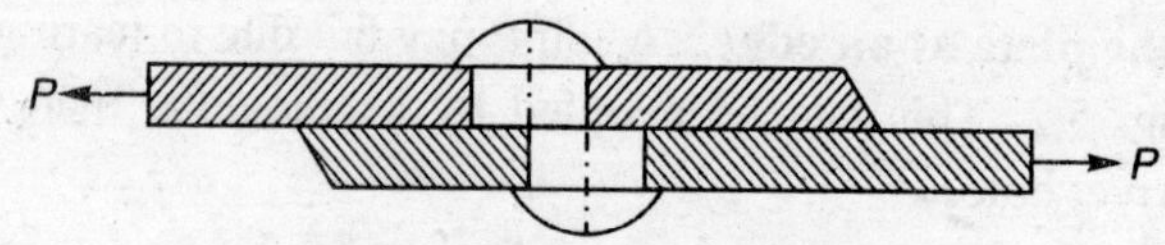

(a) Shearing off rivet in lap joint

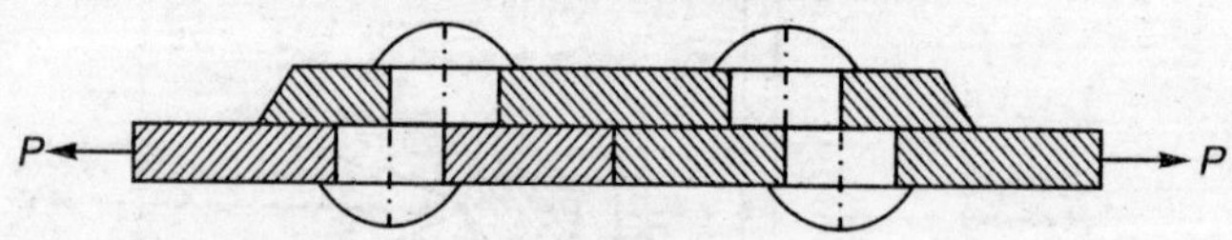

(b) Shearing off rivet in a single strap butt joint

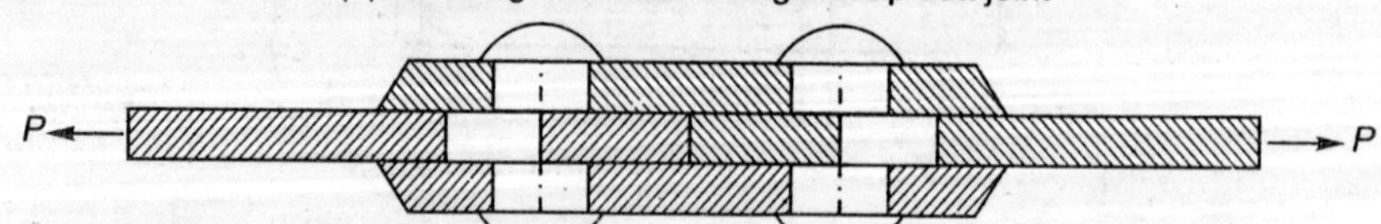

(c) Shearing off rivet in a double strap butt joint

Fig. 5.4 Shearing of rivets.

Let d_0 = diameter of rivet hole

n = number of rivets per pitch length

τ = allowable shear stress for rivets material

z = number of shears

Then Shearing area, $A_s = n \times z \times \frac{\pi}{4} \times d_0^2$ ...(5.2)

Where $z = 1$ for single shear

$= 2$ for double shear

$= 1.875$ for double shear as per Indian Boiler Regulations (I.B.R.)

Shearing strength of rivets, $P_s = A_s \times \tau$ ...(5.3)

The joint shall fail if the applied load P is more then P_s.

4. **Crushing of the plate or rivets :** The crushing of the rivets is shown in Fig. 5.5 as a result of which the hole becomes of oval shape and the joint becomes loose. This type of failure of rivet is called crushing or bearing. The resistance offered by the rivets is called crushing or bearing strength. If σ_c is the allowable crushing stress for the rivets, then

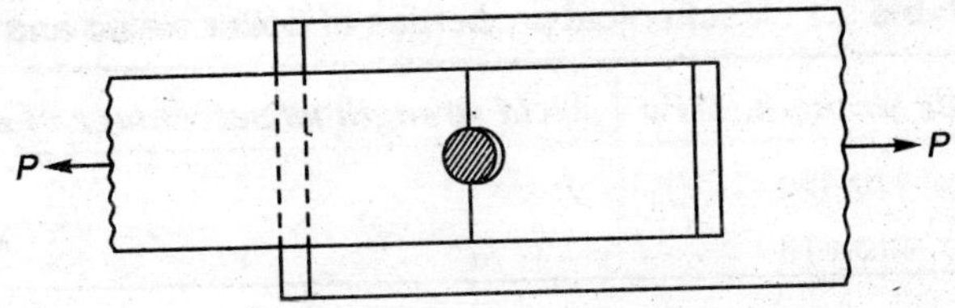

Fig. 5.5 Crushing of a rivet.

Crushing or projected area of n rivets, $A_c = n\,dt$

Crushing strength of rivets, $P_c = A_c \times \sigma_c$ ...(5.4)

The joint shall fail when the applied load P is more than P_c.

5. **Strength of the joint :** The strength of the joint is defined as the maximum value of the load it can transmit. It is the smallest value of P_t, P_s and P_c.

6. **Strength of the solid plate :** The strength of the solid plate per pitch length is calculated after ignoring the rivet hole.

Strength of solid plate, $P = p \times t \times \sigma_t$ ...(5.5)

7. **Efficiency of the joint :** It is defined as the ratio of the strength of riveted joint to the strength of the solid plate.

Efficiency, $\eta = \frac{\text{Least value of } P_t, P_s \text{ and } P_c}{P}$ ...(5.6)

5.2.4 Design of Boller Joints

The boiler is fabricated by making a riveted longitudinal joints and circumferential joints. The longitudinal joint is a butt joint with two cover straps to maintain the circular shape of the boiler. The circumferential joint is a lap joint which is required to obtain the longitudinal joints are staggered. The circumferential joint is a lap joint which is required to obtain the desired length of the boiler. For this purpose one ring overlapping the other alternately is used.

1. Longitudinal Butt Joint

(i) Plate thickness : The thickness of boiler shell plate is given by the thin cylindrical pressure vessel theory under internal pressure as :

$$t = \frac{p_s D}{2\sigma_t \eta_l} + c \quad \text{...(5.7}$$

where p_s = steam pressure, MPa gauge

D = shell diameter, mm

σ_t = permissible tensile stress for shell plates material at operating temperature, MPa (see Table 5.1 for tensile strength of boiler plates and rivets.

η_l = efficiency of longitudinal joint (see Table 5.3)

c = corrosion allowance ≈ 1 mm

The boiler plate thickness should not be less than 7 mm. For standard sizes of plates, refer to Table 5.5.

Table 5.1 : Mechanical properties of boiler plates and rivets.

Grode	*Tensile strength, MPa*	*Yield strength as percentage of minimum tensile strength*
1	410-490	55
2	510-610	50

Table 5.2 : Recommended type of joint for pressure vessels

Diameter of shell, mm	*Thickness of shell, mm*	*Type of joint*
600-1800	6-13	Double riveted
900-2100	13-25	Triple riveted
1500-2700	16-40	Quadruple riveted

Table 5.3 : Range of expected efficiencies for riveted joints

Type of joint	*Efficiency, %*	*Type of joint*	*Efficiency, %*
Lap of joints		Butt joints	
Single	45-65	Single	60-70
Double	63-77	Double	75-85
Triple	75-85	Triple	80-90
		Quadruple	85-95

Table 5.4 : Recommended factor of safety for boiler joints

Type of joint	*Factor of safety based on ultimate strength*	
	Hand riveting	*Machine riveting*
Lap joint	4.75	4.50
Single strap butt joint	4.75	4.50
Single riveted butt joint with two equal cover strap	4.75	4.50
Double riveted butt joint with two equal cover straps	4.25	4.00

Table 5.5 : Standard nominal size of steel plates

Thickness, mm	5, 6, 8, 12, 14, 16, 18, 20, 22, 25, 28, 32, 36, 40, 45, 50, 56 and 63
Width, mm	900, 1000, 1100, 1200, 1250, 1400, 1500, 1600, 1800, 2000, 2200 and 2500
Length, mm	220, 2500, 2800, 3200, 3600, 4000, 4500, 5000, 5600, 6300, 7100, 8000, 9000, 10000, 11000 and 12000

(ii) Rivet hole diameter : Rivet hole diameter is given by Unwin's formula,

$$d_0 = 6\sqrt{t} \text{ for } t \geq 8\text{mm} \qquad \text{...(5.8a)}$$

For $t \leq 8$ mm, calculate rivet hole diameter as follows :

$$d_0 = \sqrt{50t} - 4 \text{ mm for lap joints} \qquad \text{....(5.8b)}$$

$$d_0 = \sqrt{50t} - 5 \text{ mm for single riveted butt joints} \qquad \text{...(5.8c)}$$

$$d_0 = \sqrt{50t} - 6 \text{ mm for double riveted butt joints} \qquad \text{...(5.8d)}$$

(iii) Diameter of rivets : Diameter of rivet,

$$d = d_0 - 1 \text{ mm for } 12 \text{ mm} \leq d \leq 24 \text{ mm}$$

$$= d_0 - 1.5 \text{ mm for } 27 \text{ mm} \leq d \leq 36 \text{ mm}$$

$$= d_0 - 2 \text{ mm for } 39 \text{ mm} \leq d \leq 48 \text{ mm}$$

Standard diameter of rivets for boilers are : 12, 14, 16, 18, 20, 22, 24, 27, 30, 33, 36, 39, 42 and 48 mm.

(iv) Pitch of rivets : The pitch of rivets should not be less than $2d_0$ to enable the rivet heads to be formed. The maximum pitch of rivets allowed in the longitudinal joint prescribes by IBR is given by :

$$p = C \times t + 40 \text{ mm} \qquad \text{...(5.9)}$$

where $C =$ coefficient as given in Table 5.6.

Table 5.6 : Values of coefficient C.

Number of rivets	*Lap joint per pitch*	*Single strap butt joint*	*Double strap butt joint*
1	1.31	1.53	1.75
2	2.62	3.06	3.50
3	3.47	4.05	4.63
4	4.17	–	5.52
5	–	–	6.00

(v) Transverse pitch : Transverse pitch,

$$p_t = 2d_0 \text{ for chain riveting} \qquad \text{...(5.10a)}$$

$$= 0.33\, p + 0.67\, d_0 \text{ for zig-zag riveting} \qquad \text{...(5.10b)}$$

In case of zig-zag riveting, if the outermost row has half the number of rivets of inner rows, the distance of outermost row from the inner row should not be less than $(0.2p + 1.15\, d_0)$. The transverse pitch of inner rows having same number of rivets should not be less than $(0.165p + 0.67\, d_0)$. In all the above distances, p is the pitch of rivets in the outer rows.

(vi) Thickness of butt strap, t_1.

(a) Minimum thickness of butt strap = 10 mm

(b) $t_1 = 1.125t$ for ordinary chain riveting with single but strap. ...(5.11a)

$= 1.125t\left[\dfrac{p-d_0}{p-2d_0}\right]$, for single butt straps, every alternate rivet in ...(5.11b)

outer rows being omitted.

$s = 0.625t$, for double butt straps of equal widths having chain riveting ...(5.11c)

$= 0.625t\left(\dfrac{p-d_0}{p-2d_0}\right)$, for double butt straps of equal width having ...(5.11d)

every alternate rivet in outer rows being omitted.

(c) For unequal width of butt straps, the thickness of butt strap will be

$t = 0.75t$ for widestrap on the inside, and ...(5.12a)

$= 0.625t$, for narrow strap on the outside ...(5.12b)

(vii) Margin : $m = 1.5d_0$ ...(5.13)

2. Circumferential Lap Joint

(i) Thickness of boiler shell and diameter of rivets : It shall be same as for the **longitudinal** joint.

(ii) Number of rivets : Shearing resistance of the rivets in single shear,

$$P_s = n \times \frac{\pi}{4} d_0^2 \times \tau$$

Total shearing load due to steam pressure,

$$P_s = \frac{\pi}{4} D^2 p_s$$

$$\pi \times \frac{\pi}{4} d_0^2 \times \tau = \frac{\pi}{4} D^2 p_s$$

$$n = \left(\frac{D}{d_0}\right)\left(\frac{p_s}{\tau}\right) \quad ...(5.14)$$

(iii) Pitch of rivets : Let η_c = efficiency of circumferential joint = 0.5 η_l

Then
$$\eta_c = \frac{p_c - d_0}{p_c}$$

(iv) Number of rows : Number of rows of rivets,

$$N = \frac{p(D+t)}{p_c} \quad ...(5.15)$$

(v) Overlap : Overlap = $(N-1)\,p_b + m$

■ **Example 5.1** A steam boiler of 1.5 m inside diameter is to be designed for a steam pressure of 2.5 MPa. Design the longitudinal and circumferential joints assuming the following allowable stresses in steel plates and rivets :

Tension = 80 MPa, shear = 60 MPa, crushing = 120 MPa

Solution : Given : $p_s = 2.5$ MPa, $D = 1.5$ m, $\sigma_t = 80$ MPa, $\tau = 60$ MPa, $\sigma_c = 120$ MPa

1. Longitudinal joint

(i) **Plate thickness :** From Table 5.2 we observe that the joint shall be triple riveted diamond butt joint with double strap unequal cover plates and efficiency, $\eta_l = 80\%$. The thickness of boiler shell plate :

$$t = \frac{p_s D}{2\sigma_t \eta_l} + c = \frac{2.5 \times 1.5 \times 10^3}{2 \times 80 \times 0.8} + 1$$

$$= 30.29 \approx 32 \text{ mm}$$

(ii) Diameter of rivet hole for $t > 8$ mm, $d_0 = 6\sqrt{t} = 6\sqrt{32} = 33.94 \approx 34.5$ mm

Diameter of rivet, $d = d_0 - 1.5 = 34.5 - 1.5 = 33$ mm

(iii) **Pitch of rivets :** Tearing resistance of plate per pitch length,

$$p_t = (p - d_0)t\sigma_t = (p - 34.5) \times 32 \times 80 = 2560\,(p - 34.5)\ N$$

Number of rivets per pitch length are these out of which two are in double shear and one in single shear

Shearing resistance of rivets,

$$P_s = (2 \times 1.875 + 1) \times \frac{\pi}{4} \times (34.5)^2 \times 60 = 266424 \text{ N}$$

For $P_t = P_s$, we have

$$2560(p - 34.5) = 266424$$

$$p = 138.6 \text{ mm}$$

As per IBR, $p_{max} = Ct + 40$

For double strap butt joint with 3 rivets per pitch length, (Table 5.6), $C = 4.63$

$$p_{max} = 4.63 \times 32 + 40 = 188 \text{ mm}$$

Adopt $p = 140$ mm

Pitch of rivets in the inner row, $p_i = \frac{p}{2} = 70$ mm

(iv) **Distance between rows of rivets :** According to IBR, the distance between the outer row and adjacent row,

$$= 0.2p + 1.15d_0$$

$$= 0.2 \times 140 + 1.15 \times 34.5 = 67.7 \approx 70 \text{ mm}$$

Distance between the inner rows for zig-zag riveting,

$$= 0.165p + 0.67d_0$$

$$= 0.165 \times 140 + 0.67 \times 34.5 = 46.2 \approx 50 \text{ mm}$$

(v) **Thickness of butt straps :** Thickness of wide butt strap,

$$t = 0.75t = 0.75 \times 32 = 24 \text{ mm}$$

Thickness of narrow butt strap, $t_2 = 0.625t = 0.625 \times 32 = 20$ mm

Wide butt strap is put on the inside of boiler shell.

(vi) **Margin,** $m = 1.5d_0 = 1.5 \times 34.5 = 51.75 \approx 52$ mm

(vii) **Efficiency of joint :** Tearing resistance of plate in the outer row,

$$P_t = (p - d_0)\, t\sigma_t = (140 - 34.5) \times 32 \times 80 = 405120\ N$$

Shearing resistance of rivets,

$$P_s = 266424 \text{ N}$$

Crushing resistance of rivets,

$$P_c = nd_0 t\sigma_c = 3 \times 34.5 \times 32 \times 120 = 397440 \text{ N}$$

Considering tearing of the plate between the rivets in the second row when the rivets in the outermost row shears, we have

Combined tearing and shearing resistance

$$= (p - 2d_0)\, t\sigma_t + \frac{\pi}{4} \times d_0^2 \times \tau$$

$$= (140 - 2x34.5)\, 32 \times 80 + \frac{\pi}{4} \times (\times 34.5)^2 \times 60 = 237849 \text{ N}$$

Strength of the joint = 237849 N

Strength of solid plate, $P = pt\sigma = 140 \times 32 \times 80 = 358400$ N

Efficiency of joint,

$$\eta_l = \frac{237849}{358400} = 0.6636 \text{ or } 66.36\% \approx = 80\% \text{ as assumed}$$

2. Circumferential Joint

The overlapping of longitudinal and circumferential joints is shown in Fig. 5.6.

(i) $t = 32$ mm, $d_0 = 34.5$ mm, $d = 33$ mm

(ii) Number of rivets.

$$n = \left(\frac{D}{d_0}\right)^2 \left(\frac{p_s}{\tau}\right)$$

$$= \left(\frac{1500}{34.5}\right)^2 \left(\frac{2.5}{60}\right) = 78.7 \approx 80$$

(iii) **Pitch of rivets :** Assuming double riveted lap joint with zig-zag riveting, number of rivets per row

$$= \frac{n}{2} = \frac{80}{2} = 40$$

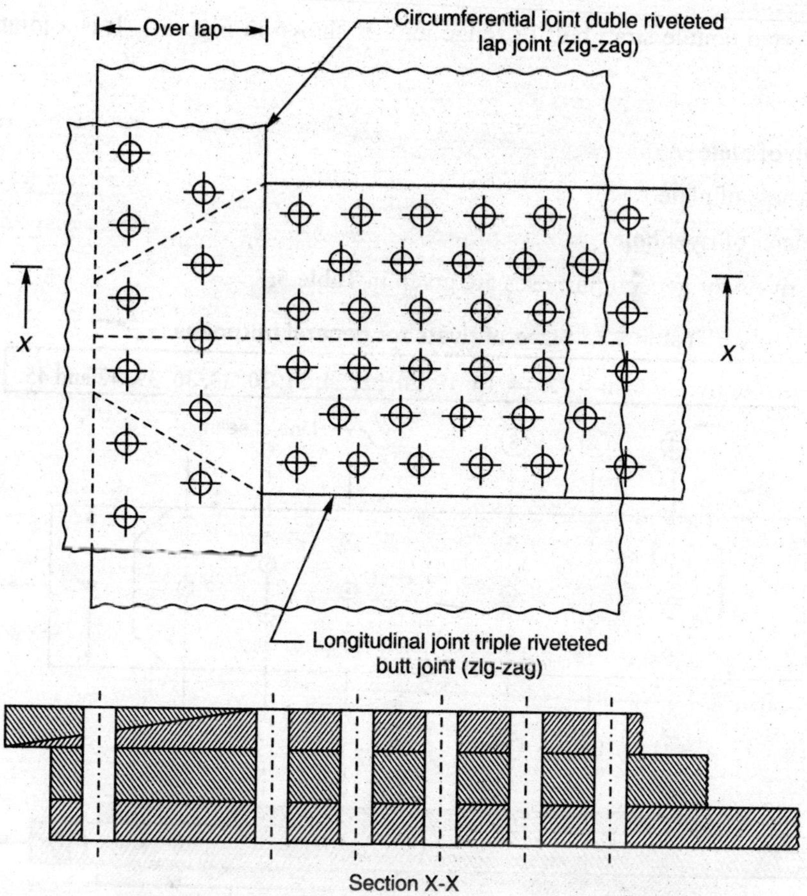

Fig. 5.6 Longitudinal and circumferential joints of a boiler.

Pitch of rivets,

$$p_c = \frac{\pi(D+t)}{\text{Number of rivets per row}}$$

$$= \frac{\pi(1500+32)}{40} = 120.32 \approx 120 \text{ mm}$$

(iv) Efficiency of joint,

$$\eta_c = \frac{p_c - d_0}{p_c} = \frac{120-34.5}{120} = 0.7125 \text{ or } 71.25\%$$

(v) Distance between rows of rivets : For zig-zag riveting, distance between the rows of rivets

$$= 0.33\,p_c + 0.67\,d_0$$

$$= 0.33 \times 120 = 0.67 \times 34.5 = 62.7 \approx 65 \text{ mm}$$

(vi) Margin, $m = 1.5\,d_0 = 1.5 \times 34.5 = 51.75 \approx 52$ mm

5.2.5 Structural Riveted Joint-Lozenge Joint

Structural riveted joints are those which are used for the fabrication of roof trusses, brideges, etc. A Lozenge joint can be lap joint or butt joint with single or double cover straps. The number of rivets in a row decreases from near the line of seam towards the edge of the plates.

The triple riveted double strap butt Lozenge joint is shown in Fig,. 5.7. It is a joint of uniform strength.

Let

w = width of plate

t = thickness of plate

d_0 = diameter of rivet hole.

The zises of rivets for general purposes are given in Table 5.7.

Table 5.7 : Sizes of rivets for general purposes.

Diameter of rivet, d, mm	12, 14, 16, 18, 20, 22, 24, 27, 30, 33, 36, 39, 42 and 45

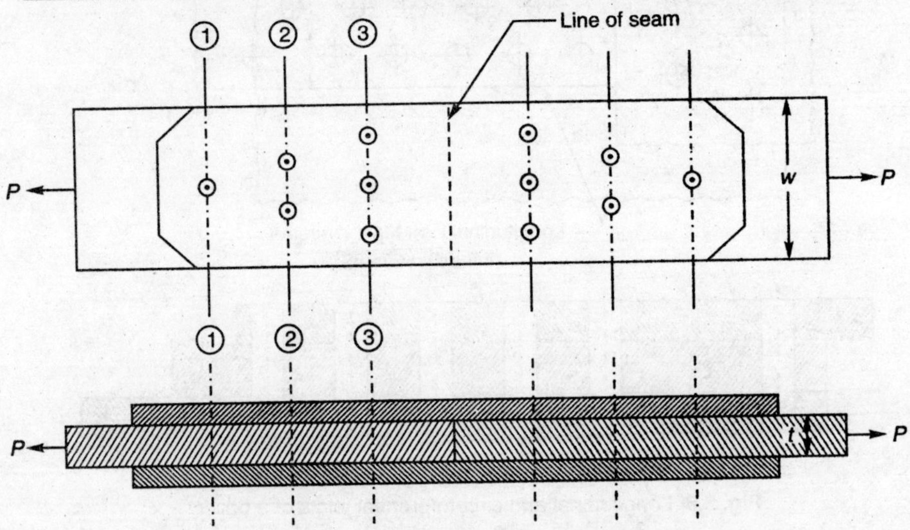

Fig. 5.7 Lozenge structural riveted joint.

Diameter of rivet hole, mm :

$$d_0 = d + 1.5 \text{ mm for } 12 \text{ mm} \leq d \leq 24 \text{ mm} \qquad ...(5.16)$$

$$= d + 2 \text{ mm for } 27 \text{ mm} \leq d \leq 50 \text{ mm}$$

(i) Diameter of rivets :

$$d_0 = 6\sqrt{t} \qquad ...(5.17)$$

Diameter rivets, $d = d_0 -$ clearance

(ii) Number of rivets : Tearing resistance of the plate in the outermost row,

$$P_t = (w - d_0)\, t\sigma_t,$$

Shearing resistance of rivets,

$$P_s = 1.75 \times \frac{\pi}{4} \times d_0^2 \times \tau$$

Crushing resistance of one rivet, $P_c = dt\sigma_c$

Number of rivets, $$n = \frac{P_t}{\text{Least of } P_c \text{ and } P_s} \qquad ...(5.18)$$

(iii) Thickness of butt straps.

$$t_1 = 1.25t \text{ for single cover strap}$$

$$= 0.75 \text{ for double cover strap} \quad ...(5.19)$$

(iv) Efficiency of joint : At section 1-1, there is only one rivet hole.

Resistance of the joint against tearing along 1-1,

$$P_1 = (w - d_0)t\sigma_t$$

Resistance of joint along 2-2,

P_2 = tearing resistance along 2-2 + shearing resistance of one rivet at 1-1

$$= (w - 2d_0)t\sigma_l + 1.75 \times \frac{\pi}{4} \times d_0^2 \times \tau$$

Resistance of joint along 3-3,

P_3 = tearing resistance along 3-3+ shearing resistance of two rivets at 2-2 and one rivet at 1-1

$$= (w - 3d_0)\sigma_t + 3 \times 1.75 \times \frac{\pi}{4} \times d_0^2 \times \tau$$

Strength of solid plate, $P = wt\sigma_l$

Shearing resistance of all rivets,

$$P_s = n \times 1.75 \times \frac{\pi}{4} \times d_0^2 \times \tau$$

Crushing resistance of all rivets,

$$P_c = ndt\sigma_c$$

Efficiency of joint, $$\eta = \frac{\text{Least of } P_1, P_2, P_3, P_s \text{ and } P_c}{P} \quad ...(5.20)$$

(v) Pitch of rivets. $p = 3d_0 + 5 \text{ mm}$ ...(5.21)

(vi) Margin, $m = 1.5d_0$ ...(5.22)

(vii) Distance between rows of rivets $= 2.5d_0$ ...(5.23)

■ **Example 5.2** A tie bar consisting of flat plate 400 mm wide and 20 mm thick is connected to a gusset plate of the same thickness by a double cover butt joint. Design an economical joint assuming the following permissible stresses :

Tension = 90 MPa, Shear = 60 MPa and crushing = 140 Mpa.

Solution : Given : $w = 400$ mm, $t = 20$ mm, $\sigma_t = 90$ MPa, $\tau = 60$ MPa, $\sigma_c = 140$ MPa

For the design to be economical, the Lozenge joint is chosen.

1. **Diameter of rivet :**

$$d_0 = 6\sqrt{t} = 6\sqrt{20} = 26.83 \text{ mm}$$

The standard diameter of rivet hole, $d_0 = 29$ mm and the corresponding diameter of rivet, $d = 27$ mm

2. **Number of rivets :** Maximum tearing strength of plate,

$$P_t = (w - d_0)t\sigma_t = (400 - 29) \times 20 \times 90 = 667800 \text{ N}$$

Shearing resistance of one rivet in double shear,

$$P_s = 1.75 \times \frac{\pi}{4} \times d_0^2 \times \tau$$

$$= 1.75 \times \frac{\pi}{4} \times 29^2 \times 60 = 69355 \text{ N}$$

Number of rivets, $n = \frac{P_t}{P_s} = \frac{667800}{69355} = 9.628 \approx 10$

3. The arrangement of rivets is shown in Fig. 5.8.

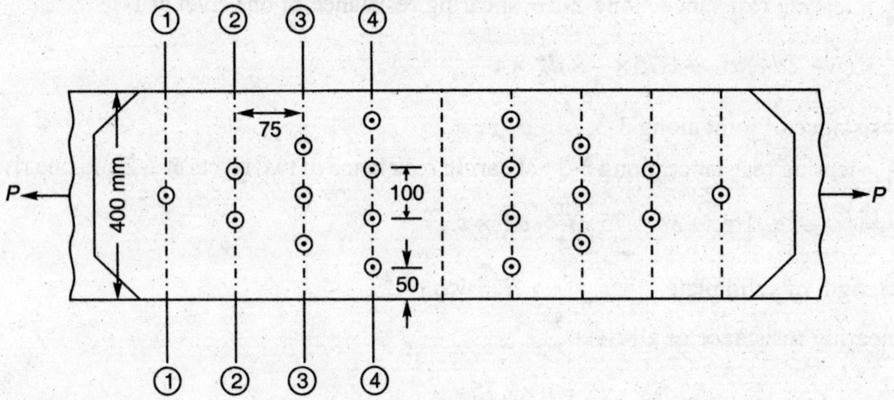

Fig. 5.8

4. Thickness of butt straps.

Inner strap, $t_1 = 0.75t = 0.75 \times 20 = 15$ mm

Outer strap, $t_2 = t_1 = 15$ mm

5. **Efficiency of joint.**

Section 1-1 :

$$P_1 = (w - d_0)t\sigma_c = (400 - 29) \times 20 \times 90 = 667800 \text{ N}$$

Section 2-2 :

$$P_2 = (w - 2d_2)t\sigma_t + \frac{\pi}{4} \times d_0^2 \times \tau$$

$$= (400 - 2 \times 29) \times 20 \times 90 + 1.75 \times \frac{\pi}{4} \times 29^2 \times 60 = 684955 \text{ N}$$

Section 3-3 :

$$P_3 = (400 - 3 \times 29) \times 20 \times 90 + 3 \times 69355 = 771465 \text{ N}$$

Section 4-4 :

$$P_4 = (400 - 29) \times 20 \times 90 + 6 \times 69355 = 927330 \text{ N}$$

Shearing resistance of all the rivets, $P_s = 10 \times 69355 = 693550$ N

Crushing resistance of all the rivets,

$$P_c = 10 \times 29 \times 20 \times 140 = 812000 \text{ N}$$

Strength of the joint $= P_1 = 667800$ N

Strength of solid plate,

$$P = wt\sigma_t = 400 \times 20 \times 90 = 720000 \text{ N}$$

Efficiency of joint,

$$\eta = \frac{667800}{720000} = 0.9275 \text{ or } 92.75\%$$

6. Pitch of riverts, $p = 3d_0 + 5 = 3 \times 29 + 5 = 92 \approx 100$ mm
7. Margin, $m = 1.5\, d_0 = 1.5 \times 29 = 43.5 \approx 50$ mm

$$w = 3p + 2m = 3 \times 100 + 2 \times 50 = 400 \text{ mm}$$

8. Distance between the rows $= 2.5d_0 = 2.5 \times 29 = 72.5 \approx 75$ mm

5.2.6 Eccentrically Loaded Riveted Joint

Consider the eccentrically loaded riveted joint, as shown in Fig. 5.9(a). Let

P = eccentric load

e = eccentricity of load

n = number of rivets in the group

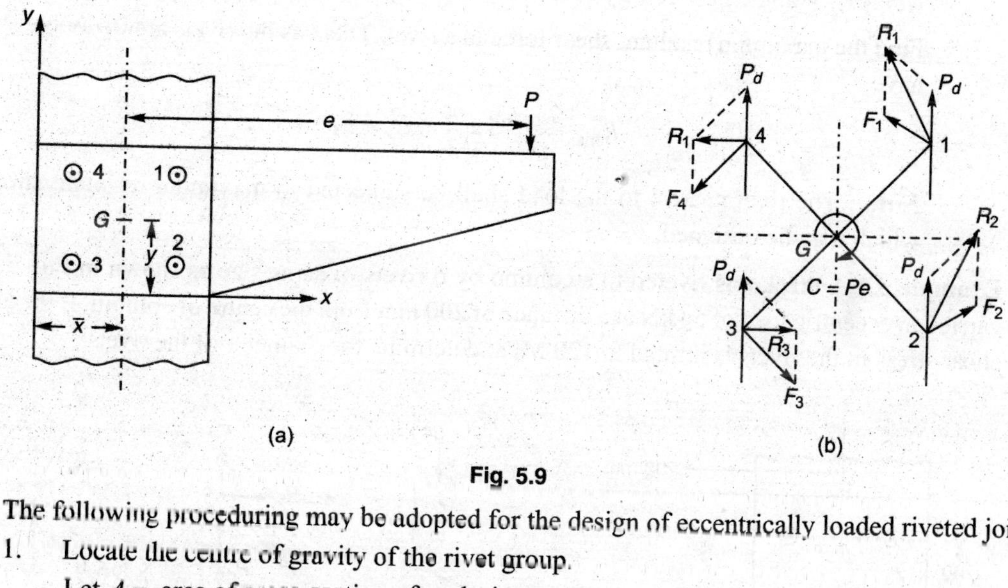

Fig. 5.9

The following proceduring may be adopted for the design of eccentrically loaded riveted joint :

1. Locate the centre of gravity of the rivet group.

   Let A = area of cross-section of each rivet.

   x_i = distance of ith rivet from y-axis

   y_i – distance of ith rivet from x-axis

   Then
$$x = \frac{A\Sigma x_i}{nA}$$

$$\bar{y} = \frac{A\Sigma y_i}{nA}$$

2. Direct primary shear force in each rivet, $P_d = \frac{P}{n}$

3. Couple about C.G., $C = P \times e$ [see Fig. 5.9(b)]

4. Let F_i = secondary shear force in each rivet

   l_i = perpendicular distance of F, from C.G.

   The secondary shear force in each rivet is proportional to the perpendicular distance from the C.G. Therefore, we have

$$\frac{F_1}{l_1} = \frac{F_2}{l_2} = \ldots = \frac{F_i}{l_i} \quad \ldots(5.24)$$

5. Equate the couple to the resisting moment of all the rivets.

$$C = \Sigma F_i l_i \left(\frac{F_1}{l_1}\right) \Sigma l_i^2 \quad \ldots(5.25)$$

6. Calculate the resultant shear force in each rivet due to the primary shear and secondary hear.

$$R_i = [P_d^2 + F_i^2 + 2P_d F_i \cos\theta_i]^{1/2} \quad \ldots(5.26)$$

   where θ_i = angle between P_d and F_i

7. Find the maximum resultant shear force in a rivet. Then, as the rivets are in double shear, we have

$$R_{max} = 1.75 \times \frac{\pi}{4} \times d_0^2 \times \tau \quad \ldots(5.27)$$

   Note : The rivet nearest to the load shall be subjected to maximum resultant shear force, which may be designed.

■ **Example 5.3** A bracket is riveted to a column by 6 rivets of same size as shown in Fig. 5.10(a). It carries an eccentric load of 54 kN at a distance of 200 mm from the centre of column. If the allowable shear stress in the rivets is limited to 120 MPa, determine the diameter of the rivets.

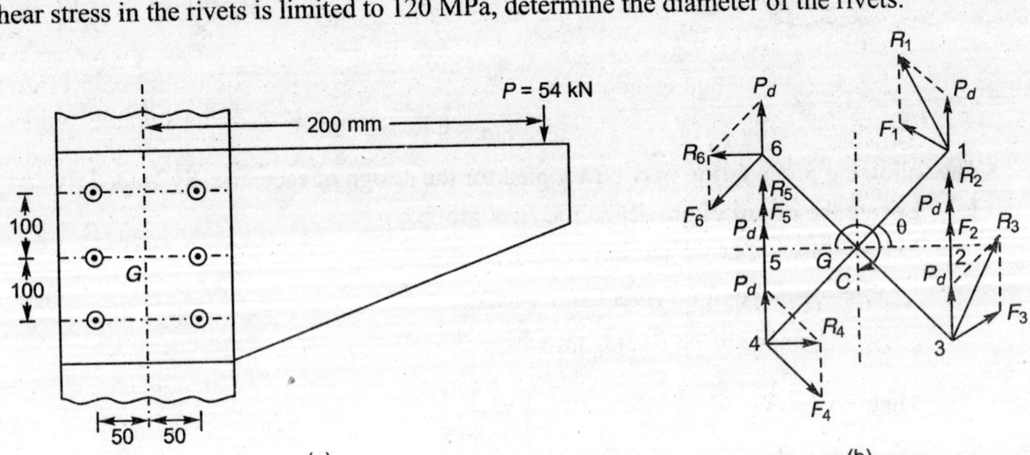

Fig. 5.10

Solution : Given : $n = 6$, $P = 54$ kN, $e = 200$ mm, $\tau = 120$ MPa

Primary or direct shear load in the rivets, $P_d = \frac{54}{6} = 9$ kN

The C.G. of the rivet group lies in the centre.

Couple, $C = P \times e = 54 \times 200 = 10800$ kNmm

$$l_1 = l_3 = l_4 = l_6 = [50^2 + 100^2]^{1/2} = 111.8 \text{ mm}$$

$$l_2 = l_5 = 50 \text{ mm}$$

$$\frac{F_1}{l_1}[4l_1^2 + 2l_2^2] = C$$

$$\left(\frac{F_1}{111.8}\right)[4 \times (111.8)^2 + 2 \times 50^2] = 10800 \times 10^3$$

$$F_1 = 21955 \text{ N}$$

Rivet 1 is nearest to the load line and shall be subjected to maximum shear load.

$$\cos\theta_1 = \frac{50}{111.8} = 0.447236$$

$$\theta_1 = 63.434°$$

Resultant force in rivet 1

$$R_1 = [P_d^2 + F_1^2 + 2P_d F_1 \cos\theta_1]^{1/2}$$

$$= [9000^2 + 21955^2 + 2 \times 9000 \times 21955 \times 0.44723]^{1/2}$$

$$= 27198 \text{ N} = \frac{\pi}{4} \times d_0^2 \times 120$$

$$d_0 = 16.98 \approx 17.5 \text{ mm}$$

$$d = 16 \text{ mm}$$

■ **Example 5.4** A bracket is attached to a column by means of five rivets as shown in Fig. 5.11 given below. Find the suitable diameter of rivets for the riveted connection. Assume suitable values of permissible stresses. What will be the change in design steps/procedure if the force P is applied in a direction perpendicular to the plane of the bracket?

Solution : Given : $n = 5$, $P = 45$, kN, $\alpha = 30°$

Primary or direct shear load in the rivets,

$$P_d = \frac{54}{5} = 9 \text{ kN}$$

The C.G. of the rivet group :

$$5\bar{x} = 3 \times 200$$

$$\bar{x} = 120 \text{ mm}$$

$$5\bar{y} = 1 \times 50 + 1 \times 100 + 1 \times 100 + 1 \times 200 = 500$$

$$\bar{y} = 100 \text{ mm}$$

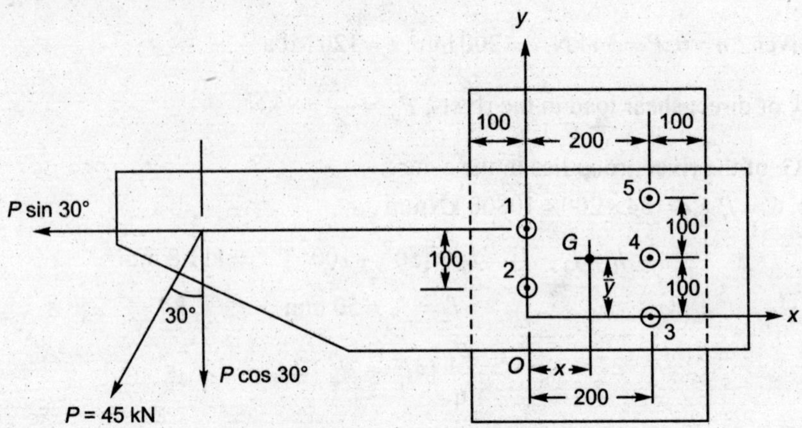

(a)

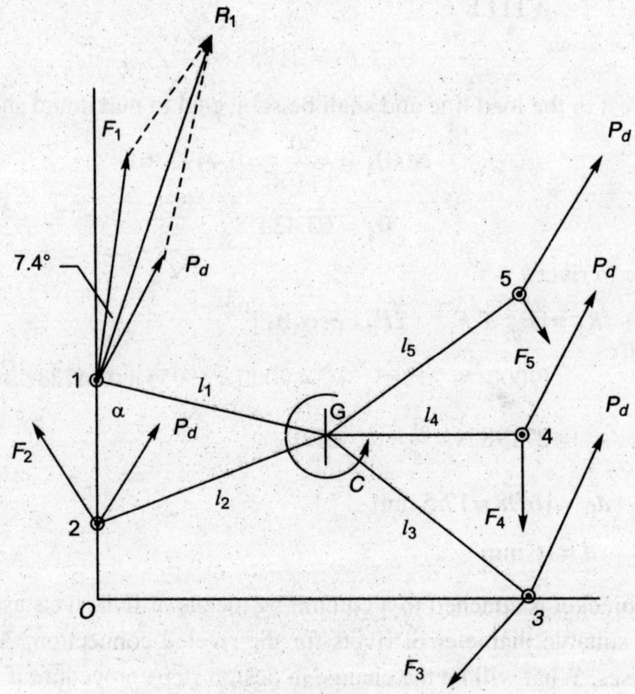

(b)

Fig. 5.11

$$e_x = 300 + 100 + 120 = 520 \text{ mm}$$

$$e_y = 100 \text{ mm}$$

$$l_1 = (120^2 + 50^2)^{1/2} = 130 \text{ mm}$$

$$l_2 = l_1$$

$$l_3 = (100^2 + 80^2)^{1/2} = 128.06 \text{ mm}$$

$$l_4 = 80 \text{ mm}$$

$$l_5 = l_3$$

$$2F_1 l_1 + 2F_3 l_3 + F_4 l_4 = P\cos\alpha \times e_x + P\sin\alpha \times e_y$$

$$(F_1/130)[2\times 130^2 + 2\times(128.06)^2 + 80^2] = [45\cos 30°\times 520 + 45\sin 30°\times 100]\times 10^3$$

$$F_1 = 40096 \text{ N}$$

Rivet 1 is nearest to the load line and shall be subjected to maximum shear load,

$$\tan\beta = \frac{120}{50} = 2.4$$

$$\beta = 67.4°$$

$$\theta_1 = \beta - (90° - \alpha) = 67.4 - (90° - 30°) = 7.4$$

$$\cos\theta_1 = \cos 7.4° = 0.99167$$

Resultant force in rivet 1,

$$R_1 = [P_d^2 + F_1^2 + 2P_d F_l \cos\theta_1]^{1/2}$$

$$= [9000^2 + 40096^2 + 2\times 9000\times 40096\times 0.99167]^{1/2} = 49035 \text{ N}$$

Let allowable shear stress in rivets, $\tau = 120$ MPa

$$49035 = \frac{\pi}{4\times d_0^2 \times 120}$$

$$d_0 = 204 \approx 21.5 \text{ mm}$$

$$d = 20 \text{ mm}$$

If the force is applied in a direction perpendicular to the plane of the bracket then the rivets shall be under the combined effect of tension and bending.

■ **Example 5.5** A double riveted lap joint with chain riveting is to be made for joining two plates of 6 mm thickness. Determine diameter of the rivets, distance between rows of rivets and pitch of rivets. Indicate how the joint will fail? Assume $\sigma_t = 220$ MPa, $\tau = 100$MPa, $\sigma_c = 150$ MPa.

Solution : Given : $t = 6$ mm, $\sigma_t = 220$ MPa, $\tau = 100$ MPa, $\sigma_c = 150$ MPa.

There are 2 rivets per pitch length and they are in single shear.

Shearing strength of rivets,

$$P_s = 2\times\frac{\pi}{4}\times d_0^2 \times \tau$$

Crushing strength of rivets,

$$P_c = 2d_0 t\sigma_c$$

Equating shearing and crushing strength, we have

$$d_0 = \left(\frac{4t}{\pi}\right)\left(\frac{\sigma_c}{t}\right)$$

$$= \left(4\times\frac{6}{\pi}\right)\left(\frac{150}{100}\right)$$

$$= 11.46 \approx 13.5 \text{ mm}$$

Diameter of rivet, $d = 12$ mm

Tearing strength of plate per pitch length,

$$P_t = (p - d_0)\, t\sigma_t = (p - 13.5) \times 6 \times 220$$

Shearing strength of rivets per pitch length,

$$P_s = 2 \times \frac{\pi}{4} \times (13.5)^2 \times 100$$

For $P_t = P_s$, we have

$$(p - 13.5) \times 6 \times 220 = 2 \times \frac{\pi}{4} \times (13.5)^2 \times 100$$

$$p = 35.2 \approx 35 \text{ mm}$$

Back pitch, $p_b = 0.6p = 0.6 \times 35 = 21$ mm

Tearing strength of plate per pitch length,

$$P_t = (35 - 13.5) \times 6 \times 220 = 28380 \text{ N}$$

Shearing strength of rivets per pitch length,

$$P_s = 28628 \text{ N}$$

Crushing strength of rivets per pitch length,

$$P_c = 2 \times 13.5 \times 6 \times 150 = 24300 \text{ N}$$

Strength of joint = 24300 N

Strength of solid plate per pitch length,

$$P = pt\sigma_t = 35 \times 6 \times 220 = 46200 \text{ N}$$

$$\text{Efficiency of joint} = \frac{24300}{46200} = 0.526 \text{ or } 52.6\%$$

The joint shall fail by crushing of rivet in the rivet hole.

■ **Example 5.6** A bracket is supported by means of 4 rivets of same size as shown in Fig. 5.12. Determine the diameter of rivets if the maximum permissible shear stress is 150 N/mm^2.

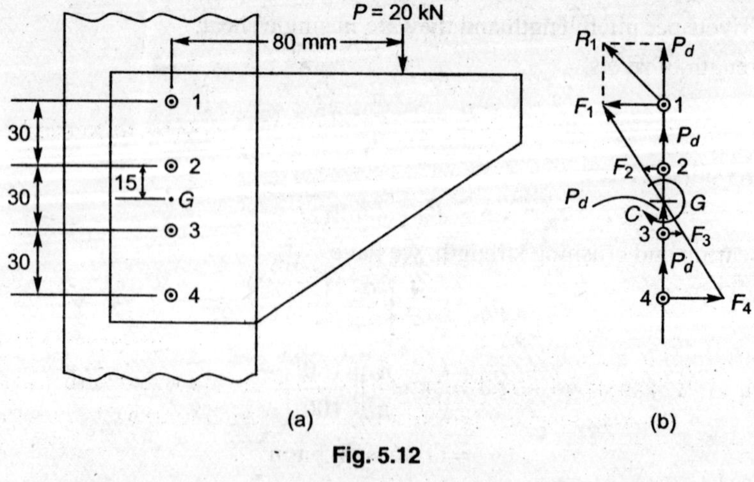

Fig. 5.12

Solution : Given : $P = 20$ kN, $e = 80$ mm, $\tau = 150$ MPa

The C.G. lies at 45 mm from top rivet.

Direct shear, force, $$P_d = \frac{P}{4} = \frac{20}{4} = 5 \text{ kN}$$

Couple, $$C = P \times e = 20 \times 80 = 1600 \text{ kN.mm}$$

$$l_1 = l_4 = 45 \text{ mm}, l_2 = l_3 = 15 \text{ mm}$$

$$\left(\frac{F_1}{45}\right)[2 \times 45^2 = 2 \times 15^2] = 1600 \times 10^3$$

$$F_1 = 16{,}000 \text{ N}$$

Rivet 1 is nearest to the load line and shall be subjected to maximum shear stress.

$$R_1 = [P_d^2 + F_1^2]^{1/2}$$

$$= [5000^2 + 16.000^2]^{1/2} = 16763 \text{ N}$$

$$= \frac{\pi}{4} \times d_0^2 \times \tau$$

$$d_0 = 11.928 \approx 13.5 \text{ mm}$$

$$d = 12 \text{ mm}$$

Part-II : WELDED JOINTS

5.3 INTRODUCTION

A welded joint is a permanent joint which is obtained by the fusion of the edges of the two parts to be jointed together, with or without the application of pressure and a filler metal. The advantages and disadvantages of welded joints over riveted joints are :

Advantages :

1. Lighter in weight
2. Easier to make alterations and additions in existing structures.
3. Higher efficiency.
4. Gives pleasing look, being smoother in appearance.
5. No weakening effect due to drilling of holds.
6. Higher strength and rigidity.
7. Faster process.

Disadvantages :

1. Introduces distortion and residual stresses due to heating.
2. Requires highly skilled labour and supervision.
3. Possibility of cracks due to excessive expansion.
4. More difficult process.

5.3.1 Types of Welded Joints

There are two types of welded joints as explained below :

Lap (or fillet) joint : It is obtained by overlapping the plates and then welding the edges of the plates.

The cross-section of the fillet weld is approximately a triangle. If h is the side of the weld triangle then throat thickness, $t = 0.705h$ in a simple weld and $0.85t$ in a reinforced weld. t is called the size of the weld.

Butt joint : It is obtained by placing the plates edge to edge and then welding. The plate edges may be square, single -V, double -V, single -U and double - U

The various types of welded joints are shown in Fig. 5.13.

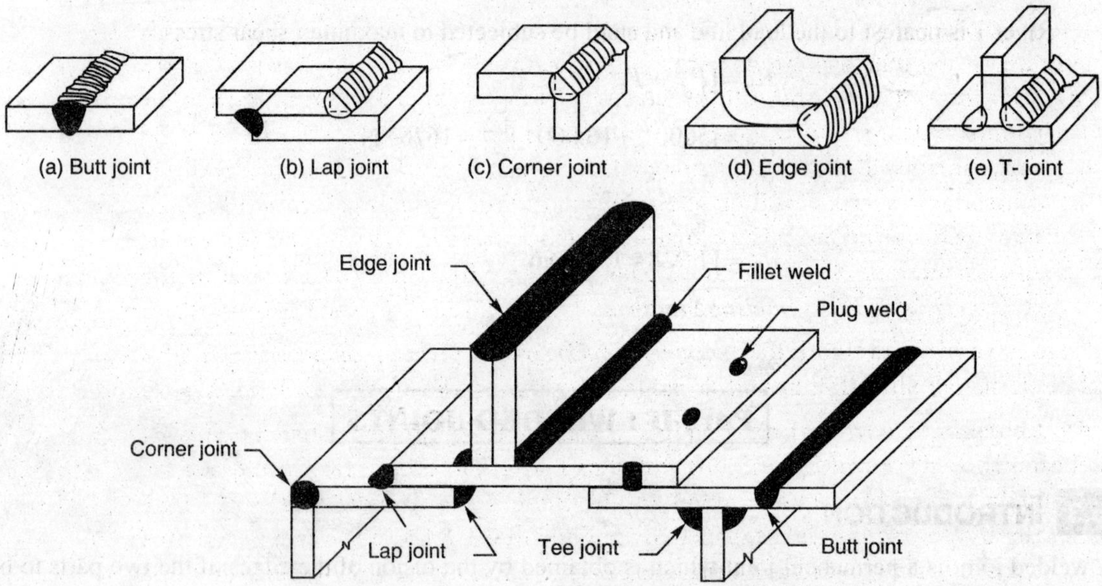

Fig. 5.13 Types of welded joints.

5.3.2 Streses in Welded Joints

The stresses for welded joints are given in Table 5.8.

Table 5.8 : Stresses for welded joints.

Type of weld	*Bare electrode*		*Coated electrode*	
	Steady load MPa	*Fatigue load MPa*	*Steady load MPa*	*Fatigue load MPa*
Fillet weld (all types)	80	21	98	35
Butt welds :				
Tension	90	35	110	55
Compression	100	35	125	55
Shear	55	21	70	35

5.3.3 Design of Welded Joints

Assumptions : The following assumptions are made in the design of welded joints :

1. The load is distributed uniformly over the entire length of the weld.
2. The cross-section of the fillet weld is a right angled triangle with equal sides of length h.
3. The minimum section is the hypotenuses with throat thickness, $t = 0.707\ h$.
4. The area which resists the applied load is always the throat area.

(a) Fillet Welds

1. **Transverse Fillet Welds :** The transverse fillet welded joints as shown in Fig. 5.14 are designed under tension.

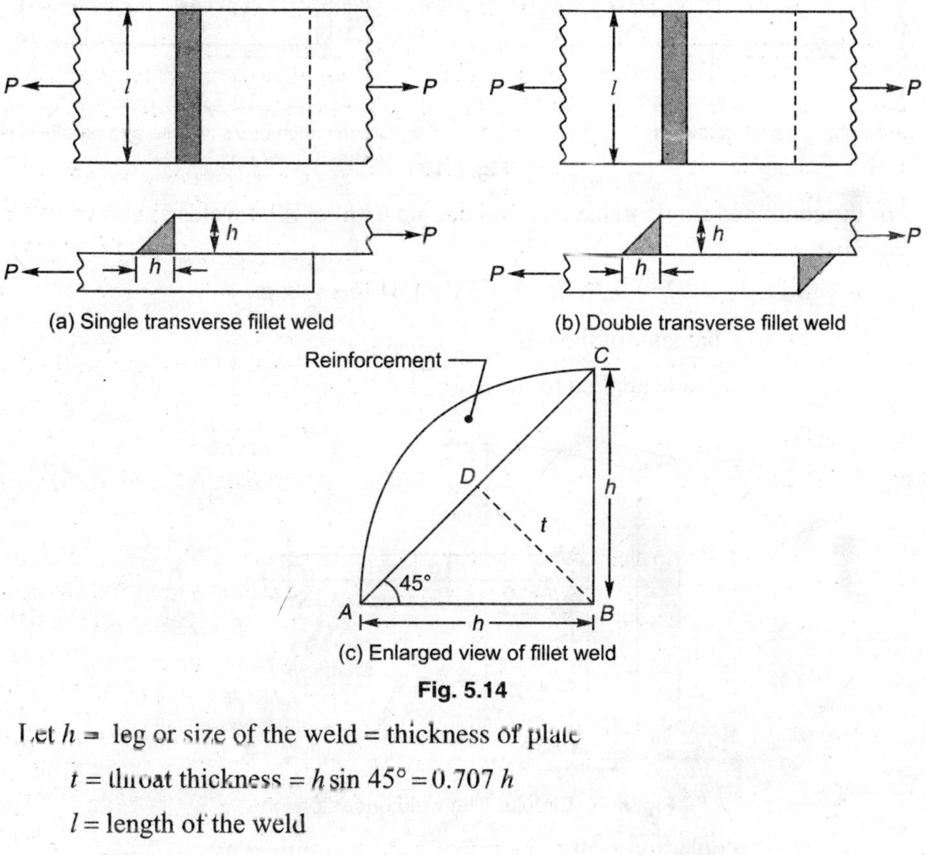

(a) Single transverse fillet weld

(b) Double transverse fillet weld

(c) Enlarged view of fillet weld

Fig. 5.14

Let h = leg or size of the weld = thickness of plate

t = throat thickness = $h \sin 45° = 0.707\ h$

l = length of the weld

σ_t = allowable tensile stress in weld

P = applied tensile force

Minimum area of weld,

$$A = t \times l = 0.707\,hl$$

Tensile strength of single fillet weld (Fig. 5.14a),

$$P = A\sigma_t = 0.707\,h \times l \times \sigma_t \qquad ...(5.28a)$$

For double fillet weld (Fig. 5.14b),

$$P = 2A\sigma_t = 1.414h \times l \times \sigma_t \qquad \text{...(5.28b)}$$

Fig. 5.14(c) shows the reinforced fillet weld.

2. **Parallel Fillet Welds :** The parallel fillet welded joints as shown in Fig. 5.15(a) are designed under shear. For the double parallel fillet welds, we have

$$P = 2A\sigma_t = 1.414h \times l \times \sigma_t \qquad \text{...(5.29)}$$

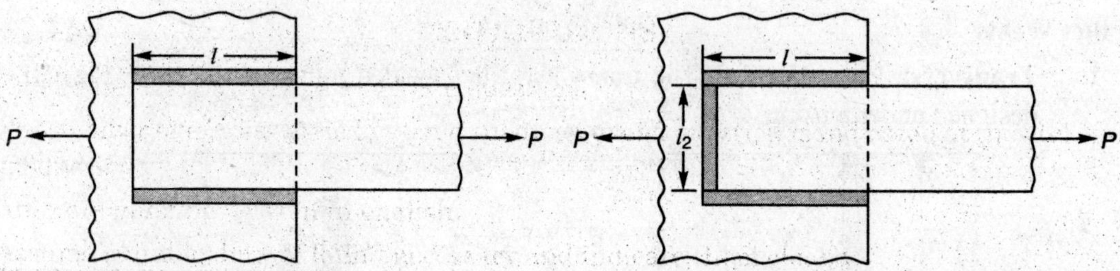

(a) Double parallel fillet weld (a) Combination of transverse and parallel fillet weld

Fig. 5.15

For the combined single transverse and double parallel fillet welds as shown in Fig. 5.15(b), we have

$$P = 0.707h \times l_1 \times \sigma_t + 1.414h \times l_2 \times \pi \qquad \text{...(5.30)}$$

d = diameter of the rod

T = torque applied to the rod

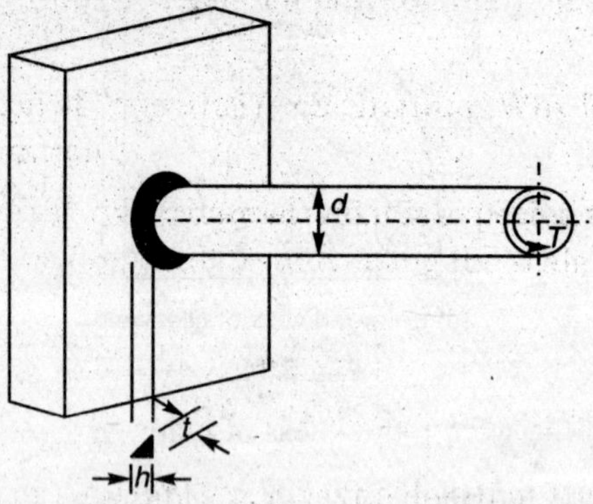

Fig. 5.16 Circular fillet weld under torsion.

J = polar moment of inertia of weld section

$$= \frac{\pi[(d+2t)^4 - d^4]}{32}$$

$$= \frac{\pi[d+2t)^2 + d^2(d \mp 2t)^2 - d^2]}{32}$$

$$= \frac{\pi[2d^2 + 4t^2 + 4dt)(4t^2 + 4dt)]}{32}$$

$$= \frac{\pi[(2d^2 + 4dt)(4dt)]}{32}, \text{ neglecting } t^2 \text{ terms}$$

$$= \frac{\pi[8d^3 t]}{32} \text{ neglecting } t^2 \text{ terms} = \frac{\pi t d^3}{4}$$

$$z_p = \frac{2J}{d} = \frac{\pi t d^2}{2}$$

Shear stress induced in the weld material,

$$\tau = \frac{T}{z_p} = \frac{2T}{\pi t d^2} = \frac{2T}{\pi \times 0.707\, h \times d^2} = \frac{2.83T}{\pi \times h \times d^2} \quad ...(5.31)$$

4. **Circular Fillet Weld under Bending Moment :** A circular fillet weld under bending is shown in Fig. 5.17. Let

d = diameter of the rod

M = bending moment applied to the rod

$= Pl$

z = secton modulus of weld section $= \frac{\pi t d^2}{4}$

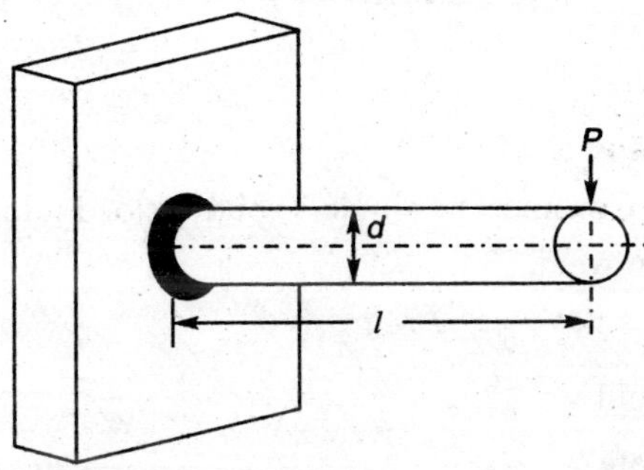

Fig. 5.17 Circular fillet weld under bending.

Bending stress induced in the rod material,

$$\sigma_b = \frac{M}{z} = \frac{4M}{\pi t d^2} = \frac{4M}{\pi \times t \times d^2}$$

$$= \frac{4M}{\pi \times 0.707 h \times d^2} = \frac{5.66M}{\pi \times h \times d^2} \quad ...(5.32)$$

5. **Long Fillet Weld under Torsion :** Consider a vertical plate welded to a horizontal plate by two identical fillet welds as shown in Fig. 5.18 and subjected to torque T.

Polar moment of inertia of weld section,

$$J = \frac{2 \times t \times l^3}{12} = \frac{t \times l^3}{6}$$

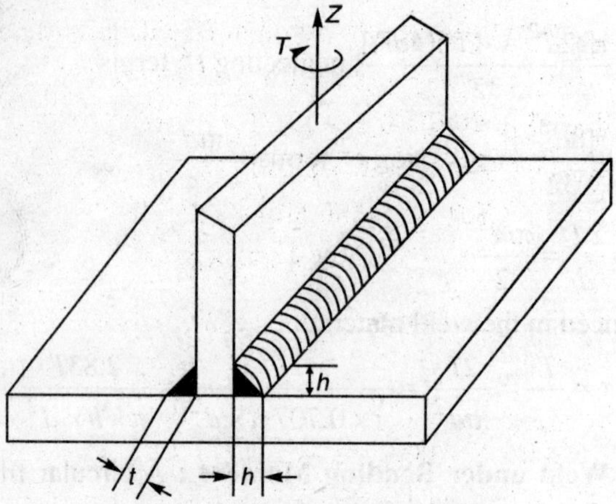

Fig. 5.18 Long fillet weld under torsion.

Shears stress in weld,

$$\tau = \frac{T \times (l/2)}{t \times (l^3/6)}$$

$$= \frac{3T}{t \times l^2} = \frac{3T}{0.707 h \times l^2} = \frac{4.242T}{h \times l^2} \qquad ...(5.33)$$

(b) Butt Welds

1. Butt Welds under Tension :

(i) Single V-butt joint : Consider the single V-butt welded joint as shown in Fig. 5.19(a).

Tensile strength of the weld,

$$P = t \times l \times \sigma_t \qquad ...(5.34)$$

where l = length of weld

= width of plate

(ii) Double V-butt joint : Tensile strength of double V-butt joint as shown in Fig. 5.19(b),

$$P = (t_1 + t_2)\, l \times \sigma_t \qquad ...(5.35)$$

where t_1, t_2 = throat thicknesses at top and bottom respectively.

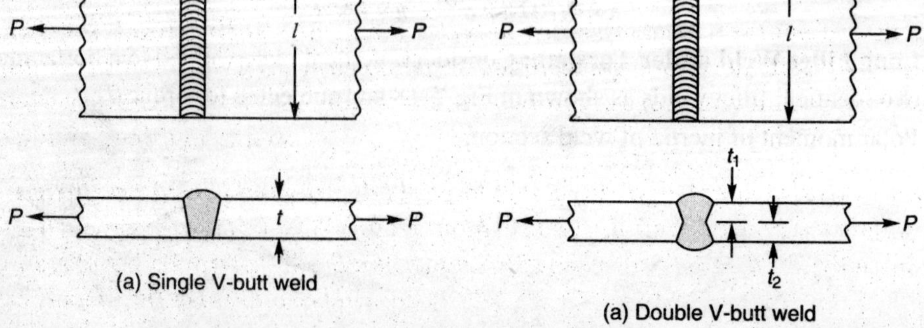

(a) Single V-butt weld

(a) Double V-butt weld

Fig. 5.19 Butt welds under tension.

2. Butt Weld under Shear : Consider a butt weld under shear as shown in Fig. 5.20.

Shear stress in weld, $$\tau = \frac{P}{hl} \quad ...(5.36)$$

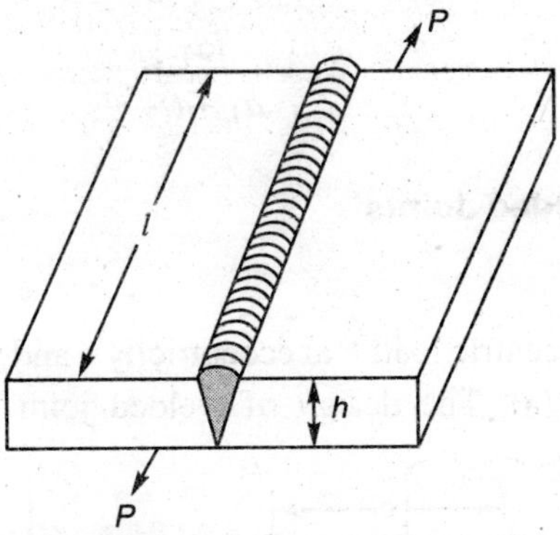

Fig. 5.20 Butt welds under shear.

5.3.4 Axially Loaded Unsymmetrical Welded Sections

Consider unsymmetrical sections like angle, channel, T-sections welded on the flange edges and loaded axially as shown in Fig. 5.21.

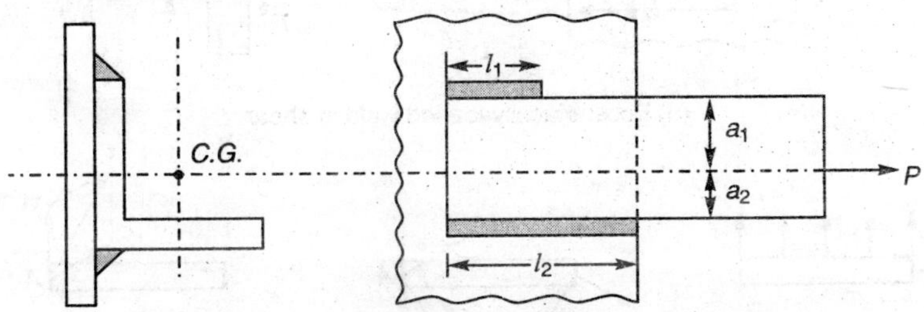

Fig. 5.21 Axially loaded unsymmetrical welded section.

Let

l_1 = length of weld at top

l_2 = length of weld at the bottom

$l = l_1 + l_2$ = total length of weld

a_1 = distance of top weld from C.G. axis.

a_2 = distance of bottom weld from C.G. axis

P = axial load

f = force of resistance offered by the weld per unit length

Moment of the top weld about the C.G. axis $= f \times l_1 \times a_1$

Moment of the bottom weld about the C.G. axis $= f \times l_2 \times a_2$

For equilibrium of the weld,

$$f \times l_1 \times a_1 = f \times l_2 \times a_2$$

or $$l_1 \times a_1 = l_2 \times a_2$$

Also $$l = l_1 + t_2$$

Thus $$l_1 = \frac{la_2}{a_1 + a_2} \quad \text{...(5.37a)}$$

$$l_2 = \frac{la_1}{a_1 + a_2} \quad \text{...(5.37b)}$$

5.3.5 Eccentrically Loaded Welded Joints

1. Welded Joint under Shear

A bracket subjecteds to an eccentric load P at eccentricity e and welded to the support by means of two welds is shown in Fig. 5.22(a). The design of welded joint under eccentric load consists of

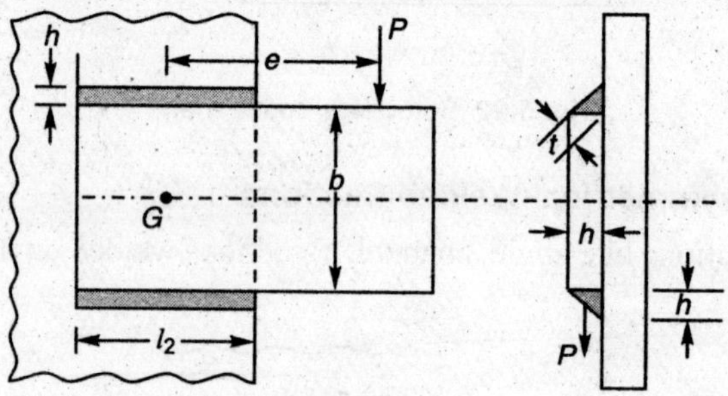

(a) Eccentriacally loaded weld in shear

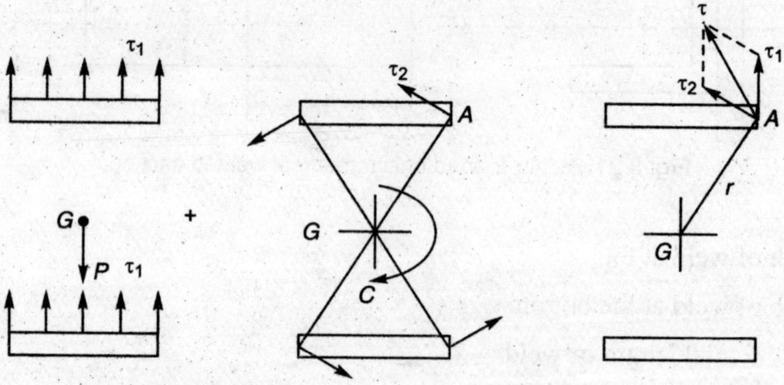

(b) Primary and secondary shear stresses

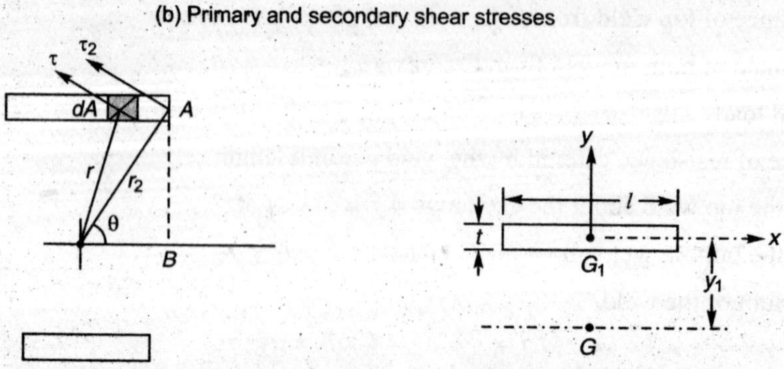

(c)

Fig. 5.22 Eccentrically loaded welded joint under shear.

determination of primary and secondary shear stresses. In such problems, the first step is to locate the C.G. of the welds, treating the weld as a line. Let G be the C.G. of two welds. The eccentric load P is equivalent to a load P of the same magnitude and sense as the eccentric load and a couple $C = Pe(cw)$ lying the same plane. The load P and G produces primary shear stress,

$$\tau_1 = \frac{P}{A}$$

where A = throat area of weld

The couple C produces secondary shear stress,

$$\tau_2 = \frac{Cr}{J} = \frac{\text{Per}}{J}$$

where r = distance of a point in weld from G

J = polar moment of inertia of all welds about G.

The primary and secondary shear stresses along with the resultant shear stress are shwon in Fig. 5.22(b).

The secndary shear stress at any point in the weld is proportaional to the distance from the C.G. Thus,

$$\frac{\tau_2}{r_2} = \frac{\tau}{r} = \text{constant}$$

where τ_2 = shear stress at the farther most point A.

$$\tau = \frac{\tau_2 \times r}{r_2}$$

Consider an elementary section of the weld having area dA at a distance r from G as shown in Fig. 5.22(c).

Shear force on this elementary section = $\tau \times dA$

Turning moment of this shear force about G,

$$dT = \tau \times dA \times r$$

$$= \left(\tau_2 \times \frac{r}{r_2}\right) \times dA \times r$$

$$= \left(\frac{\tau_2}{r_2}\right) \times dA \times r^2$$

Total turning moment over the whole weld area,

$$T = \int \left(\frac{\tau_2}{r_2}\right) \times dA \times r^2$$

$$= \left(\frac{\tau_2}{r_2}\right) \int dA \times r^2 = \left(\frac{\tau_2}{r_2}\right) \times J$$

where J = polar moment of inertia of the throat area about G.

Now $$T = C = P \times e$$

$$\sigma_2 = P \times e \times \frac{r_2}{J}$$

The polar moment of inertia of the throat area about the C.G. can be obtained by the use of parallel axis theorem as shown in Fig. 5.22(d).

$$I_{xx} = \frac{lt^3}{12} \text{ and } I_{yy} = \frac{t \times l^3}{12}, I_{zz} = I_{xx} + I_{yy}$$

Now $I_{xx} I_{yy}$ as t is small.

Thus $$J = I_{yy}$$

$$J = \frac{tl^3}{12} = \frac{tl(l^2)}{12} = \frac{Al^2}{12}$$

$$J_G = J + Ay_1^2 = A\left[\frac{l^2}{12} + y_1^2\right]$$

If there are number of welds with polar moment of inertias $J_1, J_2, J_3, \ldots$ about the C.G., then total polar moment of inertia about C.G. is given by,

$$J = J_1 + J_2 + J_3 + \ldots$$

For the welded joint shear in Fig. 5.23(a)

$$J = 2A\left[\frac{l^2}{12} + \frac{b^2}{4}\right]$$

Resultant shear stress, $$\tau = [\tau_1^2 + \tau_2^2 + 2\tau_1\tau_2 \cos\theta]^{1/2} \quad \text{...(5.38)}$$

where θ = angle between τ_1 and τ_2.

$$\cos\theta = \frac{GB}{AG} = \frac{\left(\frac{l}{2}\right)}{\left[\left(\frac{l}{2}\right)^2 + \left(\frac{b}{2}\right)^2\right]^{1/2}}$$

2. Welded Joint under Bending

Consider a cantilever beam of rectangular section welded to a support by means of two fillet welds as shown in Fig. 5.23 and subjected to bending load P at an eccentricity e.

Primary shear stress $$\tau_1 = \frac{P}{A}$$

where A = throat area of welds

Bending moment on the welds, $M = Pe$

Bending stres in the welds, $\sigma_b = \dfrac{My}{I}$

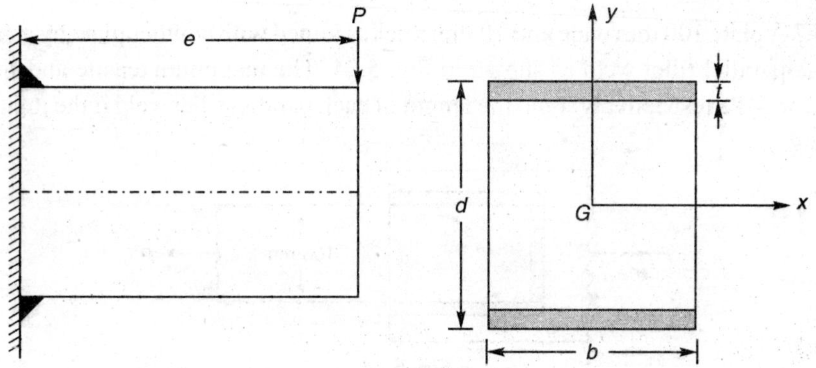

Fig. 5.23 Welded joint subjected to bending moment.

where y = distance of the point in the weld from the neutral axis

I = moment of inertia of all welds based on throat area

$$I_{xx} = \frac{bt^3}{12} + bt \times \left(\frac{d}{2}\right)^2$$

The first term is negligible. Thus

$$I_{xx} = \frac{btd^2}{2}$$

$$I = 2I_{xx} = \frac{btd^2}{2}$$

$$\sigma_b = \frac{\left(\frac{Ped}{2}\right)}{\left(\frac{btd^2}{2}\right)} = \frac{Pe}{btd}$$

Resultant shear stress in the welds

$$\tau = \left[\left(\frac{\sigma_b}{2}\right)^2 + \tau_1^2\right]^{1/2} \qquad ...(5.39)$$

5.3.6 Welded Joint Subjected to Fluctuating Forces

Stress-concentration is an important consideration while designing welded joints under fatigue loading. The fatigue stress-concentration factors K_f for welded joints are given in Table 5.9.

Table 5.9 : Fatigue stress-concentration factors for welded joints.

Type of joint	*Fatigue stress-concentration factor, K_f*
Reinforced butt welds	1.2
Toe of transverse fillet weld	1.5
End of parallel fillet weld	2.7
T-butt joint with sharp corner	2.0

The design procedufre as explained in chapter 4 may be adopted to solve a problem.

Example 5.7 A plate 100 mm wide and 10 mm thick is joined with another plate by asingle transverse and a double parallel fillet weld as shown in Fig. 5.24. The maximum tensile and shear streses are 70 MPa and 56 MPa respectively. Find the length of each parallel fillet weld if the joint is subjected to static loading.

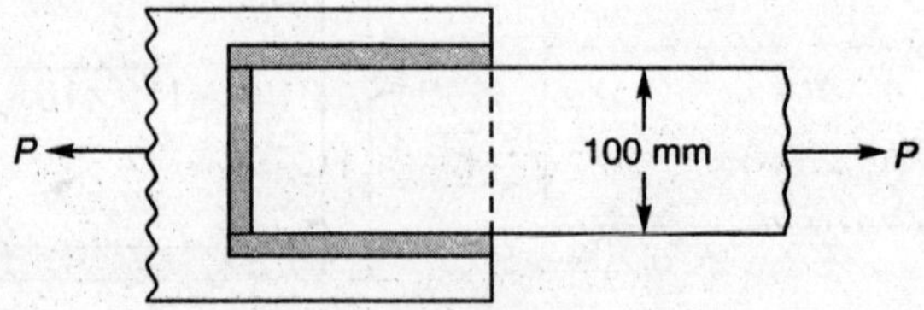

Fig. 5.24

Solution : Given : $b = 100$ mm, $h = 10$ mm, $\sigma_t = 70$ MPa, $\tau = 56$ MPa

Static loading.

Effecitve length of transverse weld, $l_1 = b - h = 100 - 10 = 90$ mm

Let l_2 = length of each parallel fillet weld

Maximum load the plate can carry,

$$P = \sigma_1 bh = 70 \times 100 \times 10 = 70{,}000 \text{ N}$$

Load carried by single transverse weld,

$$P_1 = 0.707 h l_1 \sigma_1 = 0.707 \times 10 \times 90 \times 70 = 44541 \text{ N}$$

Load carried by double paralle fillet weld,

$$P_2 = 1.414 h l_2 \tau = 1.414 \times 10 \times l_2 \times 56 = 791.84 l_2 \text{N}$$

Now
$$P = P_1 + P_2$$

$$70{,}000 = 44541 + 791.84\, l_2$$

$$l_2 = 32.2 \text{ mm}$$

Adding 10 mm for starting and stopping of weld run, we have

$$l_2 = 32.2 + \ = 42.2 \text{ mm}$$

Example 5.8 A $200 \times 100 \times 10$ mm angle is to be welded to a steel plate by fillet welds as shown in Fig. 5.25. If the angle is subjected to a static load 200 kN, find the length of welds at the top and bottom. The allowable shear stress may be taken as 60 MPa.

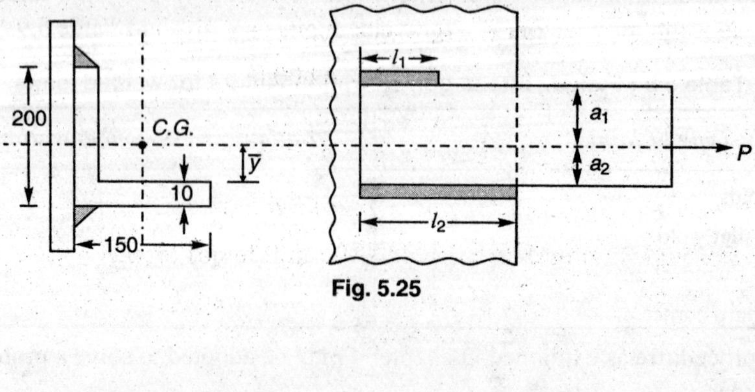

Fig. 5.25

Solution : Given : $a_1 + a_2 = 200$ mm, $P = 200$ kN, $\tau = 60$ MPa

Let l_1 = length of weld at the top

l_2 = length of weld at the bottom

$l = l_1 + l_2$ = total length of weld

$$(190 \times 10 + 150 \times 10)\bar{y} = 190 \times 10 \times 105 + 150 \times 10 \times 5$$

$$\bar{y} = 60.88 \text{ mm}$$

$$a_1 = 200 - 60.88 = 139.12 \text{ mm}$$

$$a_2 = 60.88 \text{ mm}$$

$$h = 10 \text{ mm}$$

$$P = 0.707\, hl\tau$$

$$200{,}000 = 0.707 \times 10 \times l \times 60$$

$$l = 4715.5 \text{ mm}$$

$$l_1 = \frac{la_2}{a_1 + a_2} = \frac{471.5 \times 60.88}{200} = 143.5 \text{ mm}$$

$$l_2 = 471.5 - 143.5 = 328 \text{ mm}$$

■ **Example 5.9** A circular shaft 50 mm in diameter is welded to a support by means of a fillet weld as shown in Fig. 5.26. Determine the size of the weld if the permissible shear stess in the weld is limited to 80 MPa.

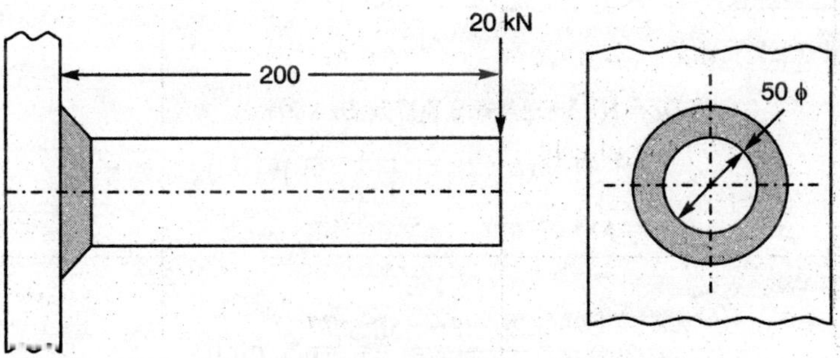

Fig. 5.26

Solution : Given : $d = 50$ mm, $\tau = 80$ MPa

Primary shear stress, $\tau_1 = \dfrac{P}{A} = \dfrac{P}{\pi dt}$

$$= \frac{10{,}0000}{\pi \times 50 \times t} = \frac{63.662}{t} \text{N/mm}^2$$

$$I_{xx} = \pi t r^3 = \pi \times t \times 25^3 = 49087.4t \text{ mm}^3$$

Bending stress, $\sigma_b = \dfrac{My}{I_{xx}} = \left[\dfrac{10{,}000 \times 200 \times 25}{49087.4t}\right.$

$$= \frac{1018.6}{t} \text{ N/mm}^2$$

Maximum shear stress,

$$\tau = \left(\frac{1}{t}\right)\left[\left(\frac{1018.6}{2}\right)^2 + (63.662)^2\right]^{1/2} = \frac{513.26}{t} = 80$$

$$t = 6.4 \text{ mm}$$

$$h = \frac{6.4}{0.707} = 9.07 \approx \text{mm}$$

■ **Example 5.10** A rectangular cross-section bar is welded to a support by means of fillet welds as shown in Fig. 5.27. Determine the size of the welds if the permissible shear stress in the weld is limited to 70 MPa.

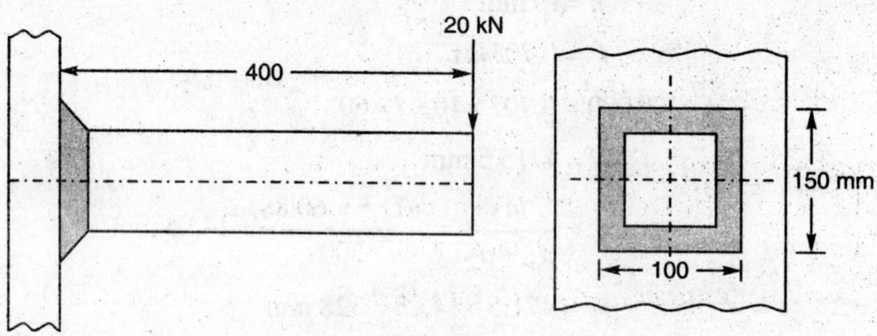

Fig. 5.27

Solution : Given : $P = 20$ kN, $\tau = 70$ MPa, $e = 400$ mm, $l = 100$ mm, $b = 150$ mm

Let h = size of welds

t = throat thickness

$$A = t(2b + 2l) = 0.707\,h(2b + 2l)$$

$$= 1.414h(b + l) = 1.414\,(150 + 100)$$

$$h = 353.5h \text{ mm}^2$$

Direct shear stress,

$$\tau = \frac{P}{A} = \frac{20{,}000}{353.5h} = \frac{56.577}{h} \text{ N/mm}^2$$

Bending moment,

$$M = Pe = 20{,}000 \times 400 = 8 \times 10^6 \text{ N.mm}$$

Section modulus,

$$z = t\left(bl + \frac{b^2}{3}\right)$$

$$= 0.707\,h\left[150 \times 100 + \frac{150^2}{3}\right] = 15907.5h \text{ mm}^3$$

Bending stress, $$\sigma_b = \frac{M}{z} = \frac{8 \times 10^6}{15907.5\,h} = \frac{502.9}{h} \text{ N/mm}^2$$

Maximum shear stress, $$\tau = \left[\left(\frac{\sigma^2}{2}\right)^2 + \tau^2\right]^{1/2}$$

$$= \frac{\left[\left(\frac{502.9}{2}\right)^2 + (56.577)^2\right]^{1/2}}{h}$$

$$= \frac{257.75}{h} = 70$$

$$h = 3.68 \approx 4 \text{ mm}$$

■ **Example 5.11** A bracket carrying a load of 20 kN is to be welded as shown in Fig. 5.28. Find the size of weld required if the allowable shear stress is not to exceed 75 MPa.

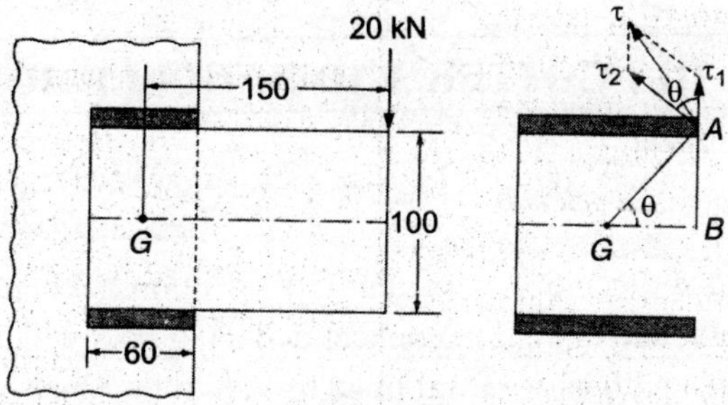

Fig. 5.28

Solution : Given : $P = 20$ kN, $e = 150$ mm, $\tau = 75$ MPa, $b = 100$ mm, $l = 60$ mm

Throat area, $A = 2tl = 2 \times 0.707 \times h \times 60 = 84.84h$ mm

Primary shear stress,

$$\tau_1 = \frac{P}{A} = \frac{20{,}000}{84.84\,h} = \frac{235.74}{h} \text{ N/mm}^2$$

Polar moment of inertia of throat about weld C.G.,

$$J = \frac{tl(3b^2 + l^2)}{6}$$

$$= \frac{0.0707h \times 60 \times [3 \times 100^2 + 60^2]}{6} = 237552h \text{ mm}^4$$

Maximum radius of the weld,

$$r_2 = \left[\left(\frac{l}{2}\right)^2 + \left(\frac{b}{2}\right)^2\right]^{1/2}$$

$$= [30^2 + 50^2]^{1/2} = 58.3 \text{ mm}$$

Secondary shear stress, $\quad z_2 = \dfrac{Per_2}{J} = \dfrac{20{,}000 \times 150 \times 58.3}{23755\, h} = \dfrac{736.26}{h}\ \text{N/mm}^2$

$$\cos\theta = \frac{l}{2r_2} = \frac{30}{58.3} = 0.5146$$

Resultant shear stress,

$$\tau = [\tau_1^2 + \tau_2^2 + 2\tau_1\tau_2 \cos\theta]^{1/2}$$

$$= \frac{[(235.74)^2 + (736.26)^2 + 2 \times 235.74 \times 736.26 \times 0.5146]^{1/2}}{h}$$

$$= \frac{881.07}{h} = 75$$

$$h = 11.74 \approx 12 \text{ mm}$$

Part-III : COTTER AND KNUCKLE JOINTS

5.4 INTRODUCTION

A cotter is a flat wedge-shaped piece of rectangular cross-section having a tapering width. A cotter is used to make a temporary joint and is used to connect two axial rods which may be either in tension or compression. The taper on the width varies from 1 : 48 to 1 : 8. If the taper is more than 1 : 8. It is called a **self-locking** cotter. A taper of 1 : 24 is generally provided. The tendency of the cotter to slacked back can be decreased by providing a gib either on one side or both sides of the cotter.

A knuckle joint is used to connect two rods which are in tension only. The two rods should lie in the same plane but may not be necessarily coaxial.

5.5 SPIGOT AND SOCKET COTTER-JOINT

A spigot and socket cotter joint is shown in Fig. 5.29. The end of the rod which goes into the socket is called the spigot. A rectangular hole is made in the socket and spigot through which the cotter is inserted. The load applied is usually axial but it may be tensile or compressive. The compressive load is taken by the collar on the spigot. The various parts may be designed as follows :

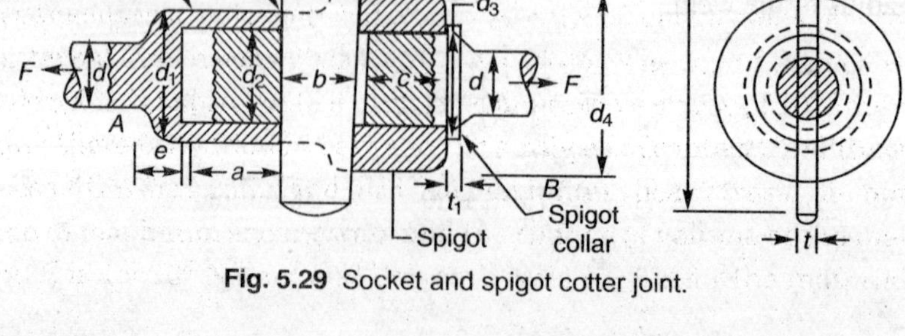

Fig. 5.29 Socket and spigot cotter joint.

1. Diameter of Rods

Let F = force in the rods in tension or compression

d = diameter of the rods

$\sigma_1 = \sigma_c$ = permissible tensile or compressive stress in the rods

Considering failure of the rods in tension or compression, we have

$$\left(\frac{\pi}{4}\right) d^2 \sigma_t = F$$

$$d = \left[\frac{4F}{\pi \sigma_t}\right]^{1/2} \quad ...(1)$$

2. Diameter of Spigot

The weakest section of the spigot is where the cotter passes through it, as shown in Fig. 5.30. Considering failure of the spigot in tension, we have

$$\left[\left(\frac{\pi}{4}\right) d_2^2 - d_2 t\right] \sigma_t = F \quad ...(2)$$

Where d_2 = diameter of spigot

t = thickness of cotter

σ_t = permissible tension stress in spigot.

Generally, $$t = \frac{d_2}{4}$$

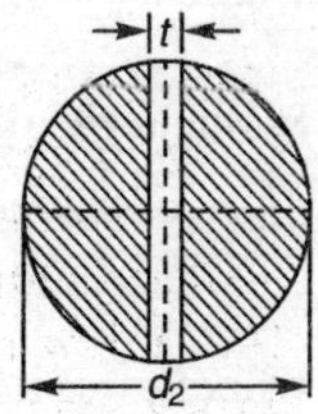

Fig. 5.30

Thus, d_2 can be determined.

3. Check for Crushing failure of Rod or Cotter

Let σ_c = allowable compressive stress for rod or cotter.

Then $$d_2 t \sigma_c = F$$

$$\sigma_c = \frac{F}{d_2 t}$$

4. Outside Diameter of Socket

Considering the failure of socket across the slot in tension or compression, as shown in Fig. 5.31, we have

$$\left[\left(\frac{\pi}{4}\right)(d_1^2 - d_2^2) - (d_1 - d_2)\, t\right] \sigma_t = F \quad ...(4)$$

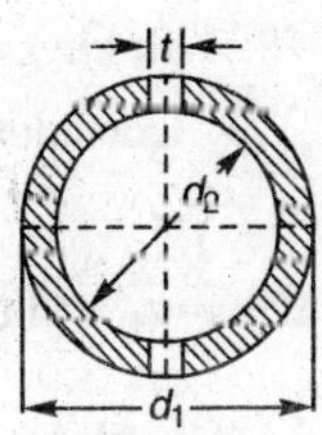

Fig. 5.31

where d_1 = outside diameter of socket.

5. Width of Cotter

Considering double shear of cotter, as shown in Fig. 5.32,

We have $$2bt\tau = F$$

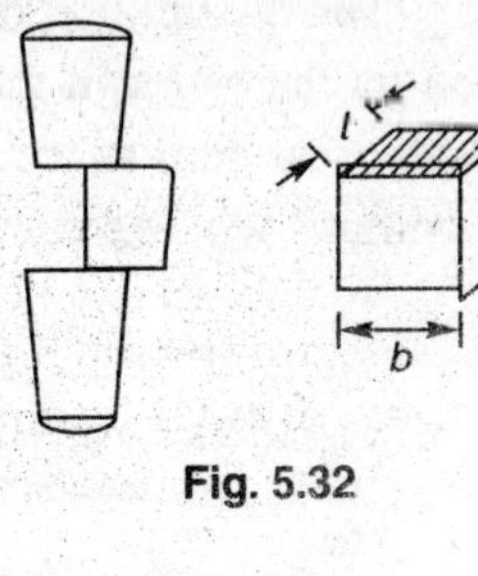

Fig. 5.32

Width of cotter,
$$b = \frac{F}{2t\tau} \quad ...(5)$$

where τ = permissible shear stress in cotter.

6. Length of Spigot from slot to the end.

Considering the failure of the socket end in double shear, we have

$$2ad_2\tau = F$$

Length of spigot from cotter slot to end.

$$a = \frac{F}{2d_2\tau} \quad ...(6)$$

7. Socket Collar Diameter

Considering failure of the socket collar in crushing, as shown in Fig. 5.33, we have

$$(d_4 - d_2)t\sigma_c = F \quad ...(7)$$

where d_4 = diameter of socket collar.

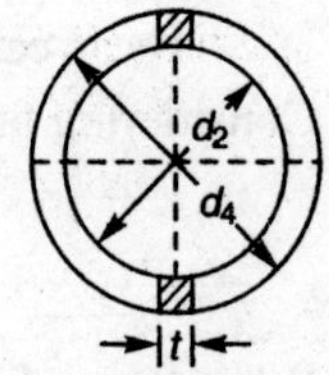

Fig. 5.33

8. Socket Collar Thickness

Considering the failure of socket collar in double shear, we have

$$2(d_4 - d_2)c\tau = F \quad ...(8)$$

where c = thickness of socket collar.

9. Diameter of Spigot Collar

Considering the failure of spigot collar in crushing, as shown in Fig. 5.34, we have

$$\left(\frac{\pi}{4}\right)(d^2 - d_2^2)\sigma_c = F \quad ...(9)$$

where d_3 = diameter of spigot collar.

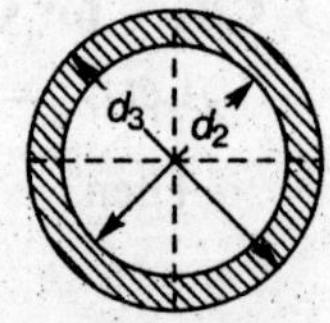

Fig. 5.34

10. Thickness of Spigot Collar

Considering the failure of spigot collar in shear, as shown in Fig. 5.35, we have

$$\pi d_2 t_1 \tau = F \quad ...(10)$$

where t_1 = thickness of spigot collar.

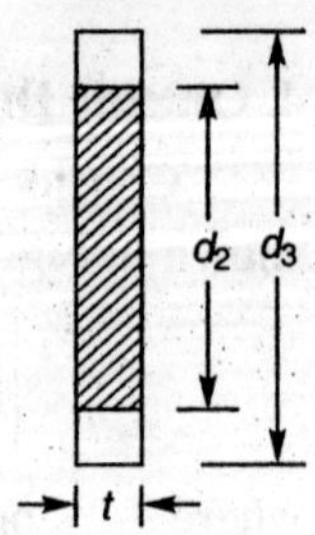

Fig. 5.35

11. Check for Cotter in Bending

The distribution of load on the cotter is shown in Fig. 5.36. It is assumed that the load on the cotter in the spigot end is uniformly distributed while in the socket end it varies from zero at the outer diameter d_4 and maximum at the inner diameter d_2. The maximum bending moment occurs at the centre of the cotter and is given by,

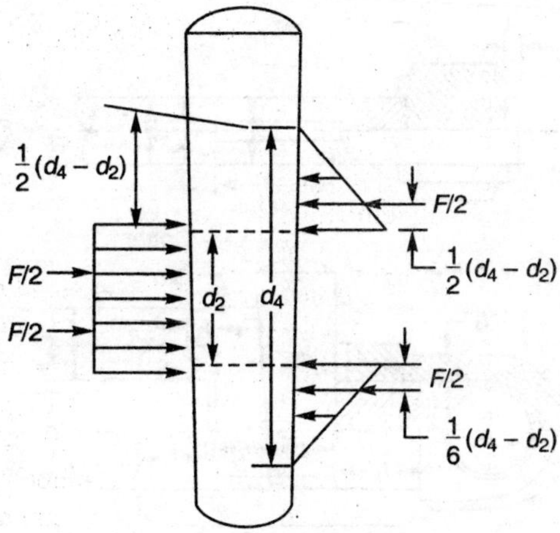

Fig. 5.36 Load distribution on the cotter.

$$M_{max} = \left(\frac{F}{2}\right)\left[\frac{(d_4 - d_2)}{6} + \frac{d_2}{2}\right] - \left(\frac{F}{2}\right) \times \left(\frac{d_2}{4}\right)$$

$$= \left(\frac{F}{24}\right)[2d_4 + d_2]$$

Section modulus of cotter,

$$z = \frac{tb^2}{6}$$

Bending stress in cotter,

$$\sigma_b = \frac{M_{max}}{z}$$

$$= \left[\frac{F}{4tb^2}\right](2d_4 + d_2) \qquad ...(11)$$

The bending stress should be less than the permissible bending stress of cotter material.

5.6 GIB AND COTTER JOINT

5.6.1 Strap End of Connecting Rod

The gib and cotter joint for the strap end (or big end) of a connecting rod is shown in Fig. 5.37. The various parts may be designed as follows :

1. Diameter of Rod

The connecting rod is subjected to tensile and compressive load. Considering the failure of rod of diameter d under tensile load F, we have

$$\left(\frac{\pi}{4}\right)d^2\sigma_t = F$$

$$d = \left[\frac{4F}{\pi\sigma_t}\right]^{1/2} \qquad ...(1)$$

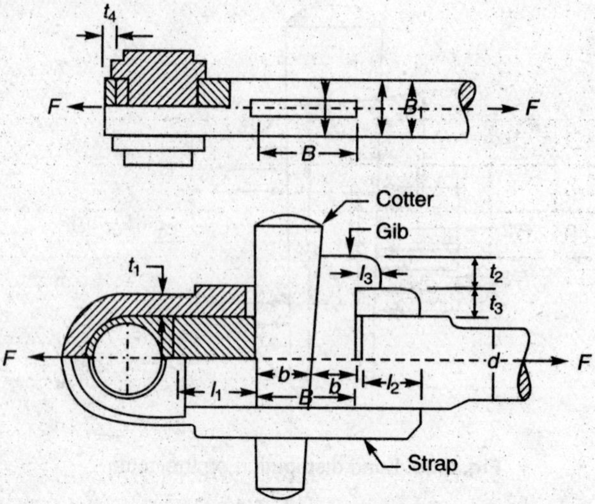

Fig. 5.37 Gib and cotter joint for strap end of a connecting rod.

2. Width of Strap

Generally, width of strap, $B_1 = d$ ...(2)

3. Thickness of Cotter

$$t = \frac{B_1}{4} \quad ...(3)$$

4. Thickness of Gib

It is kept equal to thickness of cotter $= t = \dfrac{B_1}{4}$ ...(4)

5. Height and Length of Gib Head

Height of gib head, $t_2 = t$ ...(5)

Length of gib head, $l_3 = t$ ...(6)

6. Thickness of Strap

Considering tearing of the strap at the thinnest section, we have

$$2B_1 t_1 \sigma_t = F$$

Thickness of Strap, $t_1 = \dfrac{F}{2B_1\sigma_t}$ ...(7)

Considering tearing strength of strap at the cotter shoulder to be equal to that at the thinnest section,

We have

$$2t_3(B_1 - t) = 2t_1 B_1$$

$$t_3 = \frac{B_1 t_1}{B_1 - t} \quad ...(8)$$

7. Total Width of Gib and Cotter

Width of Gib, $b_1 = 0.55\,B$ ...(9)

Width of cotter, $b = 0.45B$ ...(10)

8. Other Dimension

$$l_1 = 2t_1 \text{ and } l_2 = 2.5t_1 \quad ...(11)$$

$$t_4 = 91.15 \text{ to } 1.5)\, t_1 \quad ...(12)$$

5.6.2 For Square Rods

The gib and cotter joint for a square rod is shown in Fig. 5.38. The various parts can be deigned as follows :

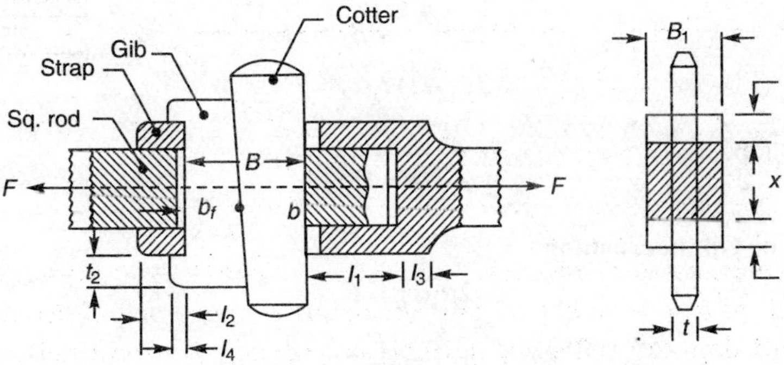

Fig. 5.38 Gib and cotter joint for square rods.

1. Size of Rod

Considering the failure of square rod of cross-sectional area x^2 under tensile force F, we have

$$\sigma_t \times x^2 = F$$

$$x = \left(\frac{F}{\sigma_t}\right)^{1/2} \quad ...(1)$$

where σ_t – permissible tensile stress in rod.

The other dimension are fixed as follows :

Width of strap, $B_1 = x$

Thickness of cotter, $t = \frac{B_l}{4}$

Thickness of gib = t

Height of gib, $t_2 = t$

Length of gib head, $l_4 = t$

2. Width of Gib and Cotter

Considering the double shear of gib and cotter, we have

$$2Bt\tau = F$$

$$B = \frac{F}{2t\tau} \qquad ...(2)$$

Width of gib, $b_1 = 0.55B$, for a single gib ...(3a)

$= 0.30B$, for a double gib ...(3b)

Width of cotter, $b = 0.45B$, for a single gib ...(4a)

$= 0.40b$, for a double gib ...(4b)

3. Thickness of Strap

Considering the failure of strap end in tension at the location of gib and cotter, as shown in Fig. 5.39, we have

$$2(B_1 t_1 - t t_1)\sigma_t = F$$

Now $B_1 = x$.

Thus $2(x - t)\, t_1 \sigma_t = F$

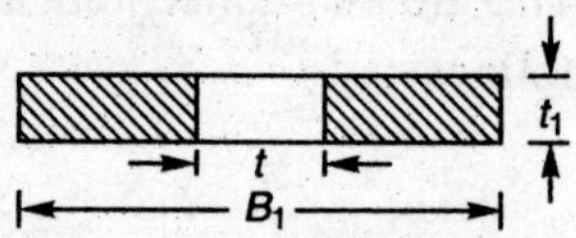

Fig. 5.39

Thickness of strap, $$t_1 = \frac{F}{2(x-t)\sigma_t} \qquad ...(5)$$

4. Check for Strap or Gib in crushing

$$2tt_1\sigma_c = F$$

$$\sigma_c = \frac{F}{2tt_1} \qquad ...(6)$$

Which should be less than the allowable compressive stress in strap or gib.

5. Length of Rod from Cotter to end

Considering the double shear of rod, as shown in Fig. 5.40, we have

$$2l_1 \times \tau = F$$

$$l_1 = \frac{F}{2 \times \tau} \qquad ...(7)$$

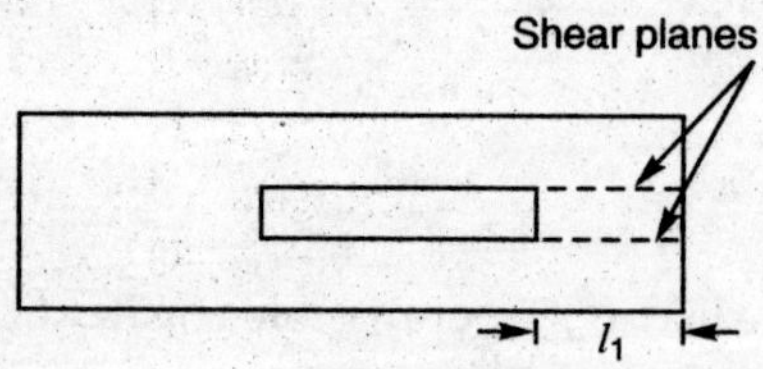

Fig. 5.40

6. Length of Rod from Gib to the end of Strap

Considering the double shear of the strap. As shown in Fig. 5.41, we have

$$2 \times 2l_2 t_1 \tau = F$$

$$l_2 = \frac{F}{4t_1\tau} \qquad ...(8)$$

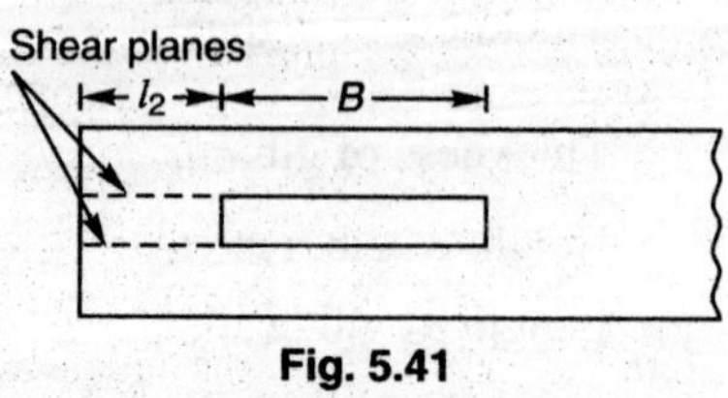

Fig. 5.41

7. Other Dimensions

$$l_3 = \frac{2x}{3} \qquad ...(9)$$

5.7 KNUCKLE JOINT

A knuckle joint is used to conect two rods subjected to tensile loads. An eye is forged at the end of one rod and a fork at the end of other rod. The eye and fork are connected by means of a pin. This joint provides flexibility in one plane and the joint can be connected and disconnected quickly. The knuckle pin is secured by means of a collar and split pin. Its rotation in the fork is prevented by means of a small stop, pin, peg or snug. The knuckle joint is made of steel or wrought iron.

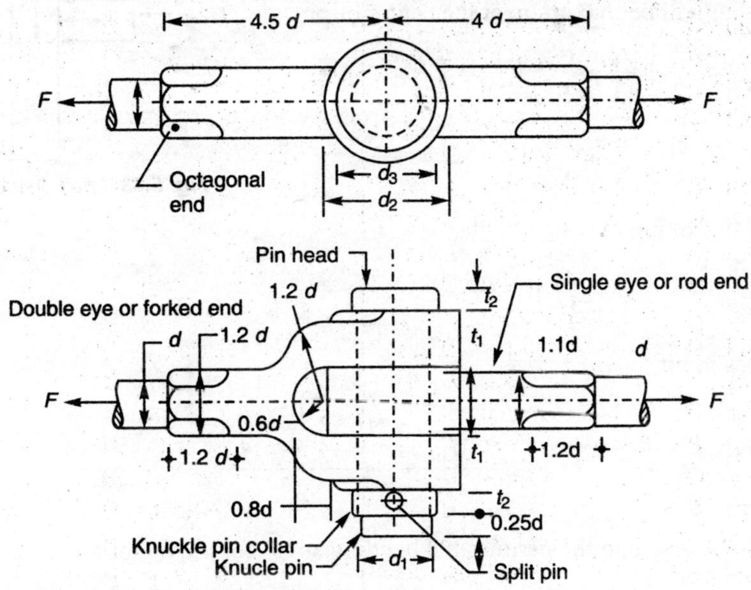

Fig. 5.42 Kunckle joint.

Two views of a knuckle joint are shown in Fig. 5.42. The various parts of the joint may be designed as follows :

1. Diameter of Rod.

Let d = diameter of the rod, and F = tensile load. Considering the failure of the rod in tension,

We have
$$\left(\frac{\pi}{4}\right) d^2 \sigma_t = F$$

$$d = \left[\frac{4F}{\pi \sigma_t}\right]^{1/2} \quad ...(1)$$

2. Diameter of the Knuckel Pin.

Considering the double shear of the pin, we have

$$2\left(\frac{\pi}{4}\right) d_1^2 \tau = F$$

$$d_1 = \left[\frac{2F}{\pi \tau}\right]^{1/2} \quad ...(2)$$

3. Check for the Knuckle Pin in Bending.

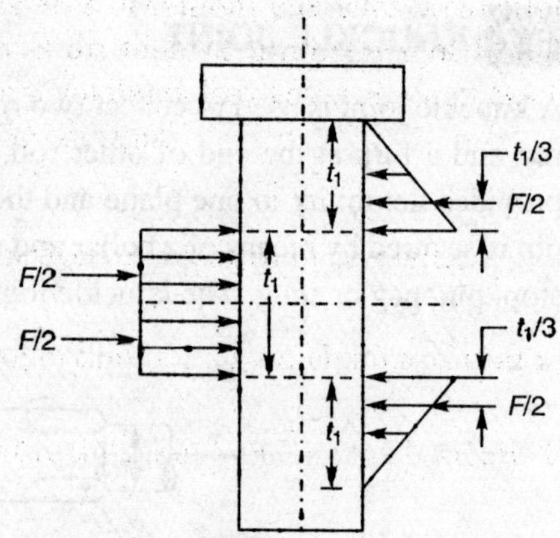

Fig. 5.43 Load distribution on the pin.

The pin fits in the eye and the fork in order to permit angular movement of eye in the fork. This introduces bending in the pin in addition to shear. To determine the bending stress in the pin, it is assumed that the load is uniformly distributed in the eye and varies uniformly over the fork, as shown in Fig. 5.43. The bending moment will be maximum at the centre of pin.

$$M_{max} = \left(\frac{F}{2}\right)\left(\frac{t_1}{3}+\frac{t}{2}\right)-\left(\frac{F}{2}\right)\times\left(\frac{t}{4}\right)$$

$$= \left(\frac{F}{2}\right)\left(\frac{t_1}{3}+\frac{t}{4}\right)$$

Section modulus of pin,

$$z = \frac{\pi d_1^3}{32}$$

Bending stress in pin,

$$\sigma_b = \frac{M_{max}}{x} = \frac{16F\left(\frac{t_1}{3}+\frac{t}{4}\right)}{\pi d_1^3}$$

Which should be test than the permissible bending stress for pin material.

4. Outer Diameter of Eye.

Considering shearing of the eye, we have

$$(d_2 - d_1)t\tau = F$$

$$d_2 = d_1 + \frac{F}{t\tau} \quad ...(4)$$

5. Check for crushing of Eye

The single eye or pin may fail in crushing due to tensile load.

$$d_1 t\sigma_c = F$$

$$\sigma_c = \frac{F}{d_1 t} \quad ...(5)$$

Which should be less than the permissible compressive stress in eye material.

6. Thickness of Fork end

Considering the tensile failure of fork end, we have

$$2(d_2 - d_1)\, t_1\sigma_c = F$$

$$t_1 = \frac{F}{2(d_2 - d_1)t_1} \quad ...(6)$$

7. Check for the Fork end in shear

The for end may fail in shear due to tensile load

$$2(d_2 - d_1)t_1\tau = F$$

$$\tau = \frac{F}{2(d_2 - d_1)t_1} \quad ...(7)$$

Which should be less than the permissible shear stress in fork.

8. Check for Fork end in crushing.

The fork end or pin may fail in crushing due to tensile load.

$$d_1 \times 2t_1 \times \sigma_c = F$$

$$\sigma_c = \frac{F}{2d_1t_1} \quad ...(8)$$

Which should be less than the permissible compressive stress in fork.

9. Empirical Dimensions.

$$d \sin 1 = d$$

$$d \sin 2 = 2d$$

$$d_3 = 1.5d$$

$$t = 1.25d$$

$$t_1 = 0.75d$$

$$t_2 = 0.5\text{d}$$

■ **Example 5.12** Design a spigot and socket cotter joint to support a load of 25 kN in tension and compression. The material used is carbon steel having following allowable stresses :

Tension and compression = 50 MPa, shear = 40 MPa, crushing = 8 MPa.

Solution : Given : $F = 25$ kN, $\sigma_t = 50$ MPa, $\tau = 40$ MPa, $\sigma_c = 80$ MPa

The spigot and socket cotter joint is shown in Fig. 5.29.

1. Diameter of Rods

Let d = diameter of the rods

Considering failure of the rods in tension or compression, we have

$$\left(\frac{\pi}{4}\right)d^2\sigma_t = F$$

$$d = \left[\frac{4F}{\pi\sigma_t}\right]^{1/2} = \left[\frac{4 \times 25 \times 10^3}{\pi \times 50}\right]^{1/2} = 25.23 \approx 26 \text{ mm}$$

2. Diameter of spigot, d_2

Considering failure of the spigot in tension, we have

$$\left[\left(\frac{\pi}{4}\right)d_2^2 - d_2t\right]\sigma_t = F$$

where d_2 = diameter of spigot

t = thickness of cotter $= \frac{d_2}{4}$

σ_t = permissible tensile stress in spigot.

$$\left[\left(\frac{\pi}{4}\right)d_2^2 - \frac{d_2 \times d_2}{4}\right]50 = 25000$$

$$d_2 = 30.56 \approx 32 \text{ mm}$$

$$t = \frac{32}{4} = 8 \text{ mm}$$

3. Check for Crushing failure of Rod or Cotter

Let σ_c = allowable compressive stress for rod or cotter.

Then

$$d_2 t \sigma_c = F$$

$$\sigma_c = \frac{F}{d_2 t} = \frac{25000}{32 \times 8} = 97.65 \text{ MPa } 80 \text{ MPa}$$

Hence not safe. We determine the value of d_2 from the safe crushing stress.

$$d_2 \times \left(\frac{d_2}{4}\right) \times 80 = 2500$$

$$d_2 = 35.35 \approx 36 \text{ mm}$$

$$t = 9 \text{ mm}$$

4. Outside Diameter of Socket, d_1

Considering the failure of socket across the slot in tension or compression,

$$\left[\left(\frac{\pi}{4}\right)(d_1^2 - d_2^2) - (d_1 - d_2)\, t\right]\sigma_t = F$$

$$\left[\left(\frac{\pi}{4}\right)(d_1^2 - 36^2) - (d_1 - 36) \times 9\right] \times 50 = 25000$$

$$\left(\frac{\pi}{4}\right)(d_1^2 - 1296) - (d_1 - 36) \times 9 = 500$$

$$d_1^2 - 1296 - 11.46 d_1 + 412.53 = 636.62$$

$$d_1^2 - 11.46 d_1 - 1920.1 = 0$$

$$d_1 = 0.5\,(11.46 \pm (11.46)^2 + 4 \times 1920.1)^{1/2}$$

$$= 49.92 \approx 50 \text{ mm}$$

5. Width of Cotter, b

Considering double shear of cotter.

$$2bt\tau = F$$

Width of cotter, $$b = \frac{F}{2t\tau} = \frac{25000}{2 \times 9 \times 40} = 34.7 \approx 35 \text{ mm}$$

6. Length of Spigot from slot to the end, a

Considering the failure of the socket end in double shear, we have

$$2ad_2\tau = F$$

Length of spigot from cotter slot to end,

$$a = \frac{F}{2d_2\tau}$$

$$= \frac{25000}{2 \times 36 \times 40} = 8.68 \approx 10 \text{ mm}$$

7. Socket Collar Diameter, d_4

Considering failure of the socket collar in crushing, as shown in Fig. 5.33, we have

$$(d_4 - d_2)\,t\sigma_c = F$$

$$(d_4 - 36) \times 9 \times 80 = 25000$$

$$d_4 = 70.7 \approx 72 \text{ mm}$$

8. Socket Collar Thickness, c

Considering the failure of socket collar in double shear, we have

$$2(d_4 - d_2)c\tau = F$$

$$2(72 - 36) \times c \times 40 = 25000$$

$$c = 10.64 \approx 12 \text{ mm}$$

9. Diameter of Spigot Collar, d_3

Considering the failure of spigot collar in crushing,

$$\left(\frac{\pi}{4}\right)(d_3^2 - d_2^2)\sigma_c = F$$

$$\left(\frac{\pi}{4}\right)(d_3^2 - 36^2) \times 80 = 25000$$

$$d_3 = 41.16 \approx 42 \text{ mm}$$

10. Thickness of Spigot Collar, t_1

Considering the failure of spigot collar in shear,

$$\pi d_2 t_1 \tau = F$$

$$t_1 = \frac{25000}{\pi \times 36 \times 40} = 5.53 \approx 6 \text{ mm}$$

11. Check for Cotter in Bending

Bending stres in cotter,

$$\sigma_b = \left[\frac{F}{4td^2}\right](2d_4 + d_2)$$

$$= \left[\frac{25000}{4 \times 9 \times 35^2}\right](2 \times 72 + 36)$$

$$= 10.2 \text{ MPa} < 80 \text{ MPa}$$

Hence safe.

12. Length of cotter

$$l = 4d = 4 \times 26 = 104 \text{ mm}$$

13. Dimension

$$e = 1.2d = 1.2 \times 36 = 31.2 \approx 32 \text{ mm}$$

■ **Example 5.13** The big end of a connecting rod, as shown in Fig. 5.37, is subjected to a maximum load of 40 kN. The diameter of the circular part of rod adjacent to the strap end is 70 mm. Design the joint assuming the following permissible stresses :

Strap : Tension = 25 MPa

Gib and cotter : Shear = 20 MPa

Solution : Given : $F = 40$ kN, $d = 70$ mm, $\sigma_t = 20$ MPa

The gib and cotter joint for the strap end (or bit end) of a connecting rod is shown in Fig. 5.37.

1. Width of Strap, B_1

Generally, width of Strap, $B_1 = d = 70$ mm

2. Thickness of Cotter, t

$$t = \frac{B_1}{4} = \frac{70}{4} = 17.5 \approx 18 \text{ mm}$$

3. Thickness of Gib

It is kept equal to thickness of cotter

$$t = \frac{B_1}{4}$$

4. Height and Length of Gib Head

Height of gib head, $t_2 = t = 18$ mm

Length of gib head, $l_3 = t = 18$ mm

5. Thickness of strap, t_1

Considering tearing of the strap at the thinnest section, we have

$$2B_1 t_1 \sigma_t = F$$

Thicknes of strap,

$$t_1 = \frac{F}{2B_1\sigma_t} = \frac{40{,}000}{2\times 70\times 25} = 11.42 \approx 12 \text{ mm}$$

Considering thearing strength of strap at the cotter shoulder to be equal to that at the thinnest section, we have

$$2t_3(B_l - t) = 2t_1B_1$$

$$t_3 = \frac{B_1t_1}{B_1 - t} = \frac{70\times 12}{70-18} = 16.15 \approx 18 \text{ mm}$$

6. Total width of Gib and Cotter, B

Considering double shear failure of gib and cotter,

$$2Bt\tau = F$$

$$B = \frac{40{,}000}{2\times 18\times 20} = 55.5 \approx 56 \text{ mm}$$

Width of gib, $b_1 = 0.55B = 0.55\times 56 = 30.8 \approx 32$ mm

Width of cotter, $b = 0.45B = 24$ mm

7. Other Dimensions

$$l_1 = 2t_1` = 2\times 12 = 24 \text{ mm}$$

$$l_2 = 2.5t_1 = 2.5\times 12 = 30 \text{ mm}$$

$$t_4 = (1.15 \text{ to} 1.5)\ t_1 = 1.25\times 12 = 15 \text{ mm}$$

Example 5.14 Design a gib and cotter joint for a square rod to carry a maximum load of 30 kN. Assume that the gib, cotter and rod are made of the same material having the following allowable streses :

Tension = 20 MPa, shear = 15 MPa, and crushing 50 MPa.

Solution : Given $F = 30$ kN, $\sigma_t = 20$ MPa, $\tau = 15$ MPa, $\sigma_c = 50$ MPa

The gib and cotter joint for a square rod is shown in Fig. 5.38.

1. Size of Rod, *x*

Considering tailure of square rod under tensile force F,

$$\sigma_t \times x^2 - F$$

$$\sigma_t \times x^2 = F$$

$$x = \left(\frac{F}{\sigma_t}\right)^{1/?} = \left(\frac{30{,}000}{20}\right)^{1/2} = 38.72 \approx 40 \text{ mm}$$

The other dimensions are fixed as follows :

Width of strap, $B_1 = x = 40$ mm

Thickness of cotter, $t = \frac{B_1}{4} = \frac{40}{4} = 10$ mm

Thickness of gib $\quad t_1 = t = 10$ mm

Height of gib, $\quad t_2 = t = 10$ mm

Length of gib head, $\quad l_4 = t = 10$ mm

2. Width of Gib and Cotter

Considering the double shear of gib and cotter,

$$2Bt\tau = F$$

$$B = \frac{F}{2t\tau} = \left(\frac{30{,}000}{2\times10\times15}\right) = 100 \text{ mm}$$

Width of gib, $\quad b_1 = 0.55B$, for a single gib

$$= 0.55\times100 = 55 \text{ mm}$$

Width of cotter, $\quad b = 0.45B$, for a single gib

$$= 0.45\times100 = 45 \times \text{mm}$$

3. Thickness of Strap

Considering the failure of strap end in tension at the location of gib and cotter,

$$2(B_1t_1 - tt_1)\sigma_t = F$$

Now $\quad B_1 = x$

Thus $\quad 2(x-t)\,t_1\sigma_1 = F$

Thickness of strap, $\quad t_1 = \dfrac{F}{2(x-t)\sigma_t}$

$$= \frac{30{,}000}{2(40-10)\times20} = 25 \text{ mm}$$

4. Check for Strap or Gib in crushing

$$2tt_1\sigma_c = F$$

$$\sigma_c = \frac{F}{2tt_1}$$

$$= \frac{30{,}000}{2\times10\times25} = 60 \text{ MPa} > 50 \text{ MPa}$$

Therefore, we calculate t_1 from crushing failure,

$$t_1 = \frac{30{,}000}{2\times10\times50} = 30 \text{ mm}$$

5. Length of Rod from Cotter to end

Considering the double shear of rod,

$$2l_1 \times \tau = F$$

$$l_1 = \frac{F}{2x\tau} = \frac{30{,}000}{2\times40\times15} = 25 \text{ mm}$$

6. Length of Rod from Gib to the end of Strap

Considering the double shear of the strap.

$$2 \times 2 l_2 t_1 \tau = F$$

$$l_1 = \frac{F}{4 t_1 r} = \frac{30{,}000}{4 \times 30 \times 15} = 16.67 \approx 18 \text{ mm}$$

7. Other dimensions

$$l_3 = \frac{2x}{3} = \frac{2 \times 40}{3} = 26.67 \approx 27 \text{ mm}$$

■ **Example 5.15** Design a knuckle joint to transmit 100 kN. The design stresses are : Tension = 80 MPa, shear = 60 MPa, and crushing = 120 MPa.

Solution : Given : $F = 100$ kN, $\sigma_t = 80$ MPa, $\tau = 60$ MPa, $\sigma_c = 120$ MPa

The knuckle joint is shown in Fig. 5.42

1. Diameter of Rod, *d*

Considering the failure of the rod in tension,

$$\left(\frac{\pi}{4}\right) d^2 \sigma_t = F$$

$$d = \left[\frac{4F}{\pi \sigma_t}\right]^{1/2}$$

$$= \left[\frac{4 \times 100{,}000}{\pi \times 80}\right]^{1/2} = 39.89 \approx 40 \text{ mm}$$

2. Diameter of the Knuckle Pin, d_1

Empirically, $d_1 = d = 40$ mm

Considering the double shear of the pin,

$$2\left(\frac{\pi}{4}\right) d_1^2 \tau = F$$

$$\tau = \frac{2F}{\pi d_1^2}$$

Hence safe $= \frac{2 \times 100{,}000}{\pi \times 40^2} = 39.8 \text{ MPa} < 60 \text{ MPa}$

3. Thickness of single eye or rod end, *t*

$$t = 1.25d = 1.25 \times 40 = 50 \text{ mm}$$

4. Thickness of fork, t_1

$$t_1 = 0.75d = 0.75 \times 40 = 30 \text{ mm}$$

5. Thickness of pin head, t_2

$$t_2 = 0.5d = 0.5 \times 40 = 20 \text{ mm}$$

6. Check for the Knuckle Pin Bending.

$$\text{Bending stress in pin,} \quad \sigma_b = 16F \frac{\left(\frac{t_1}{3} + \frac{t}{4}\right)}{\pi d_1^3}$$

$$= 16 \times 100{,}000 \frac{\left(\frac{30}{3} + \frac{50}{4}\right)}{\pi \times 40^3} = 179 \text{ MPa}$$

Which gives a lower value of factor or safety.

7. Outer Diameter of Eye.

Considering shearing of the eye, we have

$$(d_2 - d_1)\, t\tau = F$$

$$d_2 = d_1 + \frac{F}{t\tau} = 40 + \frac{100{,}000}{50 \times 60} = 73.3 \approx 75 \text{ mm}$$

8. Check for Thickness of Fork eye.

The single eye or pin may fail in crushing due to tensile load.

$$d_1 t \sigma_c = F$$

$$\sigma_c = \frac{F}{d_1 t} = \frac{100{,}000}{40 \times 50} = 50 \text{ MPa } 120 \text{ MPa}$$

Hence, safe.

9. Check for Thicknes of Fork end.

Considering the tensile failure of form end,

$$2(d_2 - d_1)\, t_1 \sigma_c = F$$

$$\sigma_t = \frac{F}{2(d_2 - d_1)t_1} = \frac{100{,}000}{2(75 - 40) \times 30}$$

$$= 47.6 \text{ MPa } 80 \text{ MPa}$$

Hence, safe.

10. Check for the Fork end in shear.

The fork end may fail in shear due to tensile load.

$$2(d_2 - d_1)\, t_1 \tau = F$$

$$\tau = \frac{F}{2(d_2 - d_1)t_1}$$

$$= \frac{100{,}000}{2(75 - 40) \times 30} = 47.6 \text{ MPa } 60 \text{ MPa}$$

Hence, safe.

11. Check for Fork end in crushing.

The fork end or pin may fail in crushing due to tensile load.

$$d_1 \times 2t_1 \times \sigma_c = F$$

$$\sigma_c = \frac{F}{2d_1 t_1} = \frac{100{,}000}{2 \times 40 \times 30} = 41.67 \text{ MPa } 120 \text{ MPa}$$

Hence safe.

SOLVED QUESTIONS FROM UNIVERSITY EXAMINATIONS

■ **Example 5.16** Design a triple riveted longitudinal, double strap butt joint with chain riveting for a boiler of 1.5 m diameter and carrying a pressure of 1.2 N/mm^2. The allowable stresses are : $\sigma_t = 105$ N/mm^2, $\tau = 62$ MPa and $\sigma_c = 120$ MPa. [MDU, Dec., 2007]

Solution : Given : $p = 1.2$ N/mm^2, $D = 1.5$ m, $\sigma_t = 105$ N / mm^2, $\sigma_c = 120$ MPa, $\tau = 62$ MPa

Triple riveted butt longitudinal joint (chain)

The joint is shown in Fig. 5.44.

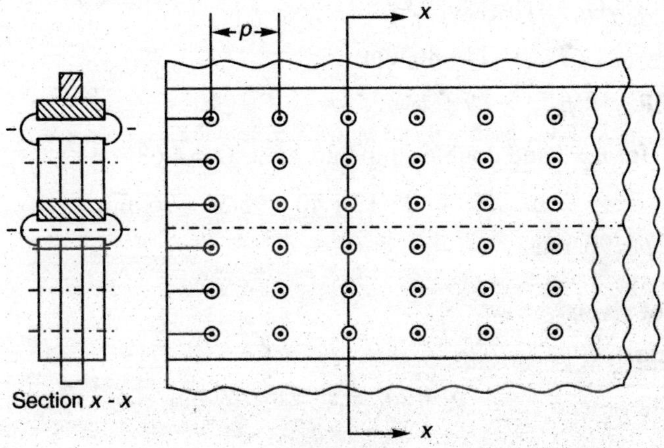

Fig. 5.44

1. Plate thickness

For a triple rivited butt joint, the efficiency from Table 5.3 is 80-90%. Taking $\eta_l = 85\%$, plate thickness is given by,

$$t = \frac{pD}{2\sigma_t \eta_l} + c$$

$$= \frac{1.2 \times 1.5 \times 10^3}{2 \times 105 \times 0.85} + 1 = 11.08 \text{ mm}$$

The standard nominal size of steel plate nearest to above value is 12 mm.

We take $t = 12$ mm.

2. Diameter of rivet

Hole for $t \geq 8$ mm, according to Unwin's formula is,

$$d_0 = 6\sqrt{t} = 6\sqrt{12} = 20.78 \approx 21 \text{ mm}$$

Diameter of rivet, $d = d_0 - 1$ mm for 12 mm $\leq d \leq$ 24 mm

$$\therefore \qquad d = 21 - 1 = 20 \text{ mm}$$

3. Pitch of rivets :

Tearing resistance of plate per pitch length,

$$P_t = (p - d_0)\, t\sigma_t = (p - 21) \times 12 \times 105$$

$$= 1260(p - 21) \text{ N}$$

Number of rivets per pitch length are 3, which are in double shear.

Shearing resistance of rivets,

$$P_s = 3 \times 1.875 \times \frac{\pi}{4}(21)^2 \times 62 = 120793 \text{ N}$$

For $\qquad P_t = P_s$

$$1260(p - 21) = 120793$$

$$p = 116.86 \text{ mm}$$

According to IBR, $\qquad p_{max} = Ct + 40$

From Table 5.6, for $n = 3$ and double strap butt joint, $C = 4.63$

$$\therefore \qquad p_{max} = 4.63 \times 12 + 40 = 95.56 \approx 96 \text{ mm}$$

Hence, we take $p = 96$ mm

4. Transverse pitch of rivets

For chain riveting,

$$p_t = 2d_0 = 2 \times 21 = 42 \text{ mm}$$

5. Thickness of buitt strap

$t_1 = 0.625t$ for double butt straps of equal width having chain riveting

$= 0.625 \times 12 = 7.5$ mm

6. Margin

$$m = 1.5d_0 = 1.5 \times 21 = 31.5 \text{ mm} \simeq 32 \text{ mm}$$

7. Tearing resistance of plate

$$P_t = (p - d_0)t\sigma_t = (96 - 21) \times 12 \times 105 = 94500 \text{ N}$$

Caushing resistance of rivets,

$$P_c = ndt\sigma_c = 3 \times 20 \times 12 \times 120 = 86400 \text{ N}$$

Shearing strength of rivets,

$$P_s = 120793 \text{ N}$$

Strength of riveted joint = 86400 N (least value out of P_t, P_c and P_s)

Strength of solid plate,

$$P = p_t \sigma_t = 96 \times 12 \times 105 = 120960 \text{ N}$$

Joint efficiency, $$\eta_l = \frac{86400}{120960} = 0.7143 \text{ or } 71.43\%$$

■ **Example 5.17** A 65 mm diameter solid rod is to be welded to a flat plate by a fillet weld around the circumference of the rod. Determine the size of the weld if the torque on the rod is 3 kN.m. The allowable shear stress in the weld is 70 MPa. [MDU, May, 2008]

Solution : Given : $d = 65$ mm, $T = 3$ kN.m, $\tau = 70$ MPa

Polar moment of inertia of weld section,

$$\tau = \frac{\pi}{4} t d^3$$

Polar section modulus of weld,

$$z_p = \frac{\pi t d^2}{2}$$

Shear stress induced in weld,

$$\tau = \frac{T}{z_p} = \frac{2T}{\pi t d^2}$$

Throat thickness, $$t = \frac{2T}{\pi d^2 \tau} = \frac{2 \times 3 \times 10^3 \times 10^3}{\pi (65)^2 \times 70}$$

$$= 6.45 = 6.5 \text{ mm}$$

Size of weld, $$h = \frac{t}{0.707} = 9.12 \simeq 10 \text{ mm}$$

■ **Example 5.18** Design the longitudinal and circumfertiol joints for a boiler whose diameter is 2.4 m and is subjected to a pressure of 1 N/mm^2. The longitudinal efficiency is about 85%. The longitudinal joint is a triple riveted butt joint with two unequal cover plates. The allowable stresses are : $\sigma_t = 7$ MPa, $\tau = 56$ MPa and $\sigma_c = 120$ MPa. [MDU, May, 2008]

Solution :

Given : $D = 2.4$ m, $p = 1 \text{ N/mm}^2$, $\eta_l \simeq 0.85$, $\sigma_t = 77$ MPa, $\tau = 56$ MPa, $\sigma_c = 120$ MPa.

1. Longitudinal joint

Plate thickness, $$t = \frac{pD}{2\sigma_t \eta_l} + 1 \text{ mm} = \frac{1 \times 2.4 \times 10^3}{2 \times 77 \times 0.85} + 1 = 19.33 \text{ mm}$$

The nearest standard nominal steel plate thickness = 20 mm

2. Rivet hole diameter

$$d_o = 6\sqrt{t} \text{ for } t \geq 8 \text{ mm}$$
$$= 6\sqrt{20} = 26.83 \simeq 27 \text{ mm}$$

3. Rivet diameter

$$d = d_o - 1.5 \text{ mm for } 27 \text{ mm} \leq d \leq 36 \text{ mm}$$
$$= 27 - 1.5 = 25.5 \text{ mm}$$

Standard diameter of boiler rivets,

$$d = 27 \text{ mm}$$

∴ Rivet hole diameter, $d_o + 27 + 1.5 = 28.5$ m

4. Pitch of rivets

The triple riveted longitudinal butt joint with two unequal cover plates in shown in Fig. 5.45. There are 4 rivets per pitch length, out of which 4 rivets are in double shear and one rivet in single shear. Assume that strength of rivets in double shear is 1.875 times that of single shear.

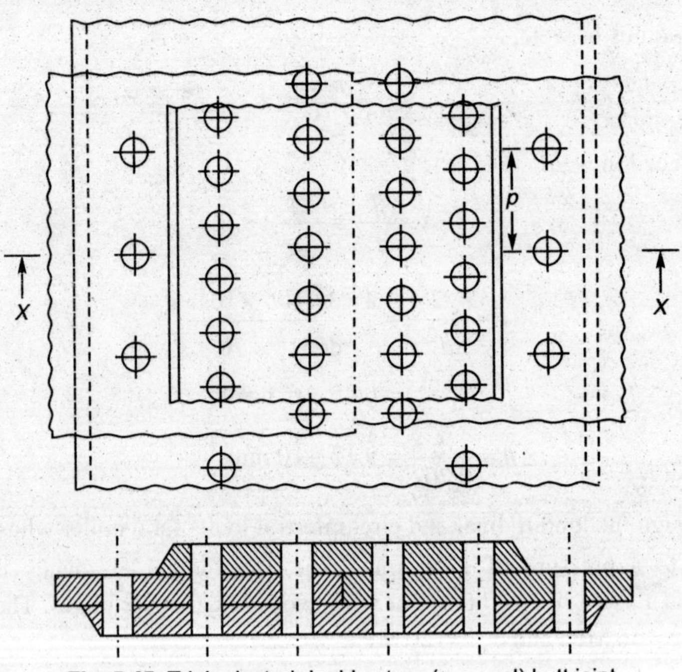

Fig. 5.45 Triple riveted double strap (unequal) butt joint.

Shearing strength of rivets per pitch length.

$$P_s = (4 \times 1.875 + 1)\frac{\pi}{4} d^2 \tau$$
$$= 8.5 \times \frac{\pi}{4} d^2 \tau$$
$$= 8.5 \times \frac{\pi}{4} (27)^2 \times 56 = 272536 \text{ N}$$

Tearing strength of plate per pitch length,

$$P_t = (p - d_o)\, t\, \sigma_t = (p - 28.5) \times 20 \times 77$$
$$= 1540\,(p - 28.5)\text{ N}$$

For $P_t = P_s$, we have

$$1540\,(p - 28.5) = 272536$$
$$p = 205.5 \simeq 206\text{ mm}$$

According to IBR, for $n = 5$, $c = 6.0$

$$p_{\max} = Ct + 40\text{ mm}$$
$$= 6 \times 20 + 40 = 160\text{ mm}$$

$p_{\max} < p$, therefore, $\quad p = p_{\max} = 160\text{ mm}$

Pitch of rivets in inner row, $\quad p_i = \dfrac{p}{2} = \dfrac{160}{2} = 80\text{ mm}$

5. Transverse pitch of rivets

According to IBR, distance of outer most row from inner row

$$= 0.2\,p + 1.15\,d_o$$
$$= 0.2 \times 60 + 1.15 \times 28.5 = 64.77 \simeq 65\text{ mm}$$

Distance between inner rows for zig-zag riveting

$$= 0.165\,p + 0.67\,d_0$$
$$= 0.165 \times 160 + 0.67 \times 28.5$$
$$= 45.5 \simeq 46\text{ mm}$$

6. Thickness of butt straps

For unequal width of butt straps, thickness of butt strap,

$$t_1 = 0.75t \text{ for wide strap on the inside}$$
$$= 0.75 \times 20 = 15\text{ mm}$$
$$= 0.625\,t \text{ for narrow strap on the outside}$$
$$= 0.625 \times 20 = 12.5\text{ mm}$$

7. Margin

$$m = 1.5d = 1.5 \times 27 = 40.5\text{ mm}$$

8. Efficiency of joint

Tearing resistance of plate in the outer row,

$$P_t = (p - d_o)\, t\sigma_t$$
$$= (160 - 28.5) \times 20 \times 77 = 202510\text{ N}$$

Shearing resistance of rivets,

$$P_s = 272536\text{ N}$$

Crushing resistance of rivets,

$$P_c = ndt\,\sigma_c = 5 \times 27 \times 20 \times 120 = 324000\text{ N}$$

Strength of joint when the single rivet in outermost row fails in shear and the plate at 2 rivets in second row tears off,

$$P_1 = (p - 2d_o)t\sigma_t + \frac{\pi}{4}d^2\tau$$

$$= (160 - 2 \times 28.5) \times 20 \times 77 + \frac{\pi}{4}(27)^2 \times 56 = 190683 \text{ N}$$

Strength of joint $= \min(P_t, P_s, P_c, P_1) = 190683$ N

Strength of solid plate,

$$P = pt\sigma_t = 160 \times 20 \times 77 = 246400 \text{ N}$$

Joint efficiency $= \dfrac{190683}{246400} = 0.7738$ or 77.38%

Circumferential joint :

(i) $t = 20$ mm, $d_o = 28.5$ mm, $d = 27$ mm

(ii) Number of rivets, $n = \left(\dfrac{D}{d_o}\right)^2\left(\dfrac{p}{\tau}\right) = \left(\dfrac{2400}{28.5}\right) \times \left(\dfrac{1}{56}\right) = 127.6 \simeq 130$

(iii) Pitch of rivets :

Assuming double riveted lap joint with zig-zag riveting, number of rivets per row $= \dfrac{n}{2} = \dfrac{130}{2} = 65$

Pitch of rivets, $p_c = \dfrac{\pi(\Delta + t)}{65} = \dfrac{\pi(2400 + 20)}{65} = 116.96 \simeq 117$ mm

(iv) Efficiency of joint, $\eta_c = \dfrac{p_c - d_o}{p_c} = \dfrac{117 - 28.5}{117} = 0.7564$ or 75.64%

(v) Distance between rows of rivets :

For zig-zag riveting, distance between rows of rivets

$$= 0.33p_c + 0.67d_o$$

$$= 0.33 \times 117 + 0.67 \times 28.5 = 57.5 \simeq 58 \text{ mm}$$

(vi) Margin, $m = 1.5d_0 = 1.5 \times 28.5 = 42.75 \simeq 43$ mm

Example 5.19 Two steel rods of equal diameter are required to join with a cotter joint. Each rod is subjected to an axial tensile force of 60 kN. Design and draw the joint specifying its main dimensions.

[MDU, Dec., 2008]

Solution : Refer to example 5.12

Example 5.20 A knuckle joint is to be used to connect two circular rods subjected to an axial tensile force of 60 kN. The rods are coaxial and a small amount of angular movement between their axes is permissible. Design and draw the joint specifying the dimensions of its components.

[MDU, May, 2009]

Solution : Refer to example 5.15

■ **Example 5.21** A bracket is supported by means of 4 rivets of same size, as shown in Fig. 5.46(a). Determine the diameter of the rivet if the maximum allowable shear stress is 100 MPa.

[MDU, Dec., 2009]

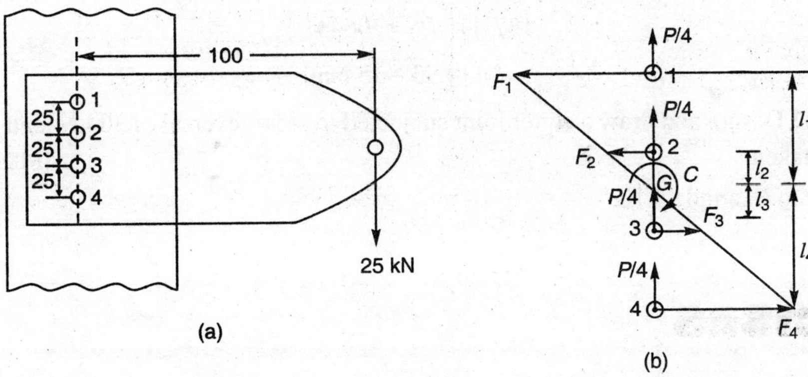

Fig. 5.46

Solution : Given : $p = 25$ kN, $e = 100$ mm

The forces on the rivets are shown in Fig. 5.46(b). The C.G.of rivet group is located at G.

$$l_1 = l_4 = 37.5 \text{ mm}, \ l_2 = l_3 = 12.5 \text{ mm}$$

Moment of force about G, $C = P \times e = 25 \times 10^3 \times 100 = 25 \times 10^5$ N-mm

$$F_1 = F_4 \text{ and } F_2 = F_3$$

$$\begin{aligned}\text{Resistance couple} &= 2(F_1 \times l_1 + F_2 \times l_2) \\ &= 2(F_1 \times 37.5 + F_2 \times 12.5) \\ &= 25(3F_1 + F_2)\end{aligned}$$

Also
$$\frac{F_1}{l_1} = \frac{F_2}{l_2}$$

$$F_2 = \frac{l_2}{l_1} \times F_1 = \left(\frac{12.5}{37.5}\right) F_1 = \frac{F_1}{3}$$

$$\therefore \text{Resisting couple} = 25\left(3F_1 + \frac{F_1}{3}\right) = 25 \times \frac{10}{3} F_1 = \frac{250}{3} F_1$$

For equilibrium,
$$\frac{250}{3} F_1 = 25 \times 10^5$$

$$F_1 = 30000 \text{ N}$$

$$\text{Direct force in each rivet} = \frac{P}{4} = \frac{25000}{4} = 6250 \text{ N}$$

Maximum resultant force occasion rivet 1 and 4.

Resultant force in rivet 1,

$$R_1 = \sqrt{\left(\frac{P}{4}\right)^2 + F_1^2}$$

$$= \sqrt{(6250)^2 + (30{,}000)^2} = 30644 \text{ N}$$

As the rivets are in double shear, strength, strength of a rivet

$$=1.75\times\frac{\pi}{4}d^2\tau=1.75\times\frac{\pi}{4}d^2\times100=1.37.445d^2$$

$$\therefore \qquad 137.445d^2=30644$$

$$d=14.93\simeq15 \text{ mm}$$

■ **Example 5.22** Design and draw a cotter joint subjected to slow reversal of 30 kN load. Clearly state the decision taken. [MDU, Dec., 2009]

Solution : Refer to Example 5.12.

EXERCISES

1. Calculate the main dimensions of the longitudinal and circumferential joints of a steam boiler whose inner diameter is 2 m and the pressure of steam is 25 bar. Assume suitable data from the Data Handbook.
2. Two steel plates 250 mm × 10 mm are connected by a diamond butt riveted joint with single strap. The plates are required to carry 300 kN load. The allowable stresses are 120 MPa in tension, 80 MPa in shear and 210 MPa in crushing. Design the joint.
3. Design an economical butt joint with two cover straps to connect two steel plates 300 mm × 12 mm. Assume suitable data from the Data Book.
4. Two plates are welded by means of fillet weld as shown in Fig. 5.47. The leg dimension of the welds is 15 mm and the permissible shear stress at the throat cross-section is 90 MPa. Determine the length of the weld. P = 40 kN.

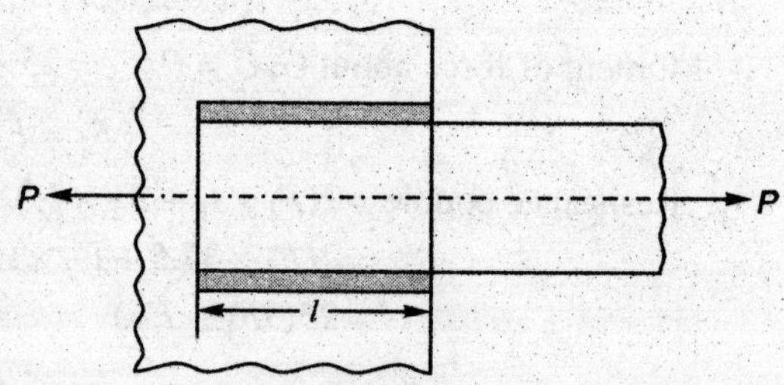

Fig. 5.47

5. A bracket as shown in Fig. 5.48 is welded to a plate with the same size welds. The permissible force per mm of weld length is 1.25 kN. Calculate the lengths l_1 and l_2.

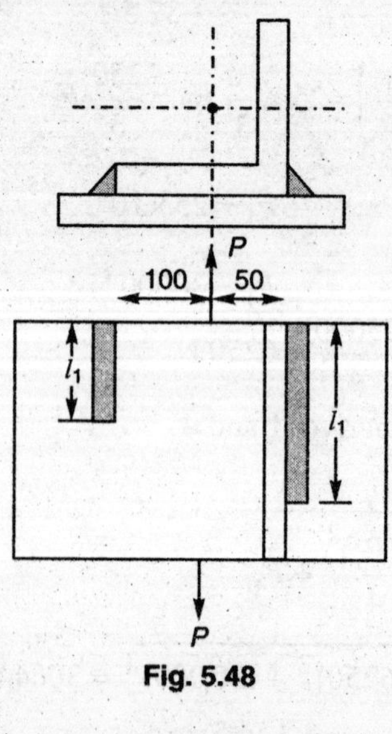

Fig. 5.48

6. A welded connection of steel plates as shown in Fig. 4.49 is subjected to an eccentric load of 15 kN. Determine the throat thickness of the weld if the permissible stress is limited to 90 MPa.

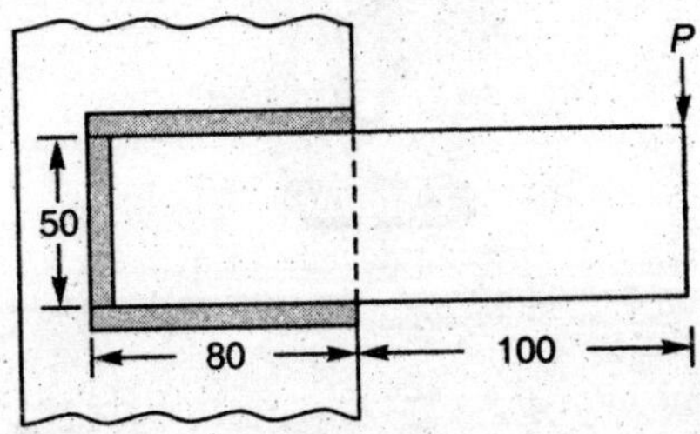

Fig. 5.49

7. A solid circular shaft 30 mm diameter is welded to a support by means of a fillet weld as shown in Fig. 5.50. Determine the leg dimensions of the weld is the permissible shear stress is 90 MPa.

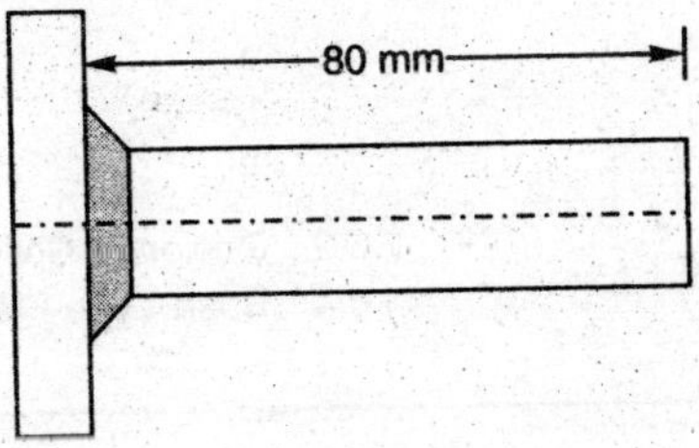

Fig. 5.50

8. Design a spigot and socket joint to connect two rods to take a pull of 60 kN. The maximum permissible stresses are : 50 MPa in in tension, 40 MPa shear and 100 MPa in crushing.

9. Design a gib and cotter joint to take an axial load of 100 kN which alternately changes from tension to compression. The allowable stresses are : 60 MPa in tension, 40 MPa in shear and 75 MPa in crushing.

10. The big end of a connecting rod is subjected to a load of 45 kN. The diameter of the circular part adjacent to the strap is 50 mm. Design the joint assuming that permissible tensile stress in strap is 35 MPa and shear stress in gib and cotter is 20 MPa.

11. Design a knuckle joint to withstand a tensile load of 30 kN. The permissible stresses are : Tension = 65 MPa, shear = 50 MPa, and crushing = 85 MPa.

6

Chapter

Keys, Couplings and Flywheel

Inside This Chapter

6.1 Keys

6.2 Design of Splines

6.3 Couplings

6.4 Bush Type Flexible Coupling

6.5 Flywheel

6.1 KEYS

A key is a piece of metal inserted between the shaft and hub of a pulley or a gear to connect them together in order to prevent relative motion between them. A keyway is a slot or recess cut in the shaft and hub to accomodate the key. It is a temporary type of fastening.

6.1.1 Types of Keys

Keys are of the following types :

1. Sunk key
2. Saddle key
3. Tangent key
4. Round key, and
5. Splines.

1. Sunk key

A sunk key is provided half in the keyway of the shaft and half in the keyway of the hub. This key is of the following types :

(i) Rectangular sunk key : This key is shown in Fig. 6.1 (a). Its usual proportions are : width $b = \frac{d}{4}$, thickness, $t = \frac{d}{6}$, where d = diameter of the shaft. Taper on the key $= 1:100$.

(ii) Square sunk key : In this case, $b = t = \frac{d}{4}$, as shown in Fig. 6.1(b).

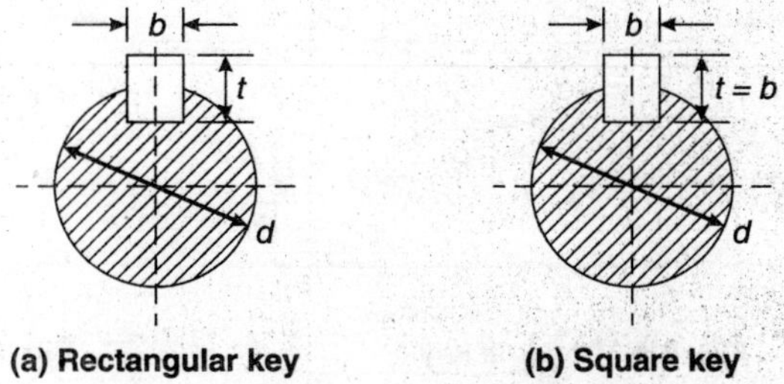

Fig. 6.1 Rectangular and square keys.

(iii) Parallel sunk key : It may be a rectangular or square key with no taper on the length of key.

(iv) Gib-head key : It is a rectangular sunk key with a head at one end known as gib head to facilitate the removal of key from the keyway, as shown in Fig. 6.2. Its usual proportions are :

$$b = \frac{d}{4} \qquad \text{...(6.1a)}$$

$$t = \frac{d}{6} \qquad \text{...(6.1b)}$$

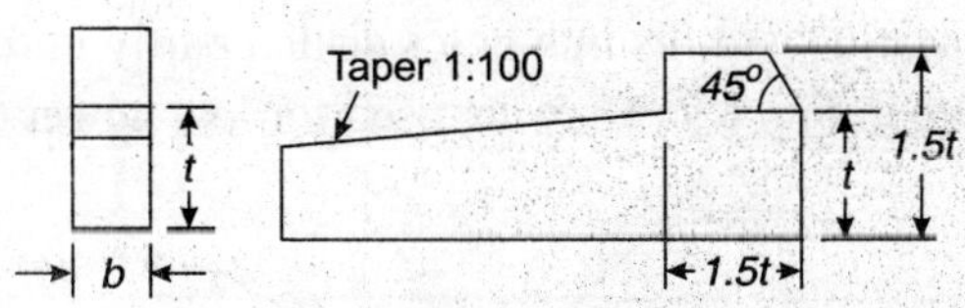

Fig. 6.2 Gib-head key.

(v) **(a) Feather key :** A feather key is used where it is essential to slide a keyed gear or pulley along the shaft axis. The key is either made a tight fit in the shaft or the hub and a sliding fit in the other. When making a tight fit in the shaft, it is screwed to the shaft, as shown in Fig. 6.3(a) and when making a tight fit in the hub, gib heads are provided on both sides of the key, as shown in Fig. 6.3(b).

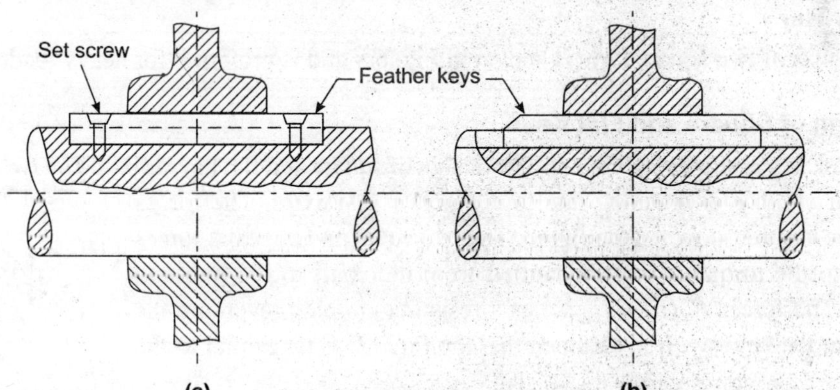

Fig. 6.3 Feather key.

(b) Woodruff key : It is an easily adjustable key cut from a cylindrical disc having segmental cross-section in front view as shown in Fig. 6.4.

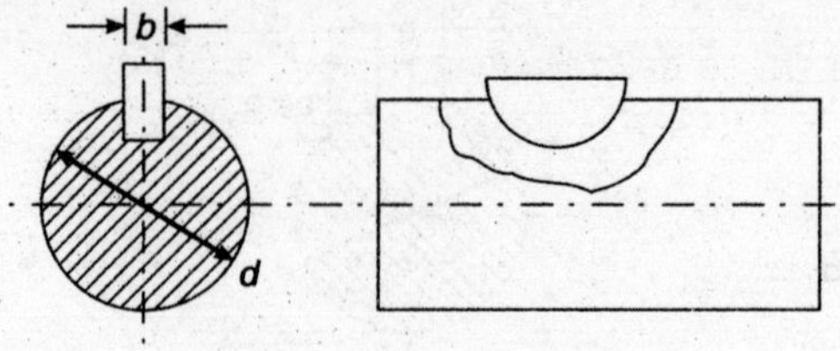

Fig. 6.4 Woodruff key.

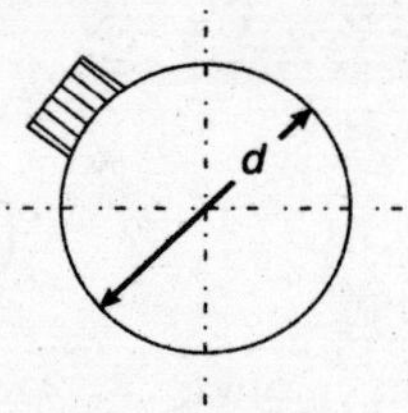

Fig. 6.5 Saddle key.

2. Saddle Key

It is similar to a flat key whose bottom surface is given a radius to match the shaft radius as shown in Fig. 6.5. They are used for light loads.

3. Tangent Key

The tangent keys are fitted in pairs at right angles as shown in Fig. 6.6. They remain in compression under load and are used for heavy loads.

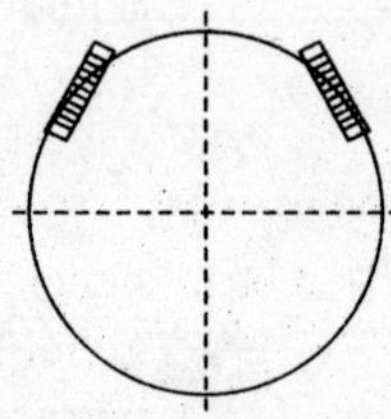

Fig. 6.6 Tangent key.

4. Round Key

This key is circular in section and fits into holes drilled partly in the shaft and partly in the hub as shown in Fig. 6.7. They are used for low power drives.

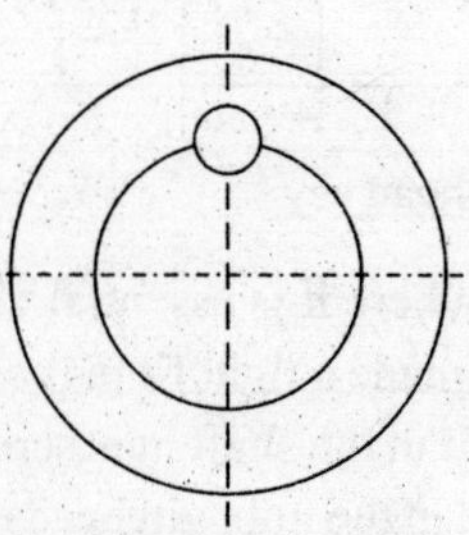

Fig. 6.7 Round key.

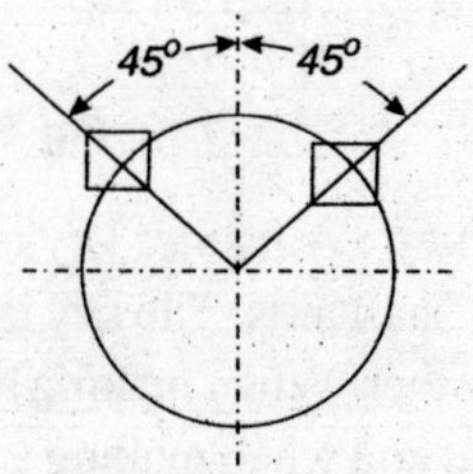

Fig. 6.8 Kennedy key.

5. Kennedy Key

It consists of two square keys is shown in Fig. 6.8 and is preferred for heavy loads.

6.1.2 Design of Square and Flat Keys

A square sunk key is shown in Fig. 6.9. The design of square and flat keys is based on two criteria-failure due to shear stresses or crushing due to compressive stresses. The forces acting on the key are shown in the figure. The two equal and opposite forces F are due to the torque being transmitted from the shaft to the hub and vice-versa. The forces F' $(F' = F)$ act as a resisting couple preventing the key to roll in the keyway. It is assumed that the force F is tangential to the shaft.

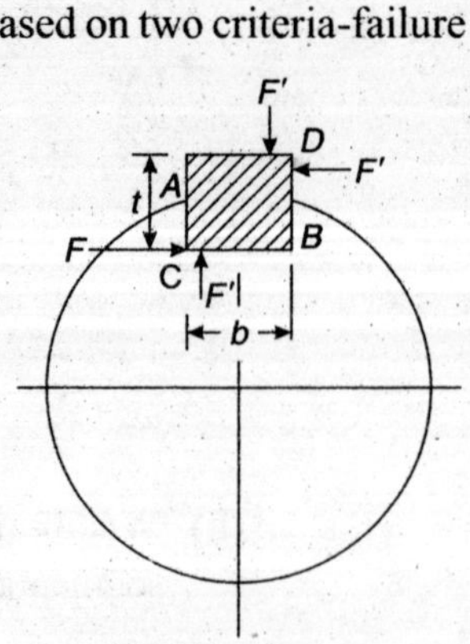

Fig. 6.9 Forces on rectangular key.

Therefore,
$$F = \frac{2T}{d} \quad \ldots(1)$$

where T = transmitted torque, N-mm

d = shaft diameter, mm

F = tangential force, N

The shear failure shall occur in plane AB so that the shear stress is given by :

$$\tau = \frac{F}{bl} \quad \text{...(2)}$$

where b = width of key

l = length of key

From Eqs. (1) and (2), we get

$$\tau = \frac{2T}{dbl}. \quad \text{...(6.2)}$$

The failure due to compressive stresses shall occur on surfaces AC and BD. Hence,

$$AC = BD = \frac{t}{2}$$

The compressive stress is given by

$$\sigma_c = \frac{F}{\frac{tl}{2}} = \frac{2F}{tl} \quad \text{...(3)}$$

where t = thickness of key

From Eqs. (1) and (3), we get

$$\sigma_c = \frac{4T}{dtl} \quad \text{...(6.3)}$$

If the key is be equally strong in shear and crushing, then

$$\frac{2T}{dbl} = \frac{4T}{dtl}$$

$$\frac{2b}{t} = \frac{\sigma_c}{\tau} \quad \text{...(6.4)}$$

Generally $\sigma_c = 2\tau$

Therefore $\frac{b}{t} = 1$, *i.e.*, a square key is the best size.

If the key is to be equally strong in shear as the shaft, then

$$\frac{2T}{dbl} = \frac{16T}{\pi d^3}$$

$$l = \frac{\pi d^2}{8b}$$

If $b = \frac{d}{4}$, then $l = \frac{\pi d}{2} \approx 1.5\,d$

6.2 DESIGN OF SPLINES

Splines are keys made integral with the shaft which fit in the keyways broached in the hub. Such shafts are called splined shafts as shown in Fig. 6.10. These shafts are stronger than shafts having a single

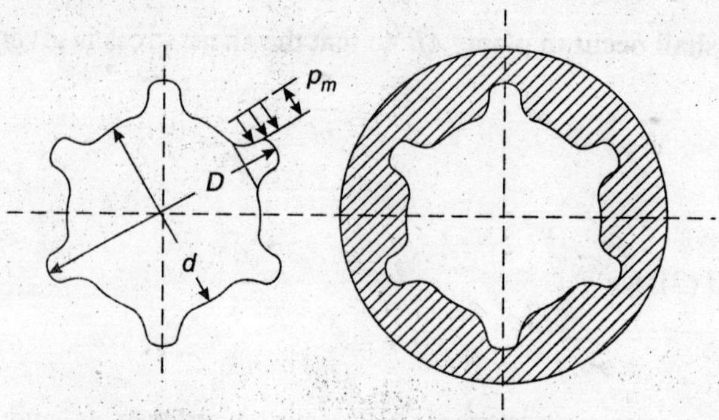

Fig. 6.10 Splines.

keyway. Splined shafts are used when the force transmitted is large and the driven member is to slide over the driving member.

Let D = major diameter of splines

d = minor diameter of splines

l = length of hub

n = number of splines

p_m = permissible pressure splines ≈ 6.5 MPa

A = total area of a splines

T = torque transmitted by the splines

Now $$A = (D - d)\frac{\ln}{2}$$

Mean radius of splines, $$R_m = \frac{D + d}{4}$$

Torque transmitting capacity of splines, $T = p_m A R_m$

$$= \left(\frac{1}{8}\right) p_m \, ln\,(D^2 - d^2) \quad \ldots(6.5)$$

■ **Example 6.1** Design the rectangular key for a shaft of 75 mm diameter. The shearing and curshing stresses for key material are 50 MPa and 75 MPa respectively.

Solution : Given : $d = 75$ mm, $\tau = 50$ MPa, $\sigma_c = 75$ MPa

For the key to be equally strong in shear and crushing

$$\frac{2b}{t} = \frac{\sigma_c}{\tau} = \frac{75}{50} = 1.5$$

$$b = \frac{d}{4} = \frac{75}{4} = 18.75 \approx 20 \text{ mm}$$

$$t = \frac{2 \times 20}{1.5} = 26.67 \approx 27 \text{ mm}$$

For key to be equally strong in shear as the shaft,

$$l = \frac{\pi d^2}{8b} = \frac{\pi \times 75^2}{8 \times 20} = 110.4 \approx 115 \text{ m.}$$

■ **Example 6.2** Design a rectangular key for a shaft of 50 mm diameter. The shearing and crushing stresses for the key and shaft are 42 MPa and 72 MPa.

Solution : Given : $d = 50$ mm, $\tau = 42$ MPa, $\sigma_c = 72$ MPa

For the key to be equally strong in shear and crushing,

$$\frac{2b}{t} = \frac{\sigma_c}{\tau} = \frac{72}{42} = 1.714$$

$$b = \frac{d}{4} = \frac{50}{4} = 12.5 \approx 15 \text{ mm}$$

$$t = \frac{2 \times 15}{1.714} = 17.5 \approx 20 \text{ mm}$$

For key to be equally strong in shear as the shaft,

$$l = \frac{\pi d^2}{8b} = \frac{\pi \times 50^2}{8 \times 15} = 65.45 \approx 60 \text{ mm}$$

■ **Example 6.3** Design a square key for fixing a gear on the shaft having 25 mm diameter. The gear rotates at 550 rpm and transmits 12 kW power to the meshing gear. The key is made of steel having yield stress in tension as 400 N/mm^2. The yield stress in compression and tension may be taken equal to each other. Assume factor of safety 2.5.

Solution : Given : $d = 25$ mm, $N = 550$ rpm, $P = 12$ kW, $\sigma_{yt} = 400$ N/mm^2 $= \sigma_{yc}$, $n = 2.5$

$$P = \frac{2\pi NT}{60 \times 10^3}$$

$$T = \frac{12 \times 60 \times 10^3}{2 \times \pi \times 550}$$

$$= 208.348 \text{ N-m}$$

According to maximum shear stress theory,

$$\tau = 0.5\,\sigma_{yt} = 0.5 \times 400 = 200 \text{ MPa}$$

Design shear stress, $$\tau_d = \frac{\tau}{n} = \frac{200}{2.5} = 80 \text{ N/mm}^2$$

Design crushing strength, $$\sigma_d = \frac{\sigma_{yc}}{n} = \frac{400}{2.5} = 160 \text{ N/mm}^2$$

Width of key, $$b = \frac{d}{4} = \frac{25}{4} = 6.25 \approx 7 \text{ mm}$$

Thickness of key, $t = 7$ mm

Length of key based on shear failure,

$$l = \frac{2T}{db\tau_d}$$

$$= \frac{2 \times 208348}{25 \times 7 \times 80} = 29.75 \approx 30 \text{ mm}$$

Length of key based on crushing failure,

$$l = \frac{4T}{dt\sigma_d}$$

$$= \frac{4 \times 208348}{25 \times 7 \times 160} = 29.75 \approx 30 \text{ mm}$$

6.3 COUPLINGS

A coupling is a machine element which can be used to connect two shafts. The applications of couplings are :

1. To provide a connection between two units manufactured separately.
2. To provide for easy disconnection during repair.
3. To provide as a connection between shaft lengths to facilitate easy transformation.
4. To provide mechanical flexibility for misalignment.
5. To minimise shock load transmission throughout the shaft length.

6.3.1 Types of Couplings

Shaft couplings are used to connect two shafts. The shafts may be either aligned or misaligned having lateral or angular misalignment. Accordingly we can have either rigid or flexible couplings. The couplings may be classified as follows :

1. **Rigid couplings :** They do not permit even minor misalignment of the axes of the connected shafts. Rigid couplings are of the following types :
   (i) Sleeve or muff coupling
   (ii) Clamp coupling
   (iii) Flange coupling, and
   (iv) Solid or forged flange coupling.
2. **Flexible couplings :** They permit slight misalignment between the axes of the connected shafts. Flexible couplings are of the following types :
   (i) Oldham's coupling
   (ii) Hooke's coupling
   (iii) Universal flexible coupling, and
   (iv) Slip coupling.

6.3.2 Sleeve or Muff Coupling

It consists of a hollow cylinder which may be keyed or pinned to the shafts as shown in Fig. 6.11. This coupling does not have any projecting part on the external surface and is very safe to use. However, this requires very close and accurate fitting and is difficult to disassemble. The sleeve may be made of cast iron or steel. The diameter of the sleeve may be determined from the torque to be transmitted by the shaft, as follows :

Torque transmitted,

$$T = \frac{\pi}{16}\left[\frac{D^4 - d^6}{D}\right] \times \tau \qquad \text{...(6.6)}$$

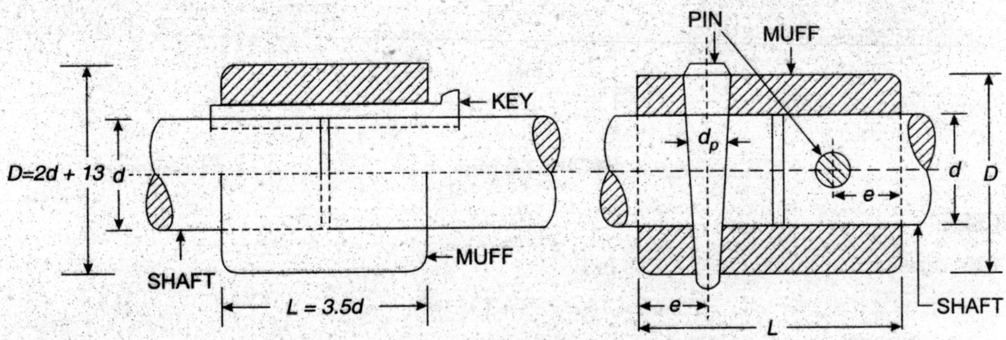

Fig. 6.11 Sleeve or muff coupling.

where D = outside diameter of muff

d = diameter of shaft

τ = permissible shear stress for muff material.

The key can be designed as explained in art. 6.1.2 sometimes two keys are used to carry haft the torque. Empirically,

$$D = 2d + 13 \text{ mm} \qquad \text{...(6.7a)}$$

Length of muff, $$L = 3.5\, d \qquad \text{...(6.7b)}$$

Edge distance, $$e = 0.75\, d \qquad \text{...(6.7c)}$$

The diameter of the pins can be determined by considering its failure in shear. Thus

$$2 \times \left(\frac{\pi}{4}\right) d_p^2 \times \tau = \frac{2T}{d} \qquad \text{...(6.8)}$$

where d_p = diameter of pin

$= (0.20 - 0.5)\, d$, empirically ...(6.9)

τ = permissible shear stress in pin material.

■ **Example 6.4** Design a muff coupling which is used to connect two steel shafts transmitting 50 kW at 300 rpm. The material for the shafts and key is plain carbon steel for which allowable shear and crushing stresses may be taken as 40 MPa and 80 MPa respectively. The material for the muff is carbon steel for which the allowable shear stress may be assumed as 15 MPa.

Solution : Given : $P = 50$ kW, $N = 300$ rpm, $\tau_s = 400$ MPa, $\sigma_{cs} = 80$ MPa, $\tau_c = 15$ MPa

Refer to Fig. 6.11.

1. Shaft Size

Power transmitted, $$P = \frac{2\pi NT}{60 \times 10^3}$$

$$T = \frac{50 \times 60 \times 10^3}{2\pi \times 300} = 1591.55 \text{ N-m}$$

Let d = diameter of shaft

$$\left(\frac{\pi}{16}\right) d^3\, \tau_s = T$$

$$d = \left[\frac{16 \times 1591.55 \times 10^3}{\pi \times 40}\right]^{\frac{1}{3}}$$

$$= 58.74 \approx 60 \text{ mm}$$

2. Muff Size

Outer diameter of muff, $D = 2d + 13$ mm

$$= 2 \times 60 + 13 = 133 \approx 135 \text{ mm}$$

Length of muff, $L = 3.5d = 3.5 \times 60 = 210$ mm

Considering the muff as a hollow shaft,

$$T = \frac{\pi}{16}\left[\frac{D^4 - d^4}{D}\right]\tau_c$$

$$1591.55 \times 10^3 = \frac{\pi}{16}\left[\frac{135^4 - 60^4}{135}\right]\tau_c$$

$$\tau_c = 3.43 \text{ MPa} < 15 \text{ MPa}$$

Hence safe,

3. Key Size

From standard tables, for shaft size 58-65 mm, the dimensions of gib-head key are :

Width, $b = 18$ mm, depth, $h = 18$ mm, length $= \frac{L}{2} = 105$ mm

(a) Check for shear stress in key.

$$T = \frac{bld\,\tau}{2}$$

Shear stress in key, $\tau = \dfrac{2 \times 1591.55 \times 10^3}{18 \times 105 \times 60}$

$$= 28 \text{ MPa} < 40 \text{ MPa}$$

(b) Check for crushing stress in key.

$$T = \frac{h}{2} \times l \times \sigma_c \times \frac{d}{2}$$

$$\sigma_c = \frac{4T}{hld}$$

$$= \frac{4 \times 1591.55 \times 10^3}{18 \times 105 \times 60}$$

$$= 56.1 \text{ MPa} < 80 \text{ MPa}$$

Hence safe.

6.3.3 Clamp or Compression Coupling

This is a modification of the muff coupling, where the muff is slip into two parts for easy assembly and disassembly of the coupling. The two halves of the coupling, as shown in Fig. 6.12, are fitted on the shaft and rigidly fastened by bolts. The torque is transmitted by frictional forces set up between the muff and the shaft. A key may also be fitted to avoid slip. The ribs add strength and serve as a protection to the workers against injury from revolving bolt heads and nuts. This coupling is used for transmitting heavy torque.

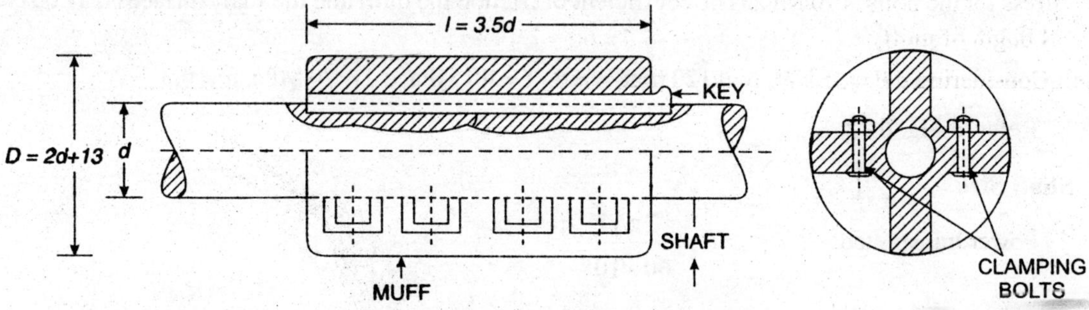

Fig. 6.12 Clamp or compression coupling.

Let,

d_b = core diameter of bolts

n = number of bolts

T = transmitted torque

d = shaft diameter

μ = coefficient of friction between shaft and sleeve materials

l = length of sleeve

σ_t = allowable tensile stress in the bolts

Clamping force on each shaft,

$$F_c = \left(\frac{n}{2}\right) \times \left(\frac{\pi}{4}\right) d_b^2 \times \sigma_t$$

Clamping area,
$$A_c = d \times \left(\frac{1}{2}\right)$$

Intensity of clamping pressure,
$$p_c = \frac{F_c}{A_c}$$

Tangential frictional force per unit area of shaft periphery, $F = \mu p_c$

Torque that can be transmitted by the coupling,

$$T = \left(\frac{d}{2}\right) \times \left(\frac{\pi d l}{2}\right) \times F$$

$$= \left(\frac{\pi^2}{16}\right) d_b^2 \, \sigma_t \mu \, n \, d \qquad \ldots(6.10)$$

Load on each bolt, $$P = \left(\frac{\pi}{4}\right) d_b^2 \, \sigma_t \quad \text{...(6.10a)}$$

$$\therefore \quad = \frac{4T}{\pi \, d \, n \, \mu} \quad \text{...(6.10b)}$$

■ **Example 6.5** Design a clamp coupling to transmit 35 kW at 120 rpm. The allowable shear stress of shaft and key is 40 MPa. The number of bolts connecting the two halves are 6. The permissible tensile stress for the bolts is 70 MPa. The coefficient of friction the muff and the shaft surfaces may be taken as 0.3.

Solution : Given : $P = 35$ kW, $N = 120$ rpm, $n = 6$, $\tau = 40$ MPa, $\sigma_t = 70$ MPa, $\mu = 0.3$

Refer to Fig. 6.12.

1. Shaft Size

Power transmitted, $$P = \frac{2\pi NT}{60 \times 10^3}$$

$$T = \frac{35 \times 60 \times 10^3}{2\pi \times 120} = 2785.2 \text{ N-m}$$

Let d = diameter of shaft

$$\left(\frac{\pi}{26}\right) d^3 \, \tau_s = T$$

$$d = \left[\frac{16 \times 2785.2 \times 10^3}{\pi \times 40}\right]^{\frac{1}{3}} = 70.78 \approx 72 \text{ mm}$$

2. Muff Size

Outer diameter of muff, $D = 2d + 13$ mm

$$= 2 \times 72 + 13 = 157 \approx 160 \text{ mm}$$

Length of muff, $L = 3.5d = 3.5 \times 72 = 252$ mm

Considering the muff as a hollow shaft,

$$T = \left(\frac{\pi}{16}\right)\left[\frac{D^4 - d^4}{D}\right] \tau_c$$

$$2785.2 \times 10^3 = \left(\frac{\pi}{16}\right)\left[\frac{160^4) - (72^4)}{135}\right] \tau_c$$

$$\tau_c = 3.04 \text{ MPa} < 15 \text{ MPa}$$

Hence safe.

3. Key Size

From standard tables, for shaft size of 72 mm, the dimensions of gib-head key are :

Width, $b = 22$ mm, depth, $h = 18$ mm, length $= \frac{L}{2} = 126$ mm

(a) Check for shear stress in key.

$$T = \frac{bld\,\tau}{2}$$

$$\text{Shear stress in key, } \tau = \frac{2 \times 27852 \times 10^3}{22 \times 126 \times 72}$$

$$= 27.9 \text{ MPa} < 40 \text{ MPa}$$

(b) Check for crushing stress in key.

$$T = \left(\frac{h}{2}\right) \times l \times \sigma_c \times \frac{d}{2}$$

$$\sigma_c = \frac{4T}{hld}$$

$$= \frac{4 \times 2785.2 \times 10^3}{18 \times 126 \times 72}$$

$$= 68.8 \text{ MPa} < 80 \text{ MPa}$$

Hence safe.

4. Size of Bolts

Let d_b = core diameter of bolts

$$T = \left(\frac{\pi^2}{16}\right) d_b^2 \sigma_t \, \mu \, n \, d$$

$$d_b = \left[\frac{16 \times 2785.2 \times 10^3}{\pi^2 \times 70 \times 0.3 \times 6 \times 72}\right]^{1/2}$$

$$= 22.3 \approx 23.32 \text{ mm for coarse threads series.}$$

6.3.4 Forged End Flange Coupling

In this coupling, the flanges are forged at the ends of the shafts as shown in Fig. 6.13. The flanges are connected together by bolts. Such type of couplings are generally used for connecting marine engine shafts.

The various dimensions of forged end coupling are :

Let d = diameter of shaft

$$d_2 = 0.2\,d + 5 \text{ mm}$$

$$D_2 = d + 2d_2 + 20 \text{ mm}$$

$$b = 0.3\,d + 5 \text{ mm}$$

$$n = \frac{d}{55} + 3$$

$$D_3 = \frac{d + d_2}{2}$$

$$b_2 = 4 \text{ to } 5 \text{ mm}$$

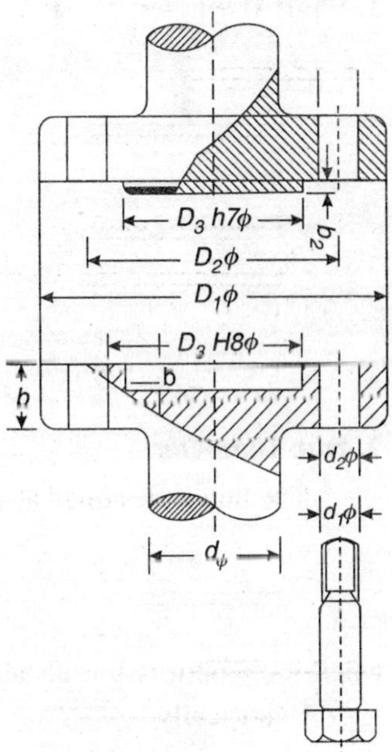

Fig. 6.13 Forged end type flange coupling.

$$b_1 = b_2 + 1 \text{ mm}$$
$$D_1 = D_2 + d_2 + 40 \text{ mm}$$

6.3.5 Rigid Protected Type Flange Coupling

It consists of two flanges keyed to the shaft ends and bolted together. To ensure proper alignment of the shafts the end of one shaft may enter into the recess provided in the flange attached to the other shaft. Protected type of flanges are often used to provided safety to the operator. These flanges project beyond the heads of the bolts and nuts. The various parts of the flange coupling are shown in Fig. 6.14. The various parts can be designed as discussed below :

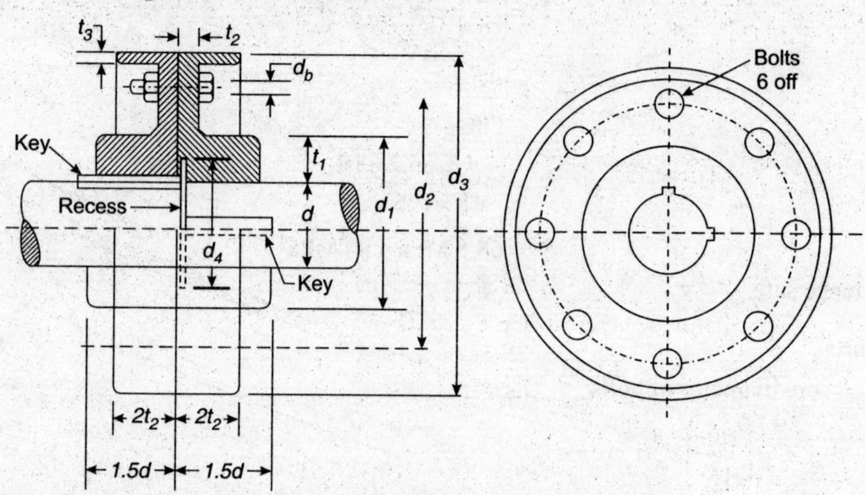

Fig. 6.14 Protected type rigid flange coupling.

1. Shaft Diameter

$$P = \frac{2\pi N T_m}{60 \times 10^3}$$

$$T_m = \frac{P \times 60 \times 10^3}{2\pi \times N}$$

$$T_{max} = (1 + \%\text{ overload factor}) \times T_m$$

$$\left(\frac{\pi}{16}\right) d^3 \tau_s = T_{max}$$

2. Hub Diameter

The hub is designed as a hollow shaft subjected to torsion. Thus

$$T = \left(\frac{\pi}{16}\right)\left[\frac{d_1^4 - d^4}{d_1}\right]\tau_h \quad \text{...(6.12)}$$

where τ_h = permissible shear stress in hub

Empirically, $d_1 = 2d$

Thickness of hub, $t_1 = \dfrac{d_1 - d}{2}$

Diameter of recess in flanges, $d_4 = d + t_1$

Outside diameter of protecting flange,

$$d_3 = 4d$$

3. Hub Length

The hub length is made equal to the key length required. The key size is determined by designed the key and the shaft to be equally strong in shear.

Width of key, $b = \dfrac{d}{4}$

Length of key, $l = 1.5d$

Generally a square or a rectangular key is used. For a rectangular key

$$t = \frac{d}{6}.$$

4. Number of Bolts

$$n = \frac{4d}{150} + 3 \quad \text{...(6.13)}$$

where d = shaft diameter, mm

5. Bolt Diameter

The bolts are in single shear and are designed to carry its own share of load. Radius of bolts circle, r_2 = radius of hub + clearance + radius of socket head wrench used for tightening the bolts

$$= \frac{d_1}{2} + 5\text{ mm} + (1.85\, d_b + 8\text{ mm}) \quad \text{...(6.14)}$$

where d_b = bolt diameter, mm

Empirically, $r_2 = 1.5\, d$

Shearing force in each bolt,

$$F = \frac{T}{r_2 n}$$

$$\left(\frac{\pi}{4}\right) d_b^2\, \tau_b = F \quad \text{...(6.15)}$$

where τ_b = permissible shear stress in bolts.

6. Bolt Diameter Based on Tensile Load

The necessity of reamed holes for bolts may be avoided if the coupling is designed on the basis that the torque is transmitted by friction between the flanges. In this case, the bolts will be in tension only. Assuming uniform pressure distribution between the flanges, the mean friction radius is given by,

$$r_f = \frac{2(r_3^3 - r_4^3)}{3(r_3^2 - r_4^2)}$$

Then $T = n \times \mu\, P_i \times r_f$

where P_i = initial tensile force in each bolt

r_3 = outer radius of the friction surface

$= \frac{d_3}{2}$

r_4 = inner radius of the friction surface

μ = coefficient of friction between flanges

Therefore
$$P_i = \left(\frac{\pi}{4}\right) d_b^2 \sigma_t \quad \text{...(6.16)}$$

where σ_t = permissible tensile stress in bolts.

7. Flange Thickness

Considering the bearing failure of flange.

$$F = n \times d_b \times t_2 \times \sigma_c \quad \text{...(6.17)}$$

Empirically,
$$t_2 = 0.5 t_1 + 6 \text{ mm}$$

where $t_1 = \frac{d_1 - d}{2}$ = hub thickness

Considering shearing of the flange at the junction with the hub,

$$2\left(\frac{\pi d_1^2}{4}\right) \times \tau \times t_2 = T \quad \text{...(6.18)}$$

8. Check for Crushing of Bolt

Considering crushing of the bolts;

$$n d_b t_2 \sigma_{cb} \times \frac{d_2}{2} = T \quad \text{...(6.19)}$$

where σ_{ab} = permissible crushing strength of bolts.

9. Thickness of Protecting Flange

$$t_3 = 0.5 t_2 \text{ (min) or } 0.25\, d$$

10. Hub overlap = 3 mm (min)

■ **Example 6.6** Design a protected type rigid cast iron flange coupling for a steel shaft transmitting 20 kW at 240 rpm and having an allowable shear stress of 45 MPa. The shearing and tensile stresses in the bolts are limited to 30 MPa and 60 MPa respectively. The material of the shaft and key is the same for which the crushing strength is twice the shear strength. The overload is 25%. The allowable shear stress for cast iron is 15 MPa and coefficient of friction is 0.15.

Solution : Given : $P = 20$ kW, $N = 240$ rpm, $\tau_s = 45$ MPa, $\tau_b = 30$ MPa, $\sigma_b = 60$ MPa, $\sigma_{cs} = 2\tau_s$, τ_{ci} = 15 MPa, $T_{max} = 1.25\, T_m$ $\mu = 0.15$

Refer to Fig. 6.14

1. Shaft Diameter

$$P = \frac{2\pi N T_m}{60 \times 10^3}$$

$$T_m = \frac{20 \times 60 \times 10^3}{2\pi \times 240} = 795.775 \text{ N-m}$$

$$T_{max} = 1.25 \times 795.775 = 994.72 \text{ N-m}$$

$$\left(\frac{\pi}{16}\right) d^3 \; \tau_s = T_{max}$$

$$\left(\frac{\pi}{16}\right) \times d^3 \times 45 = 994.72 \times 10^3$$

$$d = 48.28 \approx 50 \text{ mm}$$

2. Hub Diameter

The hub is designed as a hollow shaft subjected to torsion. Thus

$$T_{max} \left(\frac{\pi}{16}\right) \left[\frac{d_1^4 - d^4}{d_1}\right] \tau_h$$

where τ_h = permissible shear stress in hub made of C.I.
= 15 MPa

$$2\left(\frac{\pi d^2 1}{4}\right) \times \tau \times t_2 = T_{max}$$

$$2\left(\pi \times \frac{100^2}{4}\right) \times 20 \times \tau = 994720$$

$$\tau = 3.17 \text{ MPa} < 15 \text{ MPa}$$

Hence safe.

3. Check For Crushing of Bolt

Considering crushing of the bolts,

$$n d_b t_2 \sigma_{cb} \times \frac{d_2}{2} = T_{max}$$

where σ_{cb} = permissible crushing strength of bolts.

$$4 \times 12 \times 20 \, \sigma_{cb} \times \frac{150}{2} - 994720$$

$$\sigma_{cb} = 13.8 \text{ MPa} < 60 \text{ MPa}$$

Hence safe.

4. Thickness of protecting flange.

$$t_3 = 0.5 t_2 \text{ (min) or } 0.25\, d$$

$$t_3 = 0.5 \times 20 = 10 \text{ mm}$$

5. Hub overlap = 3 mm (min)

6.4 BUSH TYPE FLEXIBLE COUPLING

In a bush type coupling, motion from one-half of flange is transmitted to the other half by means of pins or bolts. The pins are rigidly bolted to one flange and fit loosely in the corresponding holes of the other flange. This type of coupling is commonly used where the driving and drives members are mounted on a monoblock. This can take up parallel misalignment of shafts upto 0.25 mm per 10 mm outside diameter of coupling and angular misalignment upto 1.5 degree and axial displacement also. The pins are covered by leather, rubber, etc. A typical bush type flexible coupling is shown in Fig. 6.15.

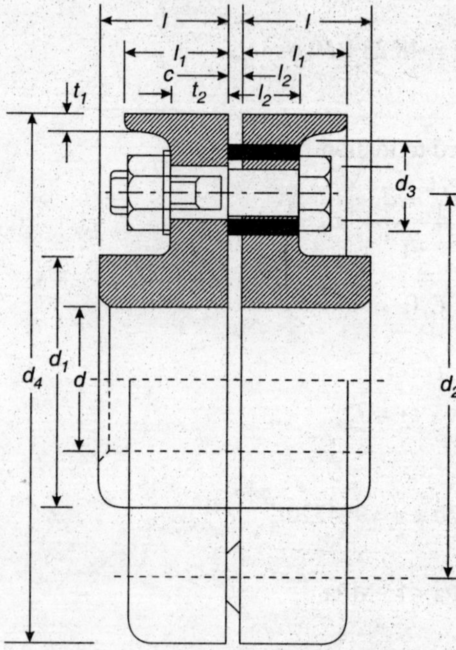

Fig. 6.15 Bush type flexible coupling.

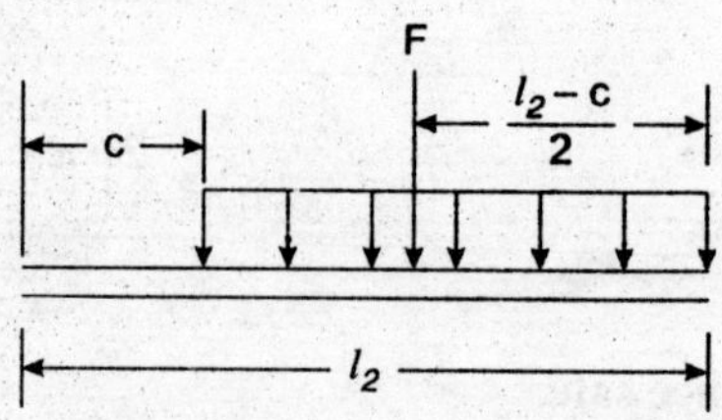

Fig. 6.16 Loading on the pin.

The various dimensions of coupling may be determined as follows :

1. Number of Bolts

$$n = 0.04d + 3 \quad \text{...(6.20)}$$

where d = diameter of shaft, mm

2. Outer Diameter of Hub

$$d_1 = 2d \quad \text{...(6.21)}$$

3. Diameter of Bolts

$$d_b = \frac{0.5d}{\sqrt{n}} \quad \text{...(6.22)}$$

Shear force in each bolt,

$$F = \frac{2T}{nd_2}$$

where d_2 = pitch circle diameter of bolts

$$\left(\frac{d}{4}\right)d_b^2\ \tau = F \qquad \ldots(6.23)$$

where τ = permissible shear stress in bolts

4. Outer Diameter of Rubber Bush

$$d_3 = \text{diameter of pin} + 2 \times \text{thickness rubber bush}$$

5. Length of Bush

Considering the bearing of bush,

$$F = l_2 d_3 p_b \qquad \ldots(6.24)$$

where p_b = bearing pressure for bush
= 2 MPa

6. Check for a Pin in Bending

Let c = clearance between two flanges. The loaded pin is shown in Fig. 6.16.

Bending moment on the pin,

$$M = Fl_3$$

where $l_3 = \dfrac{l_2 - c}{2} + c = \dfrac{l_2 + c}{2}$

$$\frac{F(l_2 + c)}{2} = \left(\frac{\pi}{32}\right)d_b^3 \times \sigma_b \qquad \ldots(6.25)$$

where σ_b = bending stress in pin

7. Maximum Shear Stress in Pin

$$\tau_{max} = \left[\left(\frac{\sigma_b}{2}\right)^2 + \tau^2\right] < 35 \text{ MPa}$$

■ **Example 6.7** Design a bushed-pin type flexible coupling to connect a compressor shaft to a motor shaft transmitting 30 kW at 750 rpm. The overload factor is 20%. The allowable material properties are as follows :

Shaft. key and pin :

Shear – 35 MPa
Crushing = 70 MPa

Cast iron :

Shear = 15 MPa

Rubber bush :

Bearing – 0.8 MPa

Solution : Given : $P = 30$ kW, $N = 750$ rpm, $T_{max} = 1.2\, T_m$, $\tau_s = \tau_k = \tau_k = \tau_b = 35$ MPa,

$\sigma_{cs} = \sigma_{ck} = \sigma_{cb} = 70$ Mpa., $\tau_{ci} = 15$ MPa, $p_b = 0.8$ MPa

Refer to Fi.g 6.16.

1. Shaft Diameter

$$P = \frac{2\pi NT_m}{60 \times 10^3}$$

$$T_m = \frac{30 \times 60 \times 10^3}{2\pi \times 750} = 381972 \text{ N-mm}$$

$$T_{max} = 1.2 \times 381972 = 458366 \text{ N-mm}$$

$$\left(\frac{\pi}{16}\right) d^3 \ \tau_s = T_{max}$$

$$\left(\frac{\pi}{16}\right) \times d^3 \times 35 = 458366$$

$$d = 40.55 \approx 42 \text{ mm}$$

2. Hub Diameter

The hub is designed as a hollow shaft subjected to torsion. Thus

$$T_{max} = \left(\frac{\pi}{16}\right)\left[\frac{d_1^4 - d^3}{d_1}\right]\tau_h$$

where τ_h = permissible shear stress in hub made of C.I.

$= 15 \text{ MPa}$

$$458366 = \left(\frac{\pi}{16}\right)\left[\frac{d_1^4 - 50^4}{d_1}\right]\tau_h$$

Empirically, outer diameter of hub, $d_1 = 2d = 84 \text{ mm}$

$$\tau_h = 4.2 \text{ MPa} < 15 \text{ MPa}$$

Hence safe. Adopt $d_1 = 84 \text{ mm}$

Thickness of hub, $t_1 = \frac{d_1 - d}{2} = \frac{84 - 42}{2} = 21 \text{ mm}$

Outside diameter of protecting flange, $d_4 = 4d = 4 \times 42 = 168 \approx 170 \text{ mm}$

3. Key Size and Hub Length

The hub length is made equal to the key length required. The key is determined by designing the key and the shaft to be equally strong in shear.

Width of key, $b = \frac{d}{4} = \frac{42}{4} = 10.5 \approx = 12 \text{ mm}$

Length of key, $l = 1.5\,d = 1.5 \times 42 = 63 \text{ mm}$

We use a square key with thickness,

$$t = b = 12 \text{ mm}$$

Hub length,

4. Number of Bolts

$$d_b = 0.04\,d + 3$$
$$= 0.04 \times 42 + 3 = 4.68 \approx 6$$

5. Diameter of Bolts

$$d_b = \frac{0.5\,d}{\sqrt{n}}$$
$$= \frac{0.5 \times 42}{\sqrt{6}} = 8.57 \text{ mm}$$

To account for bending of pin in the bush, let core diameter of pin, $d_b = 20$ mm so that shank diameter of bolt = 24 mm

6. Outer Diameter of Rubber Bush

d_3 = diameter of pin + 2 × thickness of rubber bush of 2 mm + 2 × thickness of brass bush of 6 mm

$$= 24 + 2 \times 2 + 2 \times 6 = 40 \text{ mm}$$

Pitch circle diameter of bolts,

$$d_2 = d_1 + d_3 + 2 \times 6 = 84 + 40 + 12 = 136 \text{ mm}$$

7. Check for Shear in Bolt

Shear force in each bolt,
$$F = \frac{2T_{max}}{nd_2}$$
$$F = \frac{2 \times 458366}{6 \times 136} = 1123.4 \text{ N}$$
$$\left(\frac{\pi}{4}\right) d_b^2\, \tau = F$$

where τ = permissible shear stress in bolts

$$\left(\frac{\pi}{4}\right) \times 20^2 \times \tau = 1123.4$$
$$\tau = 357 < \text{MPa} < 35 \text{ MPa}$$

Hence safe.

8. Length of bush

Considering the bearing of bush,

$$F = l_2 d_3 p_b$$

where p_b = bearing pressure for bush

$= 0.8$ MPa

l_2 = length of bush

$$l_2 = \frac{1123.4}{40\,0.8} = 35 \text{ mm}$$

9. Check for pin in bending

Let c = clearance between two flanges.

$= 5$ mm

The loaded pin is shown in Fig. 6.16.

Bending moment on the pin,

$$M = Fl_3$$

where
$$l_3 = \frac{l_2 - c}{2} + c = \frac{l_2 + c}{2}$$

$$\frac{F(l_2 + c)}{2} = \left(\frac{\pi}{21}\right) d_b^3 \times \sigma_b$$

$$1123.4 \frac{(35 - 5)}{2} = \left(\frac{\pi}{32}\right) \times 20^3 \times \sigma_b$$

Bending stress in pin, $\sigma_b = 28.6$ MPa

10. Maximum shear stress in pin

$$\tau_{max} = \left[\left(\frac{\sigma_b}{2}\right)^2\right] + [(3.75)^2]^{1/2}$$

$$= \left[\left(\frac{28.6}{2}\right)^2\right] + [(3.57)^2]^{1/2}$$

$$= 14.73 \text{ MPa} < 35 \text{ MPa}$$

Hence safe.

11. Flange thickness

Considering the bearing failure of flange,

$$F = n \times d_b \times t_2 \times \sigma_c$$

where σ_c = permissible bearing stress in flange.

Empirically, $t_2 = 0.5t_1 + 6 \text{ mm} = 0.5 \times 21 + 6 = 16.5 \approx 18 \text{ mm}$

Considering shearing of the flange at the junction with the hub,

$$2\left(\frac{\pi d_1^2}{4}\right) \times \tau \times t_2 = T_{max}$$

$$2\left(\frac{\pi \times 84^2}{4}\right) \times 15 \times \tau = 458366$$

$$\tau = 8.66 \text{ MPa} < 15 \text{ MPa}$$

Hence safe.

Thickness of protecting flange,

$$t_1 = 0.5\, t_2 = 9 \text{ mm}$$

Length of protecting flange,

$$l_1 = l_2 + 1.5d_b = 35 + 1.5 \times 20 = 65 \text{ mm}$$

6.5 FLYWHEEL

A flywheel is a heavy body rotating about its axis. It acts as a reservior of energy which is stored in the form of kinetic energy. The extra energy is stored during the idle stroke of the driven machinery and released during the working stroke. Thus the flywheel controls the fluctuation of speed during each cycle of the driven machinery. In the case of a prime mover, the flywheel absorbs the excess energy during the expansion stroke and release during the suction, compression and exhaust strokes.

6.5.1 Coefficient of Fluctuation of Speed

The difference between the maximum and minimum speeds during a cycle is called the maximum fluctuation of speed. The ratio of maximum fluctuation of speed to the mean speed is called coefficient of fluctuation of speed.

Let ω_{max} = maximum angular spccd

ω_{min} = minimum angular speed

$\omega_{mean} = \left(\frac{1}{2}\right)(\omega_{max} + \omega_{min})$ = mean angular speed

Coefficient of fluctuation of speed.

$$C_S = \frac{\omega_{max} - \omega_{min}}{\omega_{mean}} \quad \text{...(6.26)}$$

$$= \frac{N_{max} - N_{min}}{N_{mean}} \quad \text{...(6.27)}$$

where N = speed in rpm

Refer to Table 6.1 for values of C_S for various machinery.

Table 6.1 : Coefficient of fluctuation of speed, C_S

Machine type	*Coefficient of fluctuation of speed C_S*
Punching machine, pressing machine	0.2-0.07
Pumps	0.05-0.03
Machine tools	0.03-0.025
Textile and paper machinery	0.025
Spinning machinery	0.02
DC generator (lighting)	0.007
AC generator	0.003
DC generator	0.01
Belt driven DC generator	0.014
AC generator direct drive	0.006
Belt driven AC generator	0.008
Automobile (idling)	0.2
Automobile (normal speed)	0.1
Weaving machine	0.025
Flour mill machinery	0.02
Reciprocating pumps and compressors	0.04

6.5.2 Coefficient of Fluctuation of Energy

It is defined as the ratio of maximum fluctuation of energy to the work done per cycle.

Maximum fluctuation of energy.

$$\Delta E = \text{Maximum energy} - \text{Minimum energy}$$

Coefficient of fluctuation of energy,

$$C_e = \frac{\Delta E}{\text{Work done per cycle}} \quad \ldots(6.27)$$

$$\text{Work done per cycle} = T_{mean} \times \theta \quad \ldots(6.28a)$$

where T_{mean} = mean torque

$$= \frac{\text{Power transmitted in watts}}{\text{Angular speed}}$$

$$= \frac{P}{\omega}$$

$$\omega = \frac{2\pi N}{60} \text{ rad/s}$$

$$\text{Work done per cycle is also} = \frac{60\,P}{n} \quad \ldots(6.28b)$$

where n = number of working strokes per minute

= N, in case of steam engines and two stroke internal combustion engines

= $\frac{n}{2}$, in case of four stroke internal combustion engines

Fluctuation of kinetic energy,

$$\Delta E = C_e E. \quad \ldots(6.29)$$

where C_e = coefficient of fluctuation of energy (sec Table 6.2 and 6.3)

Change of kinetic energy,

$$\Delta E = \frac{1}{2}(\omega_{max}^2 - \omega_{min}^2)$$

$$= \frac{1}{2}(\omega_{max} + \omega_{min})(\omega_{max} - \omega_{min}) \quad \ldots(6.30)$$

Table 6.2 : Coefficient of fluctuation of energy for steam engines, C_e

Cut off percentage	*Coefficient of fluctuation of energy, C_e*		
	Single cylinder	*Two cylinder*	*Three cylinder*
10	0.35	0.088	0.040
20	0.33	0.082	0.037
40	0.31	0.078	0.034
60	0.29	0.072	0.032
80	0.28	0.070	0.031
100	0.27	0.068	0.030

Table 6.3 : Coefficient of fluctuation of energy for I.C. engines, C_e

Type of engine	*Number of cylinders*	*Angle between crank (s) Deg*	*Coefficient of fluctuation of energy C_e*	
			Four stroke	*Two Stroke*
Single acting	1	360	2.35-2.40	0.95-1.00
Single acting	2	180	1.50-1.60	0.20-0.25
Vertical	3	120	0.60-0.75	0.15-0.18
Vertical	4	180 and 90	0.15-0.20	0.075-0.10
Vertical	5	180 and 60	0.10-0.12	0.016-0.02

6.5.3 Energy Stored in a Flywheel

Consider a flywheel as shown in Fig. 6.17

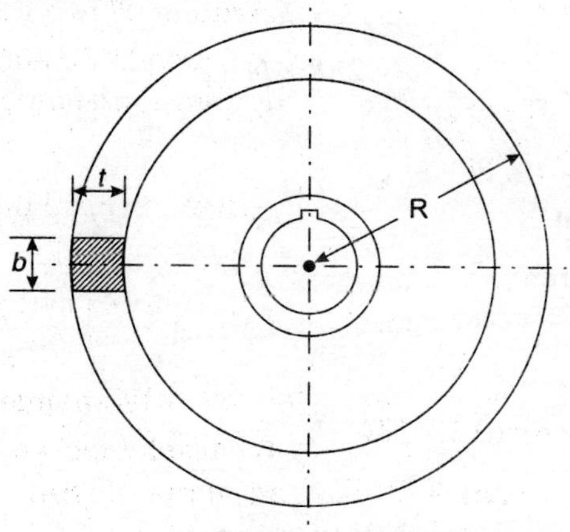

Fig. 6.17 Flywheel.

Let m = mass of flywheel, kg

K = radius of gyration of flywheel about its axis of rotation.

I = mass moment of inertia of the flywheel about is axis of rotation, kg.m^2

$= mk^2$

N_{max}, N_{min} = maximum and minimum speeds during a cycle, rpm

$\omega_{max}, \omega_{min}$ = maximum and minimum angular speeds during a cycle, rad/s

ω_{mean} = mean angular sped during a cycle

Mean kinetic energy of flywheel,

$$E = \left(\frac{1}{2}\right) I\,\omega_{mean}^2$$

$$= \left(\frac{1}{2}\right) \times mK^2 \times \omega_{mean}^2$$

Maximum fluctuation of energy as the speed changes from ω_{max} to ω_{min},

$$\Delta E = \text{Maximum K.E.} - \text{Minimum K.E.}$$

$$= \left(\frac{1}{2}\right) \times I \times [\omega_{max}^2 - \omega_{min}^2]$$

$$= \left(\frac{1}{2}\right) \times I \times (\omega_{max} + \omega_{min})(\omega_{max} - \omega_{min})$$

$$= 1 \times \omega_{mean} (\omega_{mean} - \omega_{min})$$

$$= \frac{I \times \omega_{mean}^2 (\omega_{max} - \omega_{min})}{(\omega_{mean})}$$

$$= 2EC_S \qquad \text{...(6.31)}$$

For a flywheel having very small thickness of rim as compared to the rim diameter, $K \approx R$.

Therefore, $I = mR^2$ ...(6.32)

Mass of flywheel, $m = 2\pi R \times bt \times \rho$ ...(6.33)

where b = width of rim

t = thickness of rim

ρ = density of rim material

= 7260 kg/m^3 for cast iron

= 7800 kg/m^3 for steel

Generally $= \frac{b}{t} = 2$

6.5.4 Types of Flywheels

The flywheels can be of the following types :

1. Solid disc type
2. Rimmed type with either arms or solid web.

Solid disc type flywheels are rarely used because they have less capacity of storing energy.

Rimmed types of flywheels with arms are preferred because they can store more energy.

Small rimmed type flywheels are manufactured with solid web or holes drilled in the web.

1. Solid disc type flywheel

Moment of inertial of a solid disc of radius R and mass m,

$$I = \frac{mR^2}{2} \qquad \text{...(6.34)}$$

Mass of the disc, $m = \pi R^2 t\rho$ ...(6.35)

where t = thickness of disc

ρ = density of disc material

2. Rimmed flywheel

The arms, hub and the shaft contribute only 10% of the total energy stored by a rimmed flywheel. The rim contributes approximately 90% of the required moment of inertia.

Moment of inertia of the rim,

$$I_r = \text{KI}$$

where k = a factor = 0.9

I = moment of inertia of the entire flywheel ...(6.36)

$$I_r = m_r \, R^2$$

where m_r = mass of rim

R = mean radius of rim

The limiting mean rim velocity of kept below 30 m/s for flywheels made of grey cast iron. When the velocity exceeds this limit, there is a possibility of burshing due to centrifugal force.

$$R < 30/\omega_{\text{mean}} \quad ...(6.37)$$

6.5.5 Design of Belt-Pulley Flywheel

A rimmed type flywheel is shown in Fig. 6.18

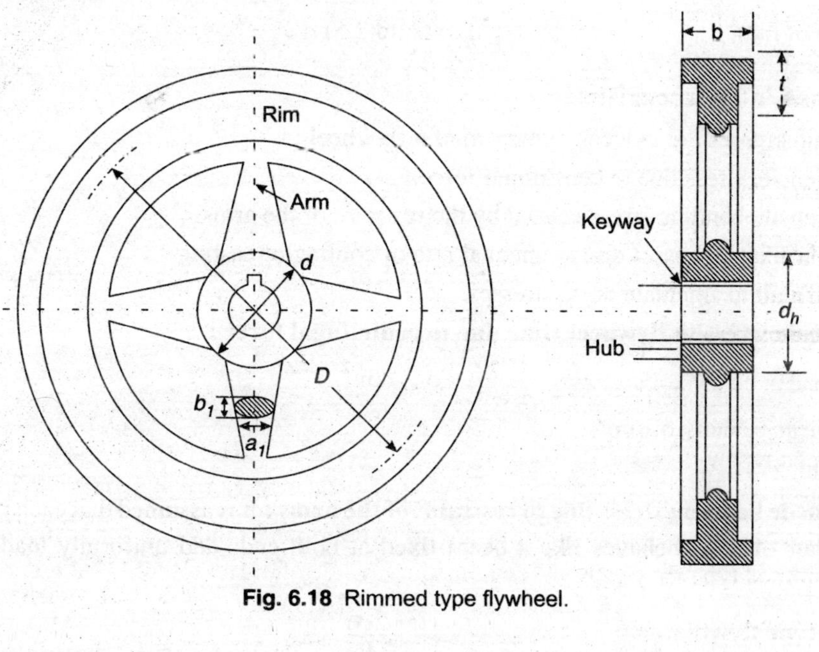

Fig. 6.18 Rimmed type flywheel.

1. Rim size

Let b = width of flywheel rim

= belt width + 40 mm

t = thickness of rim ...(6.38)

$\dfrac{b}{t} = 0.5 - 1.5$ for belt – pulley flywheel ...(6.39a)

$= 2 - 4$ for a rimmed flywheel ...(6.39b)

2. Shaft and hub size

P = power transmitted, kW

Torque on flywheel shaft, $$T = \frac{60 \times 10^3 \times P}{(2\pi N)} \text{ N-m} \qquad \text{...(6.40)}$$

where N = rpm of flywheel

M = bending moment on the flywheel shaft

= weight of flywheel × overhang ...(6.41)

Diameter of flywheel shaft,

$$d = \left[\left\{ \frac{16}{(\pi \tau_d)} \right\} \{(K_m M)^2 + (K_t T)\}^{\frac{1}{2}} \right]^{\frac{1}{3}} \qquad \text{...(6.42)}$$

where τ_d = design shear stress for shaft

K_m, K_t = shock and fatigue factors for bending and torsion respectively.

Diameter of hub, $$d_h = 2d \qquad \text{...(6.43a)}$$

Length of hub, $$l_h = (2 \text{ to } 2.5)\, d \qquad \text{...(6.43b)}$$

6.5.6 Stresses in Flywheel Rim

The following stresses are induces in the rim of a flywheel :

(i) Tensile stress due to centrifugal force.

(ii) Tensile bending stress caused by the restraint of the arms.

(iii) Shrinkage stresses due to unequal rate of cooling of casting.

It is difficult to calculate these stresses.

(i) Tensile stress in flywheel rime due to centrifugal force :

$$\sigma_t = \rho v^2 \qquad \text{...(6.44)}$$

where v = linear velocity of rim

$= R\omega$

(ii) Tensile bending stress due to restraint of the arms : It is assumed that each portion of the rim between a pair of arms behaves like a beam fixed at both ends and uniformly loaded, as shown in Fig. 6.19.

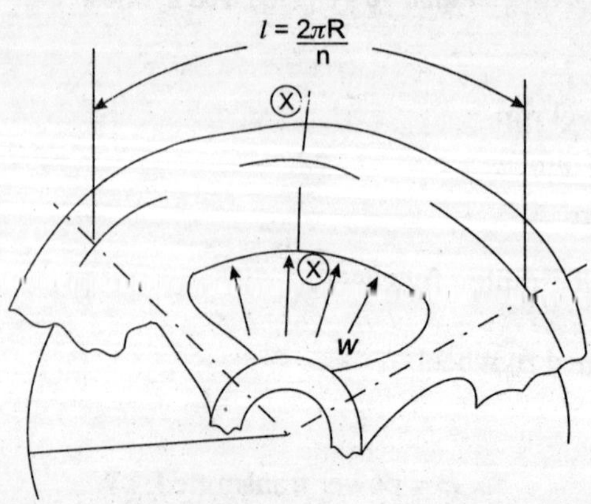

Fig. 6.19 Bending stress due to restraint of arms.

Length between fixed ends, $l = \frac{2\pi R}{n}$

where n = number of arms

The uniformly distributed load 'w' per metre length will equal to the centrifugal force between a pair of arms.

$$w = bt\rho\omega^2 R \text{ N/m}$$

Maximum bending moment.

$$M = \frac{wl^2}{12} = \left(\frac{bt\rho\omega^2 R}{12}\right)\left(\frac{2\pi R}{n}\right)^2$$

$$= \frac{bt\omega^2 R^3 \pi^2}{3n^2}$$

$$= \frac{\pi^2 bt\rho v^2 R}{3n^2}$$

Section modulus,

$$z = \frac{bt^2}{6}$$

Bending stress in flywheel rim due to restraint of arms,

$$\sigma_b = \frac{M}{z} = \frac{2\pi\rho v^2 R}{n^2 t} \qquad \text{...(6.45)}$$

where R = mean rim radius

It has been observed that the arms of a flywheel stretch about $\frac{3}{4}$th of the amount necessary for free expansion. Therefore, the total stress in the rim,

$$\sigma_{\text{total}} = \left(\frac{3}{4}\right)\sigma_t + \frac{\sigma_b}{4} \qquad \text{...(6.46a)}$$

$$= \left(\frac{3}{4}\right)\rho v^2 + \rho v^2 \left\{\frac{\pi^2 R}{(2n^2 t)}\right\}$$

$$= \rho v^2 \left[\frac{0.75 + 4.935\,R}{n^2 t}\right] \qquad \text{...(6.46b)}$$

For a safe design, $\sigma_{\text{total}} < 38$ MPa for C.I.

6.5.7 Stresses in Flywheel Arms

The following stresses are induced in the flywheel arms

(i) Tensile stress due to centrifugal force acting on the rim.

(ii) Bending stress due to torque transmitted from the rim to the shaft or vice-versa.

(iii) Shrinkage stresses due to unequal rate of cooling of casting.

It is difficult to calculate these stresses.

(i) Tensile stress due to centrifugal force : The tensile stress due to centrifugal force acting on the rim, assuming 75% effect, is given by,

$$\sigma_{t1} = \left(\frac{3}{4}\right)\sigma_t = \left(\frac{3}{4}\right)\rho v^2$$

(ii) Bending stress due to torque transmitted : Due to the torque transmitted from the rim to the shaft or vice-versa, the arms will be subjected to bending, because they are required to carry the full torque load.

Consider the arm as shown in Fig. 6.20. The arm may be assumed as a cantilever beam fixed at the hub and carrying a concentrated load at the free end at the rim.

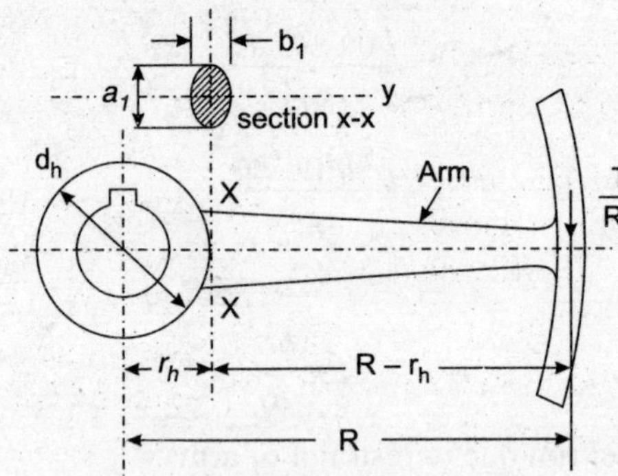

Fig. 6.20 Bending stress in arm.

Let R = mean radius of the rim

r = radius of hub $= \frac{d_h}{2}$

T = maximum torque transmitted by the shaft

Load at the mean radius of rim,

$$F = \frac{T}{R}$$

Load in each arm $= \frac{T}{\pi R}$

Maximum bending moment on the arm near the hub,

$$M = \frac{T(R-r)}{\pi R}$$

Bending stress in arm,

$$\sigma_{b1} = \frac{T(R-r)}{\pi R z_y} \quad \ldots(6.47)$$

where z_y = section modulus of arms about y-axis

Total tensile stress in the arms near the hub,

$$\sigma = \sigma_{t1} + \sigma_{b1} \quad \ldots(6.48)$$

For belt-pulley flywheel, bending stress at the hub due to belt tensions, considering that only half the number of arms are effective in transmitting belt tension,

$$\sigma_{b2} = \frac{(F_1 - F_2)(R - r)}{\left(\frac{nz_y}{2}\right)} \quad \text{...(6.49)}$$

where F_1, F_2 = belt tensions on tight and slack sides respectively

Total stress, $\sigma_{total} = \sigma_{b1} + \sigma_{b2} + \sigma_{t1}$ ...(6.50)

For a safe design, $\sigma_{total} < 20$ MPa for C.I. ...(6.51a)

< 60 MPa for steel ...(6.51b)

6.5.8 Design of Flywheel Arms

The cross-section of the arms is generally elliptical with major axis as twice the minor axis.

Let a_1 = major axis of arms near the hub

b_1 = minor axis of arms near the hub

Section modulus of arms about the y-axis, $z_y = \frac{\pi b_1 z_1^2}{32}$

For $b_1 = \frac{a_1}{2}$,

Major axis of elliptical arm section $a_1 = \left[\frac{64 z_y}{\pi}\right]^{\frac{1}{3}}$ ...(6.52)

Cross-section of arms near the rim = $\frac{2}{3}$rd of that near the hub ...(6.53)

6.5.9 Design of Shaft, Hub and Key

Size of shaft and hub may be determined as explained in art. 6.5.5. The diameter of hub is generally taken as twice the diameter of shaft. The length of hub is taken as 2 – 2.5 times the diameter of shaft or twice the rim width. A standard sunk key is used for the shaft and hub.

Example 6.8 Design a suitable flywheel for a 12 BHP, 2-cylinder vertical 4-stroke diesel engine coupled to a 7.5 kW A.C. generator. The normal speed of the engine is 900 rpm.

Solution : Given : $P = 7.5$ kW, BHP $= 12$, $N = 900$ rpm

1. For a 20-cylinder, 4-stroke, I.C. engine, $C_e = 1.5 - 1.6$ (Table 6.2)

   We take, $C_e = 1.5$

2. For A.C. generator directly coupled, $C_s = 0.006$ (Table 6.3)

3. Fluctuation of energy, $\Delta E = \frac{C_e \times 10^3 \times P \times 60}{N}$

   $= \frac{1.5 \times 10^3 \times 7.5 \times 60}{900} = 750$ N-m

4. We take $v = 25$ m/s for cast iron flywheel. ($v < 50$ m/s) for I.C. engine flywheel)

5. Mass of flywheel, $m = \frac{\Delta E}{C_e v^2} = \frac{750}{0.006 \times 25^2} = 200$ kg

6. Mass of rim, $m_r = 0.9\text{ m} = 180\text{ kg}$
7. Mass radius of flywheel,

$$R = \frac{60v}{2\pi N}$$

$$= \frac{60 \times 25}{\pi \times 900} = 0.265\text{ m}$$

8. Size of rim

$$m_r = 2\pi R b t \rho$$

$$\frac{b}{t} = 0.5 - 1.5 \text{ for pulley type of flywheel}$$

$$= 1.25$$

For C.I., $\rho = 7100\text{ kg/m}^3$

$$180 = 2\pi \times 0.265 \times 1.25 t^2 \times 7100$$

$$t = 110.4\text{ mm},\ b = 137.95\text{ mm}$$

We adopt $b = 140$ mm and $t = 112$ mm, $R = 265$ mm

9. Flywheel shaft.

Let overhang of flywheel = 150 mm

Bending moment, M = weight of flywheel × overhang

$$= 200 \times 9.81 \times 0.15 = 294.3\text{ N-m}$$

Torque, $$T = \frac{0.746 \times \text{BHP} \times 10^3 \times 60}{2\pi N}$$

$$= \frac{0.746 \times 12 \times 10^3 \times 60}{2\pi \times 900} = 94.9837\text{ N-m}$$

Let $K_m = 1.5$ and $K_t = 1.5$, $\tau_d = 40$ MPa

Diameter of shaft, $$d = \left[\left\{\frac{16}{\pi \tau_d}\right\}\{(K_m M)^2) + (K_t T)^2\}^{\frac{1}{2}}\right]^{\frac{1}{3}}$$

$$= \left[\left\{\frac{16}{\pi \times 40}\right\}\{1.5 \times 294.3)^2 + (1.5 \times 94.9837)^2\}^{\frac{1}{2}}\right]^{\frac{1}{3}} \times 10$$

$$= 38.94 \approx 40\text{ mm}$$

10. Hub diameter, $d_h = 2d = 80$ mm
11. Hub length, $l_h = 2d = 80$ mm
12. The size of the flywheel is small. We shall use web for the flywheel with thickness,

$$t_w = \frac{b}{20} = \frac{140}{20} = 7\text{ mm}$$

Example 6.9 A punching machine is required to produce 30 holes per minute of 20 mm diameter in a steel plate of 15 mm thickness having ultimate shearing strength of 340 MPa. The actual punching operation last for a period of 36° rotation of punching machine crankshaft. Mechanical efficiency of punching machine is 80%. It rotates at 240 rpm. Design a suitable flywheel for the punching machine.

Solution : Given : $d = 20$ mm, $t = 15$ mm, $\tau = 340$ MPa, $\theta = 36°$, $\eta_{mech} = 0.8$, $N = 240$ rpm

1. Punching force, $F = \pi d t \tau_u = \pi \times 20 \times 15 \times 340 = 320443$ N
2. Work done in punching one hole,

$$E_1 = 0.5 \text{ F} \times \text{distance travelled by punch}$$

$$= 0.5 \times 320443 \times 15 \times 10^{-3} = 2403.3 \text{ N-m}$$

3. Energy required per hole at flywheel shaft,

$$E = \frac{E_1}{\eta_{meech}} = \frac{2403.3}{0.8} = 3004.15 \text{ N-m}$$

4. Actual punching operation period

$$= \frac{\theta}{360} = \frac{36}{360} = 0.1 \text{ part of one revolution of crank}$$

5. Energy stored by the flywheel in the remaining period.

$$\Delta E = (1 - 0.1)\, E = 0.9 \times 3004.15 = 2703.73 \text{ N-m}$$

6. Coefficient of fluctuation of speed,

$$C_s = 0.07 - 0.2$$

We take $C_s = 0.1$

7. Let $v = 10$ m/s

Rim diameter, $D = \frac{60v}{\pi N} = \frac{60 \times 10}{\pi \times 240}$

$$= 0.796 \approx 0.8 \text{ m or } 800 \text{ mm}$$

8. Mass of flywheel, $m = \frac{\Delta E}{C_s V^2} = \frac{2703.73}{0.1 \times 10^2} = 270.4$ kg

9. Mass of rim,

$$m_r = 0.9 \text{ m} = 0.9 \times 270.4 \times 243.4 \text{ kg}$$

10. Rim size

Let $\frac{b}{t} = 1.25$

$$1.25t^2 = \frac{243.4}{\pi \times 0.8 \times 7100}$$

$$t = 104.5 \approx 105 \text{ mm}$$

$$b = 131.25 \approx 132 \text{ mm}$$

We adopt $b = 132$ mm, $t = 105$ mm, $D = 800$ mm

11. Flywheel shaft.

Energy required per minute at flywheel shaft = $E \times$ number of holes punched per minute

$$= 3004.15 \times 30 \times 90124.5 \text{ N-m/min}$$

Torque on the flywheel shaft,

$$T = \frac{90124.5}{2\pi \times 240} = 59.766 \text{ N-m}$$

Let overhang = 150 mm

Bending moment, $M = 270.4 \times 9.81 \times 0.150$

$$= 397.9 \text{ N.m}$$

Let $K_m = 1.5, K_t = 2.0, \tau_d = 40 \text{ MPa}$

$$\text{Shaft diameter, } d = \left[\left\{\frac{16}{\pi\tau_d}\right\}\{(K_m M)^2) + (K_t T)^2\}^{\frac{1}{2}}\right]^{\frac{1}{3}}$$

$$= \left[\left\{\frac{16}{\pi \times 40}\right\}\{1.5 \times 297.9)^2 + (2.0 \times 59.766)^2\}^{\frac{1}{2}}\right]^{\frac{1}{3}} \times 10$$

$$42.6 \approx 45 \text{ mm}$$

12. Hub diameter, $d_h = 2d = 90$ mm
13. Hub length, $l_h = 2.25d \approx 100$ mm
14. The size of the flywheel is small. We shall use web for the flywheel with thickness,

$$t_w = \frac{b}{20} = \frac{132}{20} = 6.6 \approx 7 \text{ mm}$$

■ **Example 6.10** Design a flywheel for a 4-stroke, 4-cylinder, 130 kW engine running at 375 rpm. The engine is directly coupled to an A.C. generator. Due to space restrictions the flywheel diameter should not be more than 1.25 m.

Solution : Given : $P = 130$ kW, $N = 275$ rpm. $D_{max} = 1.25$ m

1. Coefficient of fluctuation of energy,

$$C_e = 0.15 - 0.20$$

We take, $C_e = 0.16$

2. Fluctuation of energy

$$\Delta E = \frac{60 C_s P}{N} = \frac{60 \times 0.16 \times 130 \times 10^3}{375} = 3328 \text{ N-m}$$

3. Let $D = 1$ m, giving, $v = \dfrac{\pi \times 1 \times 375}{60} = 19.63 \text{ m/s} < 30 \text{ m/s}$.

Hence safe for C.I. flywheel.

4. For directly coupled A.C. generator,

$$C_e = 0.006$$

5. Mass of flywheel, $m = \dfrac{\Delta E}{C_e v^2} = \dfrac{3328}{0.006 \times (19.63)^2} = 1439.43$ kg

6. Mass of rim, $m_r = 0.9$ m $= 1295.5$ kg

7. Rim size,

   Let $b = 2t$, then $2t^2 = \dfrac{1295.5}{\pi \times 1 \times 7100}$

   $$t = 170 \text{ mm}, b = 340 \text{ mm}$$

   Outer diameter of flywheel,

   $$D_o = D + t = 1.17 \text{ m} < 1.25 \text{ m, the space available.}$$

8. Shaft design.

   Torque, $T = \dfrac{10^3 P \times 60}{2\pi N} = \dfrac{10^3 \times 130 \times 60}{2\pi \times 375} = 3310.4$ N-m

   Let overhang = 300 mm

   Bending moment, $M = 1439.43 \times 9.81 \times 0.3 = 42323.24$ N-m

   Let $K_m = K_t = 1.5, \tau_d = 40$ MPa

   Shaft diameter, $d = \left[\left\{\dfrac{16}{\pi \tau_d}\right\} \{(K_m M)^2 + (K_t T)^2\}^{\frac{1}{2}}\right]^{\frac{1}{3}}$

   $$\left[\left\{\frac{16}{\pi \times 40}\right\} \{(1.5 \times 42323.24)^2 + (1.5 \times 3310.4)^2\}^{\frac{1}{2}}\right]^{\frac{1}{3}} \times 10 = 101 \approx 110 \text{ mm}$$

9. Hub diameter, $d_h = 2d = 220$ mm

10. Hub length, $l_h = 2.25d = 247.5 \approx 250$ mm

11. Number of arms, $n = 4$ for $D < 450$ mm

12. Check for rim stresses.

   Total stress in rim $= 0.75\sigma_t + 0.25\sigma_b$

   $$= \rho v^2 \left[\frac{0.75 + 0.25\pi^2 D}{n^2 t}\right]$$

   $$= 7100 \times (19.63)^2 \left[\frac{0.75 + 0.25 \times \pi^2 \times 1}{16 \times 0.170}\right] \times 10^{-6}$$

   $$= 4.53 \text{ MPa} < 38 \text{ MPa}$$

   Hence safe.

13. Arms cross section.

   Required section modulus of arms,

   $$z_y = \frac{TK_t(D - d)}{nD\sigma_{b1}}$$

$$= \frac{3310.4 \times 10^3 \times 1.5\,(1000 - 110)}{4 \times 1000 \times 14}$$

$$= 78918 \text{ mm}^3$$

Major axis of arms cross-section,

$$a_1 = \left[\frac{64 z_y}{\pi}\right]^{\frac{1}{3}}$$

$$= \left[\frac{64 \times 78918}{\pi}\right]^{\frac{1}{3}} = 117 \approx 120 \text{ mm}$$

Minor axis, $b_1 = \frac{a_1}{2} = 60$ mm

14. Key size,

Rectangular key : width, $w = \frac{d}{4} = \frac{110}{4} \approx 30$ mm, thickness $= \frac{d}{6} \approx 20$ mm

■ **Example 6.11** An Otto cycle engine develops 50 kW at 140 rpm with 70 explosions per minute. The fluctuation of speed should not exceed 0.5% of the mean speed on either side. Design a suitable cast iron rim section having width four times the depth so that the hoop stress does not exceed 4.5 MPa. Assume that the flywheel stores $\frac{16}{15}$ times the energy stored by the rim and that the work done during power stroke is 1.5 times the work done during the cycle. Take density of rim material as 7200 kg/m^3.

Solution : Given : $P = 50$ kW, $N = 140$ rpm, $n = 70$, $\sigma_t = 4.5$ MPa, $\rho = 7200$ kg/m^3

1. Fluctuation of energy

Power transmitted, $P = \frac{2\pi N T_{mean}}{60 \times 10^3}$

$$T_{mean} = \frac{50 \times 60 \times 10^3}{2\pi \times 140} = 3410.5 \text{ N-m}$$

Number of explosions, $n = 70 = \frac{N}{2}$

Therefore, it is a four stroke engine. The $T - \theta$ diagram is shown in Fig. 6.21.

Work done per cycle $= T_{mean} \times \theta = 3410.5 \times 4\pi = 42857.6$ N-m

Work done during power stroke $= 1.5 \times 42857.6 = 64286.4$ N-m

The work done during power stroke is shown by the triangle ABC in which base $AC = \pi$ radius and height $FB = T_{max}$

Work done during power stroke $= \left(\frac{1}{2}\right) T_{max} \times \pi = 64286.4$

$$T_{max} = 40926 \text{ N-m}$$

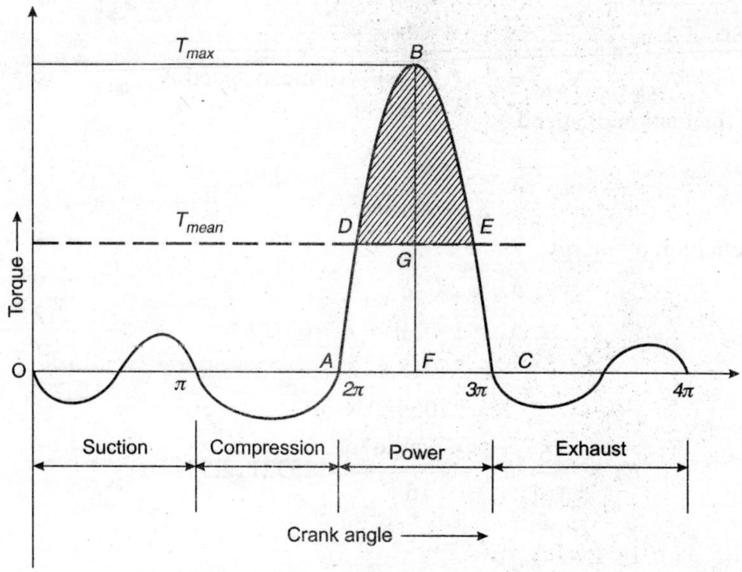

Fig. 6.21

Height above the mean torque line,

$$GB = FB - FG$$
$$= T_{max} - T_{mean}$$
$$= 40926 - 3410.5 = 37515.5 \text{ N-m}$$

Maximum fluctuation of energy, ΔE, is represented by the area *DBED*. It being a parabola, its area is proportional to the square of the height of parabola. Therefore,

$$\frac{\text{Area } DBED}{\text{Area } ABCA} = \left(\frac{GB}{FB}\right)^2$$

$$\Delta E = 64286.4 \times \left(\frac{37515.5}{40926}\right)^2 = 54018.4 \text{ N-m}$$

2. Mean diameter of flywheel

Hoop stress in flywheel rim,

$$\sigma_\theta = \rho v^2$$

$$v = \left(\frac{4.5 \times 10^6}{7200}\right)^{\frac{1}{2}} = 25 \text{ m/s}$$

$$\frac{\pi DN}{60} = v$$

$$D = \frac{25 \times 60}{\pi \times 140} = 3.41 \text{ m}$$

3. Size of rim

Width of rim, $b = 4t$

Area of cross section, $A = bt = 4t^2$

$$N_{max} - N_{min} = 2 \times 0.05 = 1\% \text{ of mean speed } N_{mean}$$

Coefficient of fluctuation of speed,

$$C_S = \frac{N_{max} - N_{min}}{N_{mean}} = 0.01$$

Maximum fluctuation of energy,

$$\Delta E = 2\,C_s E$$

$$= 2 \times 0.01 \times E = 0.02\,E$$

$$= 54018.4$$

$$E = 2700920 \text{ N-m}$$

Energy of rim, $E_r = \left(\frac{15}{16]}\right)E \frac{15 \times 2700920}{16} = 2532112.5 \text{ N-m}$

Now $E_r = \left(\frac{1}{2}\right)m_r v^2$

Mass of rim, $m_r = \frac{2 \times 2532112.5}{(25)^2} = 8102.76 \text{ kg}$

$$m_r = A \times \pi D \times \rho$$

$$= 4t^2 \times \pi \times 3.41 \times 7200$$

$$= 8102.76$$

$$t = 0.162 \text{ m or } 162 \text{ mm}$$

$$b = 4t = 648 \text{ mm}$$

■ **Example 6.12** Design a cast iron flywheel for a four-stroke I.C. Engine developing 200 kW at 240 rpm. The hoop stress developed in the rim is 5 MPa. The total fluctuation of speed is to limited to 2.5% of mean speed. The work down during the power stroke is 30% more than the average work down during the whole cycle. The maximum torque on the shaft is twice the mean torque. The density of cast iron is 7200 kg/m^3.

Solution : Given : $P = 200\text{kW}, N = 240\text{rpm}, \sigma_\theta = 5\text{ Mpa}, N_{max} - N_{min} = 0.025\,N_{mean}, T_{max} = 2T_{mean},$

$$\rho = 7200 \text{ kg/m}^3$$

1. Fluctuation of energy

$$T_{mean} = \frac{P \times 10^3 \times 60}{2\pi \times N} = \frac{200 \times 10^3 \times 60}{2\pi \times 240} = 7957.75 \text{ N-m}$$

Work done per cycle $= T_{mean} \times \theta = 5957.75 \times 4\pi = 100{,}000\ N - m$

Work done during power stroke $= 1.3 \times 100{,}000 = 130{,}000$ N-m

The $T - \theta$ diagram is shown in Fig. 6.22.

The work done during the power stroke is represented by the triangle ABC in which base

$$AC = \pi \text{ radians, and height } FB = T_{max}$$

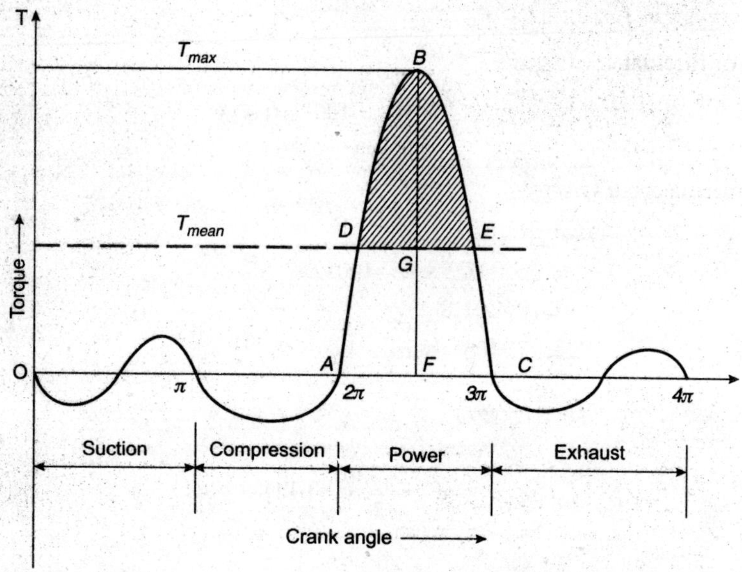

Fig. 6.22

Work done during power stroke $= \left(\frac{1}{2}\right) \times \pi \times T_{\max} = 130{,}000$

$$T_{\max} = 82760.6 \text{ N-m}$$

Height above the mean torque line,

$$GB = FB - FG$$

$$= T_{man} - T_{mean}$$

$$= 82760.6 - 7957.75 = 74802.82 \text{ N-m}$$

Maximum fluctuation of energy, ΔE, is represented by the area *DBED*. It being a parabola, its area is proportional to the square of the height of parabola. Therefore,

$$\frac{\text{Area } DBED}{\text{Area } ABCA} = \left(\frac{GB}{FB}\right)^2$$

$$\Delta E = 130{,}000 \times \left(\frac{74802.82}{82760.6}\right)^2 = 106201.8 \text{ N-m}$$

2. Mean diameter of flywheel

Hoop stress in flywheel rim,

$$\sigma_\theta = \rho v^2$$

$$v = \left(\frac{5 \times 10^6}{7200}\right)^{\frac{1}{2}} = 26.35 \text{ m/s}$$

$$\frac{\pi DN}{60} = v$$

$$D = \frac{26.35 \times 60}{\pi \times 240} = 2.097 \text{ m}$$

3. Mass of rim

Coefficient of fluctuation of speed,

$$C_s = \frac{N_{max} - N_{min}}{N_{mean}} = 0.025$$

Maximum fluctuation of energy,

$$\Delta E = 2C_s E$$
$$= 2 \times 0.025 \times E = 0.05E$$
$$= 106201.8$$
$$E = 2124036 \text{ N-m}$$

Now $$E = \left(\frac{1}{2}\right) m_r v^2$$

Mass of rim, $$m_r = \frac{2 \times 2124036}{26.35^2} = 6118.3 \text{ kg}$$

4. Rim size

Taking $$b = 2t$$

Area of rim, $$a = bt^2$$

$$m_r = A \times \pi D \times \rho = 2t^2 \times \pi \times 2.097 \times 7200$$
$$= 6118.3$$

Thickness of rim, $t = 0.254$ m or 254 mm

Width of rim, $b = 2t = 508$ mm

5. Shaft and hub size

$$T_{max} = 2T_{mean}$$
$$= 2 \times 7957.75 = 15915.5 \text{ N-m}$$
$$\left(\frac{\pi}{16}\right) d^3 \tau = T_{max}$$

Taking $$\tau = 42 \text{ MPa}$$

$$d = \left[\frac{16 \times 159155 \times 10^3}{\pi \times 42}\right]^{\frac{1}{3}}$$
$$= 124.5 \approx 125 \text{ mm}$$

Diameter of hub, $d_h = 2d = 250$ mm

Length of hub, $l_h = b = 508$ mm

6. Arms size

Number of arms for $D = 2.097$ m are 6

Bending moment on the arms,

$$M = \frac{T_{max}(D - d_h)}{Dn}$$

$$= \frac{15915.5(2.097 - 0.25)}{2.097 \times 6}$$

$$= 2336.38 \text{ N-m}$$

Section modulus, $z_y = \left(\frac{\pi}{32}\right) b_1 a_1^2 = \left(\frac{\pi}{32}\right) \times 0.5 a_1 \times a_1^2 = 0.049 a_1^3$

Taking permissible bending stress in arms = 15 MPa

$$15 = \frac{2336.38 \times 10^3}{0.049\, a_1^3}$$

Major axis of arms, $a_1 = 147 \approx 150$ mm

Minor axis of arms, $b_1 = 75$ mm

7. Key size

For a shaft of diameter 125 mm, the size of rectangular sunk key are :

Width, $w = 36$ mm

Thickness, $h = 20$ mm

Considering shearing failure of the key,

$$\frac{L \times w \times \tau \times d}{2} = T_{max}$$

$$L = \frac{15915.5 \times 10^3}{36 \times 42 \times \frac{125}{2}}$$

$$= 168.4 \approx 170 \text{ mm}$$

8. Check for maximum stress in rim

Total stress in rim, $\sigma = \rho v^2 \left[0.75 + \frac{4.935\, R}{n^2 t}\right]$

$$= 7200 \times (26.35)^2 \left[\frac{0.75 + 4.935 \times \left(\frac{2.097}{2}\right)}{6^2 \times 254}\right]$$

$$= 6.58 \text{ MPa} < 15 \text{ MPa}$$

Hence safe.

■ **Example 6.13** A punching press is used to punch 30 holes per minute in a plate using 10 kN-m of energy per hole during each revolution. Each piercing takes 42% of the time needed to make one revolution. The punch receives power through a gear reduction unit which in turn is fed by an electric motor driven belt pulley 900 mm diameter and running at 240 rpm. Find the power of the electric motor if overall efficiency of the transmission unit is 85%. Design a cast iron flywheel to be sued with the punching machine for a coefficient of steadiness of 5. Space limitations restrict the diameter to 1.2 m.

Permissible shear stress in shaft material = 45 MPa

Permissible tensile stress for cast iron = 4 MPa

Density of cast iron = 7200 kg/m^3

Solution : Given : Number of holes = 30 per minute, energy consumed per hole = 10 kN-m, $d_p = 900$ mm,

$N_p = 240$ rpm, $\eta = 85\%$, $C_s = \frac{1}{5} = 0.2$, $D_{max} = 1.2$ m $\tau = 45$ MPa, $\sigma_t = 4$ MPa, $\rho = 7200$ kg/m^3

1. Power of electric motor

Energy used for piercing holes per minute = Number of holes pierced × energy used per hole

$$= 30 \times 10000 = 300000 \text{ N-m/min}$$

Power required for the electric motor,

$$P = \frac{\text{Energy used per minute}}{60 \times \eta}$$

$$= \frac{300000}{60 \times 0.85} = 5882 \text{ W} = 5.882 \text{ kW}$$

2. Fluctuation of energy

Total energy to be supplied during each revolution,

$$E_t = \frac{10000}{0.85} = 1174.7 \text{ N-m}$$

Velocity of belt,
$$v = \frac{\pi d_p N_p}{60 \times 10^3}$$

$$= \frac{\pi \times 900 \times 240}{60 \times 10^3} = 11.3 \text{ m/s}$$

Net pull acting on the belt,
$$F = \frac{P}{v} = \frac{5882}{11.3} = 520.8 \text{ N}$$

Time required to punch a hole $= \frac{0.42}{30} = 0.014$ min

Distance moved by belt during punching a hole = velocity of belt × required to punch a hole.

$$= 11.3 \times 60 \times 0.014 = 9.492 \text{ m}$$

Energy supplied by belt during punching a hole.

$$E_b = \text{net pull in belt} \times \text{distance travelled by belt}$$
$$= 520.8 \times 9.492 = 4943.4 \text{ N-m}$$

Energy to be supplied by the flywheel,

$$\Delta E = E_t - E_b$$
$$= 11764.7 - 4943.4 = 6821.3 \text{ N-m}$$

3. Size of flywheel

Angular velocity,
$$\omega_{mean} = \frac{2\pi N}{60} = \frac{2\pi \times 240}{60} = 25.13 \text{ rad/s}$$

Diameter of flywheel rim, $D = 1.2 \text{ m}$

$$R = 0.6 \text{ m}$$

$$\Delta E = mR^2 \omega_{\text{mean}}^2 C_s$$

$$6821.3 = m \times (0.6)^2 \times (25.13)^2 \times 0.2$$

$$m = 150 \text{ kg}$$

4. Cross-section of flywheel rim

Let thickness of rim = t and width of rim, $b = 2t$

Area of cross-section of rim,

$$A = 2t^2$$

Mass of rim, $m = A \times \pi D \times \rho$

$$150 = 2t^2 \times \pi 1.2 \times 7200$$

$$t = 0.052 \text{ m or } 52 \text{ mm}$$

$$b = 104 \text{ mm}$$

5. Shaft and hub size

Mean torque, $T_{\text{mean}} = \dfrac{P \times 60}{2\pi N}$

$$= \frac{5882 \times 60}{2\pi \times 240} = 234 \text{ N-m}$$

Let $T_{\max} = 2T_{\text{mean}} = 2 \times 234 = 468 \text{ N-m}$

$$\left(\frac{\pi}{16}\right) d^3 \tau = T_{\max}$$

$$d = \left[\frac{16 \times 468 \times 10^3}{\pi \times 45}\right]^{\frac{1}{3}} = 37.56 \approx 40 \text{ mm}$$

Diameter of hub, $d_h = 2d = 80 \text{ mm}$

Length of hub, $l_h = b = 104 \text{ mm}$

6. Arms size

For $D = 1.2 \text{ m } n = 6$

Bending moment on the arms,

$$M = \frac{T_{\max}(D - d_h)}{Dn}$$

$$= \frac{468(1.2 - 0.08)}{1.2 \times 6} = 72.8 \text{ N-m}$$

Section modulus, $z_y = \left(\dfrac{\pi}{32}\right) b_1 a_1^2 = \left(\dfrac{\pi}{32}\right) \times 0.5\, a_1 \times a_1^2 = 0.049\, a_1^3$

Permissible bending stress in arms = 4 MPa

$$4 = \frac{72.8 \times 10^3}{0.049\, a_1^3}$$

Major axis of arms, $a_1 = 71.88 \approx 72$ mm

Minor axis of arms, $b_1 = 36$ mm

7. Key size

For a shaft of diameter 40 mm, the size of rectangular sunk key are :

Width, $w = 16$ mm

Thickness, $h = 10$ mm

Considering shearing failure of the key,

$$\frac{L \times w \times \tau \times d}{2} = T_{max}$$

$$L = \frac{468 \times 10^3}{16 \times 45 \times \frac{40}{2}}$$

$$= 32.5 \approx 35 \text{ mm}$$

8. Check for maximum stress in rim

Velocity of rim, $v = \frac{\pi DN}{60} = \frac{\pi \times 1.2 \times 240}{60} = 15$ m/s

Total stress in rim, $\sigma = \rho v^2 \left[\frac{0.75 + 4.935\, R}{n^2 t}\right]$

$$= 7200 \times (15)^2 \left[\frac{0.75 + 4.935 \times \left(\frac{1.2}{2}\right)}{6^2 \times 0.052}\right]$$

$$= 3.78 \text{ MPa} < 4 \text{ MPa}$$

Hence safe.

SOLVED QUESTIONS FROM UNIVERSITY EXAMINATIONS

■ **Example 6.14** Design and draw a C.I. protective flange coupling to connect two shafts to transmit 7.5 kW at 750 rpm. The permissible stresses are :

Shear stress for shaft, bolt and key material = 33 MPa

Crushing stress for bolt and key = 60 MPa

Shear stress for CI = 15 MPa [MDU, Dec., 2007]

Solution : Refer to Example 6.6.

■ **Example 6.15** Design a bushed pin type flexible coupling for connecting a motor shaft to a pump shaft for transmitting 40 kW at 1000 rpm. The allowable stress in pin is 20 MPa and bearing pressure on rubber bush is 0.45 MPa. [MDU, May, 2008]

Solution : Refer to Example 6.7.

■ **Example 6.16** Design and draw a bushed pin type flexible coupling for connecting the motor and centrifugal pump shafts. The motor delivers 25 kW power at 720 rpm. The starting torque of the motor can be assumed to be 125% of the rated torque. [MDU, Dec., 2008]

Solution : Refer to Example 6.7.

■ **Example 6.17** Design completely a flywheel, its shaft and key for a punching machine having a capacity for producing 30 holes of 19 mm diameter per minute in 16 mm thick steel plate. The material of the plate has on ultimate shear strength of 350 MN/m^2. The actual punching operation is estimated to last for $\frac{1}{5}$th of the angular rotation of the crankshaft. The crankshaft is powered by a flywheel shaft through a reduction gearing of ratio 1 : 10. Assume that the mechanical efficiency of the punching machine is 0.80 and during the punching the flywheel speed is reduced by a maximum of 10% of mean speed. The space limitations restrict the flywheel diameter to 0.9 m.

[MDU, May, 2009]

Solution : Given : Number of holes – 30 per min, $d = 19$ mm, $t = 16$ mm, $S_{ut} = 350$ MN/m^2, actual punching time $= \frac{1}{5} \times$ angular rotation of crankshaft, $i = 1{:}10$, $\eta_{mech} = 0.8, C_s = 0.1, D = 0.9$ m

Area of plate sheared, $A_s = \pi dt = \pi \times 19 \times 16 \times 955.04 \text{ mm}^2$

Force required for punching,

$$F_s = A_s \times S_{ut} = 955.04 \times 350 = 344265 \text{ N}$$

Energy required per punching hole $= \frac{1}{2} F_s \times t$

$$= \frac{1}{2} \times 334265 \times 16 \times 10^{-3}$$

$$= 2674.12 \text{ N0-m}$$

Energy required per minute = Energy required per punched hole × number of holes punched per minute

$$= 2674.12 \times 30 = 80223.6 \text{ N-m/min}$$

Maximum fluctuation of energy, $\Delta E = \left(1 - \frac{1}{5}\right) \times$ Energy per punched hole

$$= \frac{4}{5} \times 2674.12 = 2139.296 \text{ N-m}$$

Let us assume that the rotational inertial of hub and spokes

= 5% of the rotational inertia of flywheel.

Then maximum fluctuation of energy provided by the rim,

$$(\Delta E)_{min} = 0.95 \times \Delta E$$

$$= 0.95 \times 2139.296 = 2032.3 \text{ N-m}$$

Mean speed of flywheel, $N = 10 \times 30 = 300$ rpm

Angular speed, $\omega = \dfrac{2\pi N}{60} = \dfrac{2\pi \times 300}{60} = 10\pi$ rad/s

$$(\Delta E)_{\text{rim}} = mR^2\omega^2 C_s$$

$$2032.3 = m \times (0.45)^2 \times (10\pi)^2 \times 0.1$$

$$m = 101.688 \text{ kg}$$

Let density of rim = 7250 kg/m^3 and width of rim, $b = 2h$

Then $m = \pi D \times b \times h \times \rho$

$$101.88 = \pi \times 0.9 \times 2h^2 \times 7250$$

Rim thickness, $h = 0.04985$ m or $\simeq 50$ mm

$$b = 2 \times 50 = 100 \text{ mm}$$

Shaft : Motor power, $P = \dfrac{\text{Energy required per minute}}{60 \times \eta_{\text{mech}}}$

$$= \frac{80223.6}{60 \times 0.8} = 1671.3 \text{ W}$$

Now $P = \dfrac{2\pi NT}{60}$

Torque transmitted, $T = \dfrac{1671.3 \times 60}{2\pi \times 300} = 53.2$ N-m

Let d_s = diameter of shaft.

Then $T = \dfrac{\pi}{16} d_s^3 \tau$

Let $\tau = 42$ MPa for steel shaft

$\therefore$ $\dfrac{\pi}{16} d_s^3 \times 42 = 53.2 \times 10^3$

$$d_s = 18.6 \simeq 20 \text{ mm}$$

Let length of key, $L = 2 \times d_s = 2 \times 20 = 40$ mm

w = width of key

$$T = L\omega\tau = \frac{ds}{2}$$

$$53.2 \times 10^3 = 40 \times \omega \times 42 \times \frac{20}{2}$$

$$\omega = 3.2 \simeq 5 \text{ mm}$$

Depth of key = 5 mm

■ **Example 6.18** Design a pin type flexible coupling to connect two shafts of 50 mm diameter transmitting 45 kW at 1200 rpm. The bearing pressure between rubber bush and cast iron flange is 0.5 N/mm^2 and the allowable shear stress for the pin is 30 N/mm^2. Thickness of input and output flanges are taken as diameter of bush and half of shaft diameters respectively. Assume that there are six used in the coupling and the diameter of each pin is 20 mm. [MDU, Dec., 2009]

Solution : Refer to Example 6.7.

EXERCISES

1. What are the advantages of flexible coupling? How do you classify flexible coupling?
2. Design a flange coupling to transmit 135 kW power at 120 rpm from one shaft to another. Allowable shearing stress of the bolt material is 45 N/mm^2 while that of the shaft material is 55 N/mm^2. The flanges are made of C.I. for which the stresses in compression and shear are 75 N/mm^2 and 175 N/mm^2 respectively. Suggest suitable dimensions for the shaft, bolts and coupling. (Do not design the key).
3. Design a cast iron protected type flange coupling to connect two shafts of 36 mm diameter, transmitting 15 kW at 720 rpm. The overload capacity is 1.25 times the average torque. The bolts and key are made of C20 steel and the flanges are made of cast iron.
4. Design a square key for a 45 mm diameter shaft to transmit 10 kW at 250 rpm. The key and shaft are made of steel having $S_{yt} = S_{yc} = 380$ MPa. The factor of safety desired is 3.
5. Design a cast iron protected type rigid flange coupling to connect a 50 kW, 1440 rpm electric motor to a centrifugal pump. The starting torque of motor is 200% of the rated torque. Assume suitable data from the Date Book.
6. Design a bushed pin type flexible coupling for connecting a motor and a pump shaft for the following data :

Power transmitted	= 20 kW
Speed	= 900 rpm
Diameter of motor and pump shafts	= 60 mm
Allowable bearing pressure in rubber bush	= 0.35 MPa

7. Design a cast iron flywheel for a four-stroke engine to develop 100 kW at 160 rpm. The work done in the power stroke is 1.25 times the average work done during the whole cycle. The mean diameter of the flywheel is 2.5 m and the fluctuation of speed is limited to 5% of the mean speed. The material density is 7200 kg/m^3. The permissible shear stress in the shaft material is 42 MPa and the bending stress in the arms of the flywheel is 20 MPa.
8. A punch press is fitted with a flywheel which can furnish 2500 N-m of energy during quarter of a revolution near the bottom dead centre while blanking a hole in sheet metal. The maximum speed of the flywheel during the operation is 180 rpm and the speed decreases by 10% during the blanking stroke. The mean diameter of the rim is 12 m. Calculate the mass of the flywheel rim assuming that it contributes 90% of the energy requirements.
9. Design a flywheel for a punching machine having a capacity of punching 30 holes per minute of 25 mm diameter in a steel plate 20 mm thick. The ultimate strength for the material of the plate is 360 MPa. The actual punching operation takes place during 30° rotation of the punching machine crankshaft. This crankshaft is powered by a flywheel shaft through reduction gearing having a ratio of 1 : 8. Assume that the mechanical efficiency of the punching machine is 80% and during the actual punching operation the flywheel speed is reduced by 10%. The diameter of the flywheel is restricted to 0.8 m.

10. A cast iron flywheel is to be designed for a single cylinder double acting steam engine which delivers 140 kW at 100 rpm. The maximum fluctuation of energy per revolution is 10% and total fluctuation of speed is 5% of the mean speed. The mean diameter of the flywheel rim is 2.5 m Determine :
   (a) Cross-section of rim, assuming that the hub and arms provide 5% of the rotational inertia of the wheel. The density of cast iron is 7200 kg/m^3 and permissible tensile stress is 15 MPa. Width is twice the thickness.
   (b) Diameter of shaft, hub and key size, the permissible shear stress in shaft and key is 42 MPa.
   (c) Cross-section of ellipitcal arms assuming major axis to be double the minor axis. The number of arms are 6.

❑❑❑

7

Chapter

Screwed Joints and Power Screws

Inside This Chapter

7.1 Screwed Joints
7.2 Power Screws

7.1 SCREWED JOINTS

A screwed thread is formed by cutting continuous helical groove on a cylindrical surface. The screw thread may be single start or multi-start and either right hand or left hand. A screwed joint consists of a bolt and a nut assembly. It is a temporary type of fastening. The advantages and disadvantages of screwed joints are :

Advantages :

1. Highly reliable in operation.
2. Easy to assemble and disassemble.
3. Adaptable to various operating conditions.
4. Relatively cheaper to produce.

Disadvantages :

1. Stress-concentration in the threaded portion.
2. Lower strength as compared to riveted and welded joints.

7.1.1 Terminology

- **Major diameter :** It is the largest diameter of an external or internal screw thread. It is also called outside or nominal diameter.
- **Minor diameter :** It is the smallest diameter of an external or internal screw. It is also called core or root diameter.

- **Pitch diameter :** It is the diameter of an imaginary cylinder on a cylindrical screw thread the surface of which would pass through the thread at such points as to make equal the width of the thread and the width of the spaces between the threads.
- **Pitch :** It is the distance measured in an axial direction from a point on one thread to the corresponding point on the next.
- **Lead :** It is the distance between two corresponding points on the same helix.

  Lead = pitch × number of starts
- **Angle of the thread :** It is the angle included by the flanks of the thread.

7.1.2 Forms of Screw Threads

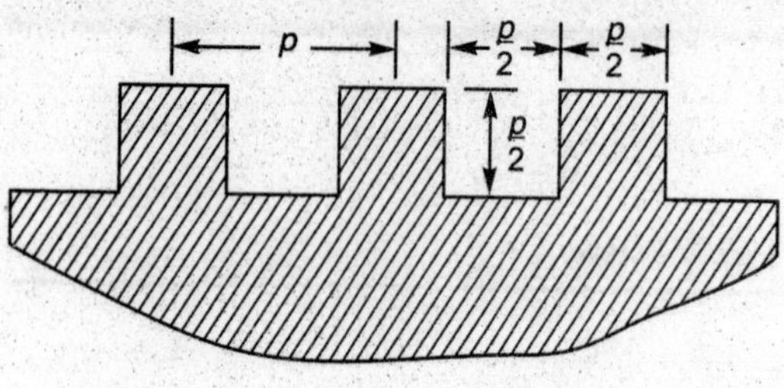

(a) Square thread

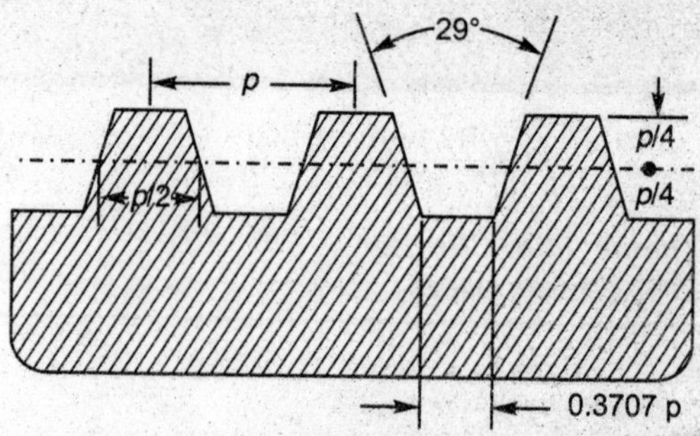

(b) Acme thread

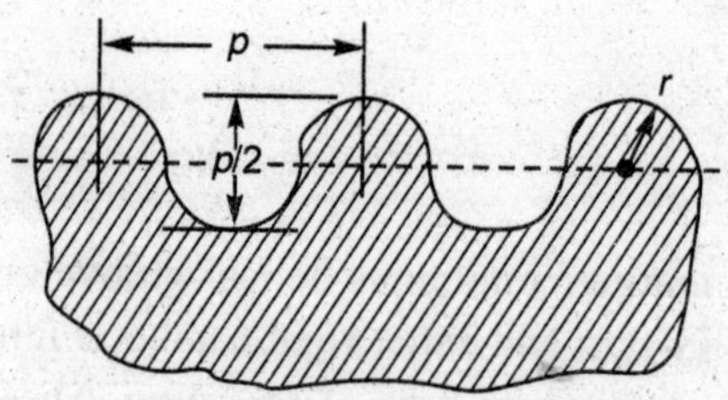

(c) Knucke thread

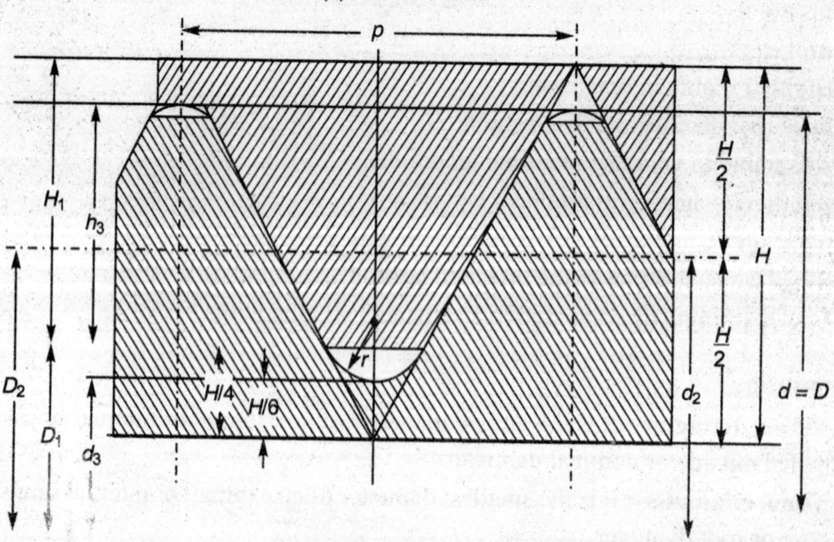

(d) Design profile of nut and bolt

7.1 Forms of screw threads.

The four types of screw threads commonly used for nuts and bolts are :

1. Square thread
2. Acme thread
3. Knuckle thread
4. Metric thread

These threads have been shown in Fig. 7.1.

7.1.3 Designation of Threaded Fasteners

Sizes of screw threads are designated by the letter M followed by diameter and pitch. The diameter and pitch being separated by the sign ×. When there is no mention of the pitch, it shall mean that a coarse pitch will be used.

Thus a fine series screw of major diameter 8 mm and pitch 1 mm shall be designated as : M8 × 1

A coarse series screw of major diameter 6 mm and pitch 1 mm shall be designed ss : M6.

To specify the tolerances on the screw threads, the following tolerance grades are used :

- '7' for fine grade
- '8' for normal (medium) grade
- '9' for coarse grade.

The tolerance position is specified as follows :

- 'H' for unit thread
- '*d*' for bolt thread with allowance, and
- '*h*' for bolt thread without allowance.

7.1.4 Standard Dimensions of Screw Threads

The dimensions of ISO metric screw threads (coarse series) used for screws, bolts and nuts are given in Table 7.1. For medium and fine series, refer to the standards in the Machine Design Data Hand Book.

Table 7.1 : Design dimensions of ISO metric screw threads, bolts and nuts.

Designation	*Nominal or major dia mm Nut and bolt, d = d*	*Pitch mm p*	*Pitch dia mm Nut and bolt $d_2 = D_2$*	*Minor or core diameter*		*Tensile thread area mm^2*
				Nut, d_3	*Bolt D_3*	
(a) Coarse series						
M4	4	0.70	3.545	3.141	3.242	8.78
M5	5	0.80	4.480	4.019	4.134	14.20
M6	6	1.00	5.350	4.773	4.917	20.10
M8	8	1.25	7.188	6.466	6.647	36.60
M10	10	1.50	9.026	8.160	8.376	58.00
M12	12	1.75	10.863	9.853	10.106	84.30
M16	16	2.00	14.701	13.546	13.835	157
M20	20	2.50	18.376	16.933	17.294	245
M24	24	3.00	22.051	20.319	20.752	353

M30	30	3.50	27.727	25.706	26.211	561
M36	36	4.00	33.402	31.093	31.670	817
M42	42	4.50	39.077	36.479	37.129	1120
M48	48	5.00	44.752	41.866	42.587	1470
M56	56	5.50	52.428	49.252	50.046	2030
M64	64	6.00	60.103	56.639	57.505	2680
M72	72	6.00	68.103	64.639	65.505	3460
M80	80	6.00	76.103	72.639	73.505	4340
M90	90	6.00	86.103	82.639	83.505	5590
M100	100	6.00	96.103	92.639	93.505	7000

7.1.5 Stresses in Screwed Joints

(a) Static Loading

The following stresses are induced in screwed joints

1. Internal stresses due to screwing up forces.
2. Stresses due to external forces.
3. Combined stresses due to (1) and (2)

(i) *Initial stresses*

1. Torsional shear stress caused by the frictional resistance of threads during tightening : Due to friction, the turning moment acts on the screw in a plane normal to the screw axis. The useful force P_i is exerted by the thread in the direction of the axis. If there were no friction, this force would exert a pressure N normal to the surface of the thread as shown in Fig. 7.2(a). The angle between P_i and N is the helix angle of the threads.

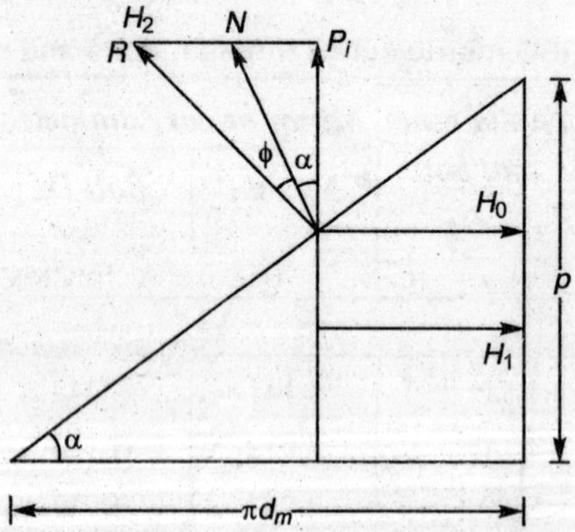

(a)

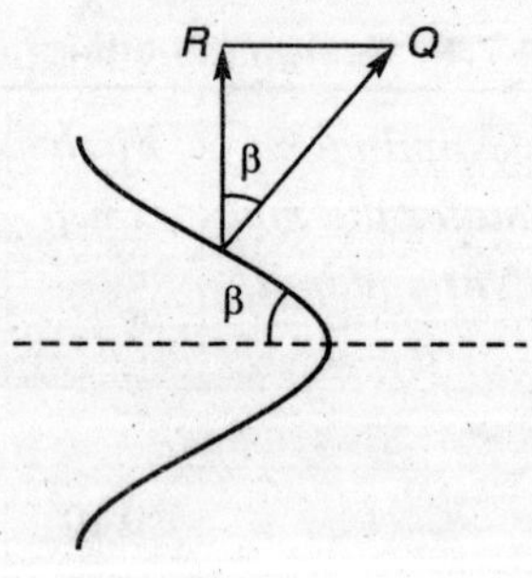

(b) Traperoidal thread

Fig. 7.2

Now $$\tan \alpha = \frac{p}{\pi d_m}$$

where p = pitch of threads

d_m = mean pitch diameter of threads.

Because of friction the resultant force R will be inclined to the normal by the amount of the angle friction ϕ, and will have a component H_1 in the place normal to the screw axis. The component H_1 is overcome and balanced by an opposite force H_o, where

$$H_1 = H_o = P_i \tan(\alpha + \phi) = P_i \left[\frac{\tan\alpha + \tan\phi}{1 - \tan\alpha \tan\phi}\right]$$

$$= P_i \left[\frac{\tan\alpha + \mu}{1 - \mu \tan\alpha}\right] \quad \text{...(7.1)}$$

where μ = coefficient of friction

For triangular threads as shown in Fig. 7.2(b),

$$Q = \frac{R}{\cos\beta}$$

where 2β = angle of threads

and the corrected tangential resistance becomes,

$$H_1 = P_i \left[\frac{\tan\alpha + \mu \sec\beta}{1 - \mu \tan\alpha \sec\beta}\right] \quad \text{...(7.2)}$$

when the nut or a screw with head is turned, the additional frictional force with which the nut or head bears is given by :

$$H_2 = \frac{P_i \mu_c d_c}{d_m} \quad \text{...(7.3)}$$

where μ_c = coefficient of friction between the collar (nut or bolt head) and the surface on which the collar bears.

d_c = mean collar diameter (nut or bolt head)

Total frictional force,

$$H = H_1 + H_2 = P_i \left[\frac{\tan\alpha + \mu \sec\beta}{1 - \mu \tan\alpha \sec\beta} + \frac{\mu_c d_c}{d_m}\right] \quad \text{...(7.4)}$$

Frictional torque, $$T_f = \frac{H d_m}{2} \quad \text{...(7.5)}$$

2. Initial tensile stresses due to screwing up forces : The initial tension in a bolt is given empirically by :

$$P_i = 2805d \text{ N, for a leak proof joint} \quad \text{...(7.6a)}$$
$$= 1420d \text{ N, for a ordinary joint} \quad \text{...(7.6b)}$$

where d = nominal diameter of bolt, mm

Initial tensile stress, $$\sigma_i = \frac{P_i}{A_s} \quad \text{...(7.7a)}$$

where A_s = stress area of bolt.

3. Torsional stress : Initial torsional shear stress due to tightening,

$$\tau_i = \frac{16 T_f}{\pi d^3} \quad \text{...(7.7b)}$$

4. Shear stress across the threads : It is given by :

$$\tau_s = \frac{P_i}{\pi d_c b n}$$

where b = width of thread section at the root

n = number of threads in engagement

5. Crushing stress on threads : The crushing stress is given by :

$$\sigma_c = \frac{P_i}{\pi(d^2 - d_c^2)n} \qquad \text{...(7.8a)}$$

where d = major diameter

6. Bending stress : The bending stress is induced when the surface under the head or nut are not perfectly perpendicular to the bolt axis. The bending stress induced in the shank of the bolt is given by :

$$\sigma_b = \frac{aE}{2l} \qquad \text{...(7.8b)}$$

where a = difference in height between the extreme corners of the nut or head

l = length of shank of the bolt

E = Young's modulus of elasticity of bolt material

(ii) *Stress due to external forces*

1. Tensile stress : This stress is induced when the load acts in the direction of the bolt axis. It is given by :

$$\sigma_t = \frac{P}{A_s} \qquad \text{...(7.9)}$$

2. Shear stress : This stress is induced in the bolt when the load acts perpendicular to the bolt axis.

$$\tau = \frac{4P}{\pi d^2 n}$$

3. Combined tension and shear stress : In such a case, the maximum shear stress l given by :

$$\tau_{max} = \left[\left(\frac{\sigma_t}{2}\right)^2 + \tau^2\right]^{1/2} \qquad \text{...(7.10)}$$

and the maximum principal stress is given by :

$$\sigma_{max} = \frac{\sigma_t}{2} + \left[\left(\frac{\sigma_t}{2}\right)^2 + \tau^2\right]^{1/2} \qquad \text{...(7.11)}$$

(b) Fluctuating Loading

In many applications, the external forces acting on the bolted assembly fluctuates between two limits. The endurance strength is the criterion of failure in these cases. The endurance limit is determined as explained in art. 4.8. The surface finish of the bolt is usually considered as equivalent to a machined surface. The fatigue stress-concentration factor K_f varies from 2 to 3. The modified Goodman diagram may be used to determine the bolt size.

7.1.6 External Load with Preload

(a) Axial Loads

Assume that a flexible gasket is used in a bolted joint as shown in Fig. 7.3. Let

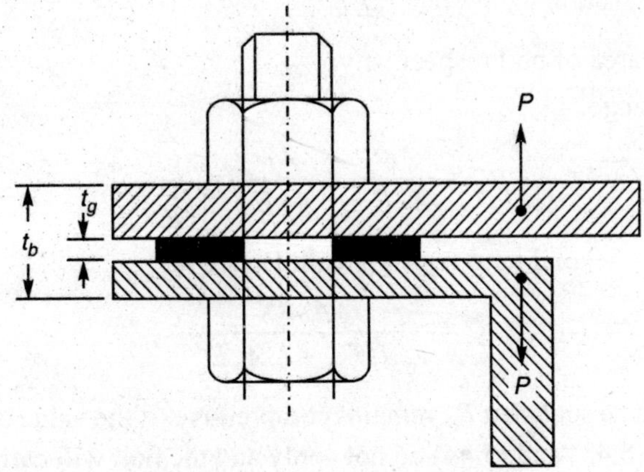

Fig. 7.3 Bolt joint with flexible gasket.

P_i = initial load due to tightening of the bolt.
P = external load
P_g = compression load on the gasket
P_b = combined load on the bolt
k_b, k_g = stiffness of bolt and gasket respectively, N/mm
δ_b = elongation of bolt, mm/N $= 1/k_b$
δ_g = compression of gasket, mm/N $= 1/k_g$
n = total number of bolts

Initial compression of gasket $= P_i\delta_g$

Initial elongation of bolt $= P_i\delta_b$

When the load P is applied, increase in length of bolt or gasket $= (P_b - P_i)\delta_b$

Final deformation of gasket,

$$P_g\delta_g = P_i\delta_g - (P_b - P_i)\delta_b$$

Final compressive load on gasket,

$$P_g = P_i - \frac{(P_b - P_i)\delta_b}{\delta_b}$$

Combined load on bolt, $$P_b = P_g + P = P_i = \frac{(P_b - P_i)\delta_b}{\delta_g} + P$$

$$= P_i + \left[\frac{\delta_b}{\delta_g + \delta_b}\right] P = P_i + CP \qquad \text{...(7.12)}$$

where $C = \dfrac{\delta_g}{(\delta_g + \delta_b)} = \dfrac{k_b}{(k_b + k_g)}$

Hence $$P_g = P_i - \left[\frac{\delta_g}{\delta_g + \delta_b}\right] P \qquad ...(7.13)$$

Now $$\delta_b = \frac{l_b}{A_b E_b} = 1/k_b$$

where l_b, A_b = length and area of bolt respectively.

For the gasket and flange,

$$g_g = \frac{t_g}{A_g E_g} + \frac{t_f}{A_f E_f} = \frac{1}{k_g}$$

For a thick gasket, the second term is very small. Hence

$$C = \frac{t_g A_b E_b}{t_g A_b E_g + l_b A_g E_g} \qquad ...(7.14)$$

The value of P should be such that P_g remains compressive. If the value of P becomes such that P_g becomes tensile then Eqs. (5.43) and (5.44) do not apply and the bolt will carry the total load P.

(b) Metal-to-metal Joint

For this case, when $E_b = E_f, t_f = l_b$ and $A_f >> A_b$, then $C \to 0$, and the external load does not affect the stress in the bolt as long as $P < P_i$. For a tight joint, $P < P_i$ by about 20%. If $P > P_i$, the joint will separate and the stress will be due to the external load alone.

(c) Gasket Joint

The values of C for a gasket joint are given in Table 7.2.

Table 7.2 : Values of C

	Type of joint	*Factor C*
1.	Soft, elastic gasket with studs	1.00
2.	Soft gasket with through bolts	0.90
3.	Copper-asbestos gasket	0.60
4.	Soft copper gasket with long through bolts	0.50
5.	Soft copper corrugated gasket	0.40
6.	Lead gasket with studs	0.10
7.	Narrow copper ring	0.01
8.	Metal-to-metal joint	0.00

7.1.7 Eccentric Loading on Bolts

1. Eccentric Load Parallel to the Axis of Bolts

Consider a bracket having a rectangular base bolted to a wall by means of four bolts as shown in Fig. 7.4. In this case the bolts are under tension. The direct load in each bolt, $P_d = \frac{P}{n}$, where n = number of bolts. The load P tends to rotate the bracket about the edge AB. Let w be the load in a bolt per unit

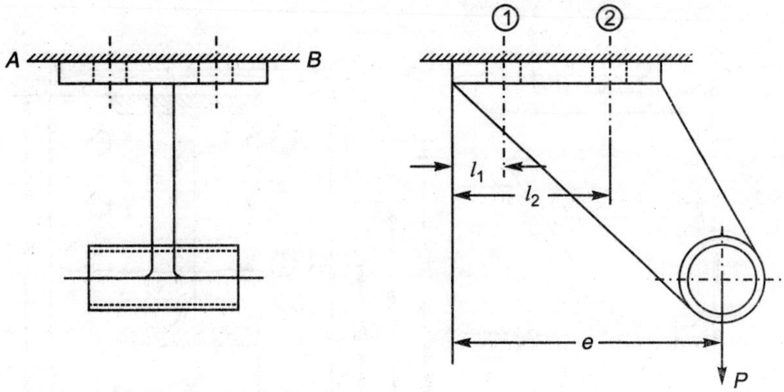

Fig. 7.4

distance due to the turning effect of the bracket. Let P_1, P_2 be the loads on each of the bolts at distances l_1 and l_2 respectively from the tilting edge. Then

Load on each bolt at distance l_1, $P_1 = wl_1$

Moment of this load about the tilting edge $= wl_1 \times l_1 \times wl_1^2$

Similarly moment of each bolt at distance $l_2 = wl_2^2$

Total moment of the load on bolts about the tilting edge $= 2w(l_1^2 + l_2^2)$

Moment due to load P about the tilting edge $= Pe$

$$2w(l_1^2 + l_2^2) = Pe$$

$$w = \frac{Pe}{2(l_1^2 + l_2^2)}$$

It may be observed that the most heavily loaded bolts are those which are located at the greatest distance from the tilting edge. Thus

$$P_2 = wl_2 = \frac{Pel2}{2(l_1^2 + l_2^2)}$$

Total tensile load on the heavily loaded bolt,

$$P_t = P_d + P_2 = \frac{P}{4} + \frac{Pel_2}{2(l_1^2 + l_2^2)} \quad ...(7.15)$$

If d_c is the core diameter of the bolt, then

$$\left(\frac{\pi}{4}\right) d_c^2 \sigma_t = P_i$$

or

$$d_c = \left[\frac{4P_i}{\pi \sigma_t}\right]^{1/2} \quad ...(7.16)$$

2. Eccentric Load Acting Perpendicular to the Axis of Bolts

Consider a wall bracket carrying an eccentric load perpendicular to the axis of the bolts on rectangular base as shown in Fig. 7.5. In this case bolts are under direct shear.

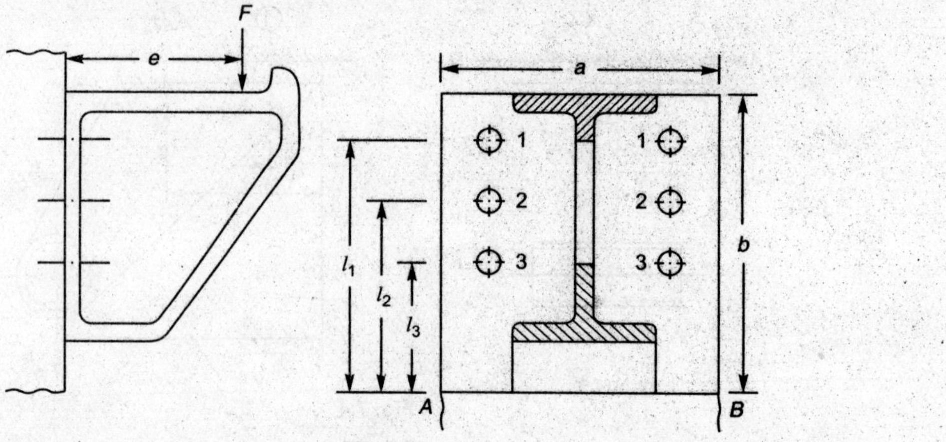

Fig. 7.5

Direct shearing load in each bolt, $P_s = \frac{P}{n}$

The load P tends to cause the bracket to rotate clockwise about the edge AB. This produces tensile load in the bolts. The moment Pe of load P must be balanced by the sum of the moments of all the bolts about the edge AB. Let P_1, P_2, P_3 be the loads on the bolts, then

$$Pe = 2(P_1 l_1 + P_2 l_2 + P_3 l_3)$$

Also, the forces in the bolts are proportional to the distance from the edge AB. Hence

$$P_2 = \frac{P_1 l_1}{l_1}, P_3 = \frac{P_1 l_3}{l_1}$$

Hence, tensile load in the most heavily loaded bolt,

$$P_1 = \frac{Pel_1}{2(l_1^2 + l_2^2 + l_3^2)}$$

The bolts are subjected to the combined action of shear and tension. The maximum tensile load in bolt 1 becomes :

$$(P_1)_{max} = \left(\frac{1}{2}\right)[P_1 + (P_1^2 + 4P_s^2)^{1/2}] \quad \text{...(7.17)}$$

and the maximum shear load,

$$(P_s)_{max} = \left(\frac{1}{2}\right)(P_1^2 + 4P_s^2)^{1/2} \quad \text{...(7.18)}$$

$$\left(\frac{\pi}{4}\right)d_c^2 \sigma_t = P_t$$

or

$$\left(\frac{\pi}{4}\right)d_c^2 \tau = (P_s)_{max}$$

3. Eccentric load on a Bracket with Circular Base

Consider round flange bearing having four bolts as shown in Fig. 7.6.

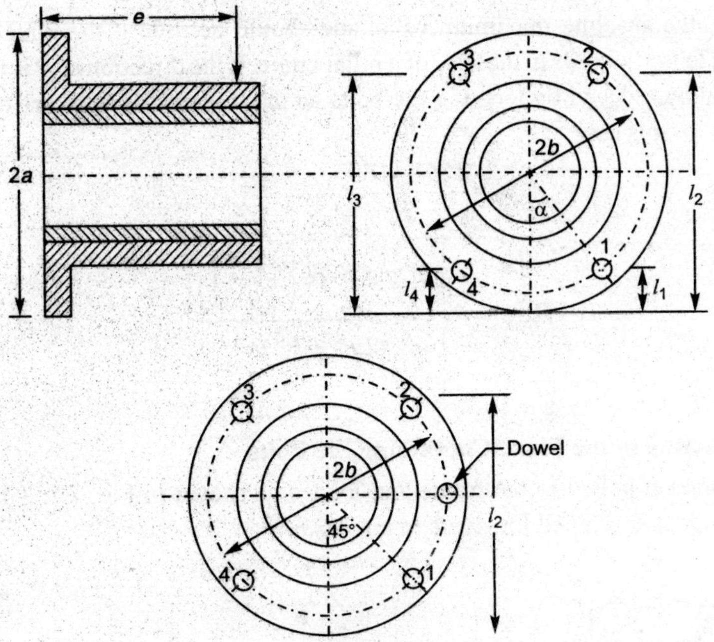

Fig. 7.6

Let

a = radius of the column flange

b = radius of the bolt pitch circle,

e = distance of the load from the lilting edge

$l_1, l_2 . l_3, l_4$ = distances of bolt centres from the tilting edge.

Then

$$P_1 = \frac{Pel_1}{l_2^2 + l_2^2 + l_3^2 + l_4^2}$$

$$l_1 = a - b\cos\alpha$$

$$l_2 = a + b\sin\alpha$$

$$l_3 = a + b\cos\alpha,$$

$$l_4 = a - b\sin\alpha$$

$$P_1 = P\left[\frac{e(a - b\cos\alpha)}{4a^2 + 2b^2}\right]$$

For n bolts, equally spaced, the general expression is,

$$P_1 = P\left[\frac{2e(a - b\cos\alpha)}{(2a^2 + 2b^2)\,n}\right] \quad \text{...(7.19)}$$

where $\alpha = \dfrac{180°}{n}$

For P_1 to be maximum, $\cos\alpha = -1 \text{ or } \alpha = 180°$

$$P_{max} = P\left[\frac{2e(a + b)}{(2a^2 + b^2)\,n}\right] \quad \text{...(7.20)}$$

Eq. 7.20 gives the absolute maximum value and should be used if the direction of load P can change with relation to the bolts, as in the case of a pillar crane. If the direction of P is fixed the maximum load on the bolts can be reduced by locating the bolts so that two of the bolts are equally stressed, as shown in Fig. 7.4(c). In this case,

$$\alpha = \frac{180° - 180°}{2n}$$

and

$$P_{max} = P\left[\frac{2\left(a + b\cos\left(\frac{180°}{n}\right)\right)}{(2a^2 + b^2)n}\right] \qquad ...(7.21)$$

4. Eccentric Load Acting in the Plane Containing the Bolts

Consider the bracket bolted to the frame and loaded as shown in Fig. 7.7. This problem can be solved in the same way as explained for rivers in art. 4.2.6.

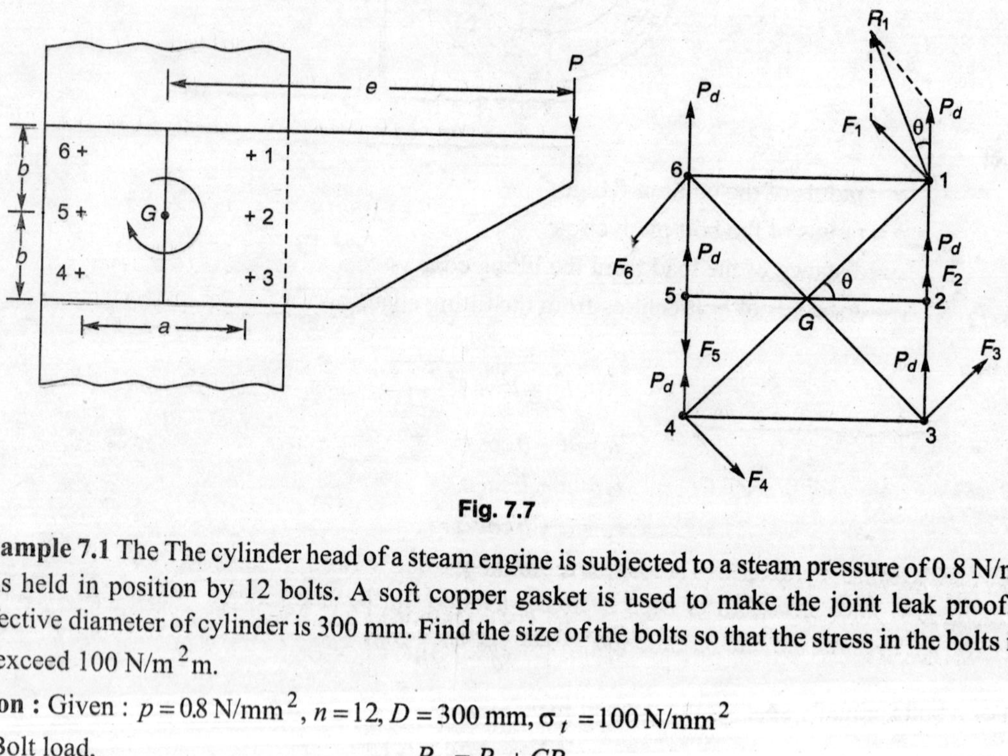

Fig. 7.7

■ **Example 7.1** The The cylinder head of a steam engine is subjected to a steam pressure of 0.8 N/mm^2. It is held in position by 12 bolts. A soft copper gasket is used to make the joint leak proof. The effective diameter of cylinder is 300 mm. Find the size of the bolts so that the stress in the bolts is not to exceed 100 N/m^{2}m.

Solution : Given : $p = 0.8$ N/mm^2, $n = 12$, $D = 300$ mm, $\sigma_t = 100$ N/mm^2

Bolt load,
$$P_b = P_i + CP$$

where P_i = initial tension due to tightening

$= 2805d$ N

$C = 0.5$ for soft copper gasket with long through bolts

$$P = \frac{\pi D^2 p}{4} = \frac{\pi \times 300^2 \times 0.8}{4} = 4712.4 \text{ N}$$

$$P_b = 2805d + 0.5 \times 4712.4 = 2805d + 2356.2$$

Stress in the bolt, $\sigma = \dfrac{P_b}{A_s}$

$$100 = 4\left[\frac{2805d + 2356.2}{\pi d^2}\right]$$

$$2805\,d + 2356.2 = 78.54\,d^2$$

or $d^2 = 35.71\,d - 30 = 0$

$$d = [35.71 + \{(35.71)^2 + 4 \times 30\}^{1/2}]$$

$$= 36.53 \approx 37 \text{ mm}$$

■ **Example 7.2** The outline of a drill press is shown in Fig. 7.8. The column is secured to the bottom block by 8 bolts as shown (two bolts on each side of column on each location). If allowable tensile stress for the bolt material is 84 MPa and the external load $P = 20$ kN acting as shown in figure, find the suitable size of the bolt.

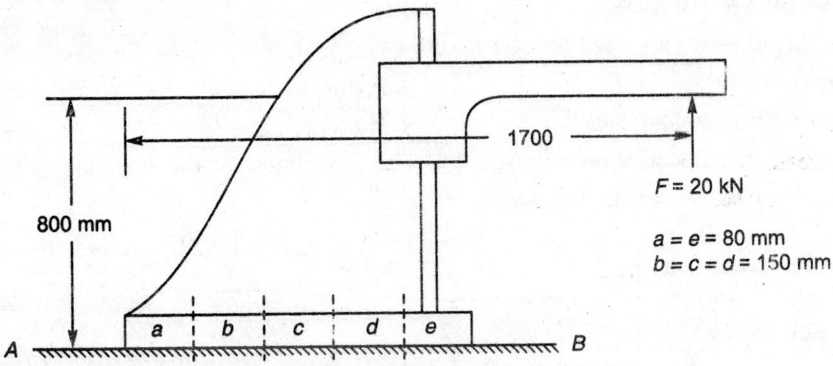

Fig. 7.8

Solution : Given : $P = 20$ kN, $e = 1700$ mm, $n = 8$, $\sigma_t = 84$ MPa

Distance of the bolts from the tilting edge are :

$$l_1 = l_2 = 80 \text{ mm}$$

$$l_3 = l_8 = 230 \text{ mm}$$

$$l_4 = l_7 = 380 \text{ mm}$$

$$l_5 = l_6 = 530 \text{ mm}$$

$$P_5 = \frac{Pel_5}{2(l_1^2 + l_3^2 + l_4^2 + l_5^2)}$$

$$= \frac{20{,}000 \times 1700 \times 530}{2(80^2 + 230^2 + 380^2 + 530^2)} = 18592 \text{ N}$$

Direct load, $P_d = \dfrac{P}{n} = \dfrac{20000}{8} = 2500$ N

Tensile load in the most heavily loaded bolt $= P_d + P_5$

$$= 2500 + 18592 = 21092 \text{ N}$$

Required stress area of bolt, $A_s = \frac{21092}{84} = 251.1 \text{ mm}^2$

M 20×2 ISO metric threads (fine series) provide, $A_s = 258 \text{ mm}^2$

Hence, we select M 20×2 bolt size.

7.2 POWER SCREWS

A power screw is a mechanical device to convert rotary motion into translational motion and for transmitting power. The main applications of power screws are :

1. To raise or lower the load, *e.g.,* screw jack.
2. To obtain accurate motion in machining operating, *e.g,* lead screw of a lathe.
3. To clamp a workpiece, *e.g.,* bench vice.
4. To load a specimen, *e.g.,* universal testing machine.

7.2.1 Types of Screw Threads

The following thread forms are used for power screws :

1. Square
2. Acme
3. ISO metric trapezoidal, and
4. Buttress

These threads have been shown in Fig. 7.9. The guidelines for the selection of a proper thread profile for the power screws are as follows :

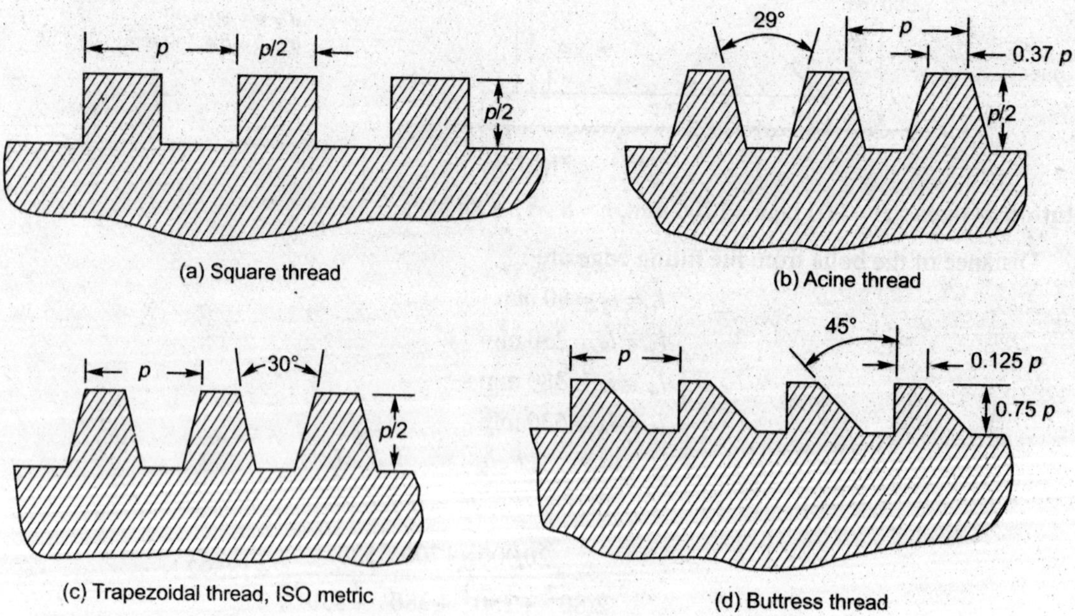

Fig. 7.9 Threads for power screw.

(i) The efficiency of square threads is more than that of other types of threads.

(ii) Square threads are difficult to manufacture. They are mainly machined on lathe. It is not possible to manufacture the threads on milling machine. On the other hand, acme and trapezoidal threads can be efficiency manufactured on thread milling machine.

(iii) The strength of acme and trapezoidal threads is more as compared to square threads due to more thread thickness.

(iv) It is not possible to compensate wear in square threads by using a split-type nut. However, it is possible to compensate axial wear in case of acme and trapezoidal threads.

(v) Buttress threads can transmit power in one direction whereas square, acme and trapezoidal threads can transmit power in both directions.

(vi) Square threads are used for screw jack and other clamping devices. Acme and trapezoidal threads for the lead screw and other power transmission devices in machine tools. Buttress threads are used in vices where force is applied only in one direction.

7.2.2 Force Analysis of Square Threads

The major dimensions of a square threaded screw are shown in Fig. 7.10(a).

Pitch, p : It is defined as the distance measured parallel to the axis from a point on one thread to the corresponding point on the adjacent thread.

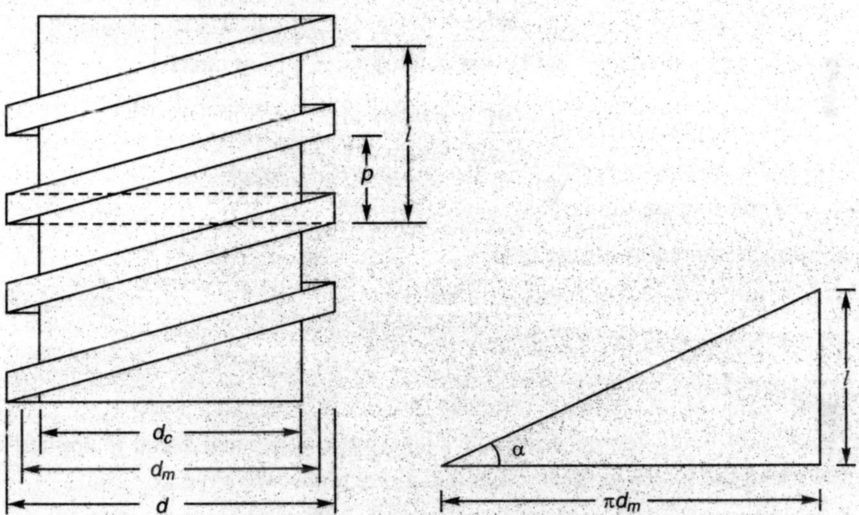

(a) Thread terminology

(b) Thread angle

Fig. 7.10

Lead, l : It is defined as the distance measured parallel to the axis that the nut will advance in one revolution of the screw. For a single-start screw lead and pitch are the same. For multi start screw, lead – number of starts × pitch.

The development of the thread is shown in Fig. 7.10(b). The helix single α of the thread is given by,

$$\tan\alpha = \frac{p}{pd_m} \text{ or } \frac{l}{\pi d_m} \qquad \ldots(7.22)$$

where d_m = mean diameter of screw

1. Lifting Load

Consider the body being lifted up the plane by a horizontal force P applied as shown in Fig. 7.11(a). Resolving the forces perpendicular and along the plane, we have

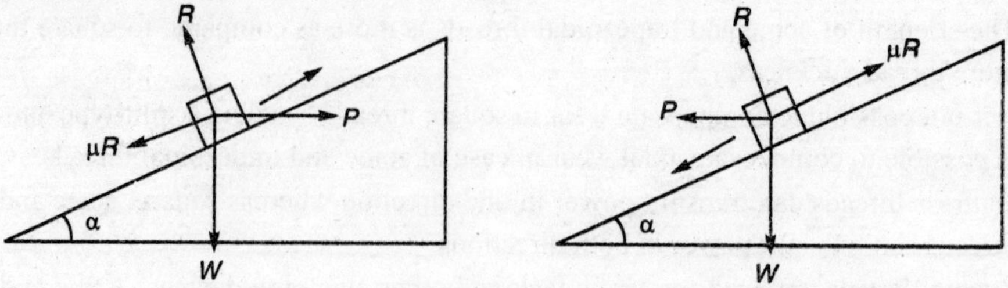

(a) Lifting load (b) Lowering load

Fig. 7.11 Force analysis of square threads.

$$P\cos\alpha = W\sin\alpha + \mu R$$

$$P\sin\alpha + W\cos\alpha = R$$

or
$$P\cos\alpha = W\sin\alpha + \tan\phi(P\sin\theta + W\cos\alpha)$$

$$P(\cos\alpha\cos\phi - \sin\phi) = W(\sin\alpha\cos\phi + \cos\alpha\sin\phi) \qquad \text{...(7.23)}$$

$$P\cos(\alpha + \phi) = W\sin(\alpha + \phi)$$

$$= W\left[\frac{\tan\alpha + \tan\phi}{1 - \tan\alpha\tan\phi}\right] = W\left[\frac{\tan\alpha + \mu}{1 - \mu\tan\alpha}\right]$$

$$= W\left[\frac{\sin\alpha + \mu\cos}{\cos\alpha - \mu\sin\alpha}\right] \qquad \text{...(7.24)}$$

where $\phi = \tan^{-1}\mu$ = angle of friction

Frictional torque required to lift the load,

$$T = \frac{Pd_m}{2} = \left(\frac{Wd_m}{2}\right)\tan(\alpha + \phi) \qquad \text{...(7.25)}$$

2. Lowering Load

Consider the body being loered down the plane by a horizontal force P applied as shown in Fig. 7.11(b). Resolving the forces perpendicular and along plane, we have

$$P\cos\alpha + W\sin\alpha = \mu R$$

$$P\sin\alpha + R = W\cos\alpha$$

or
$$P\cos\alpha + W\sin\alpha = \mu(W\cos\alpha - P\sin\alpha)$$

$$P(\cos\alpha + \mu\sin\alpha) = W(\mu\cos\alpha - \sin\alpha)$$

$$P = W\frac{(\mu\cos - \sin\alpha)}{(\cos\alpha + \mu\sin\alpha)}$$

$$= W\frac{(\mu\tan\alpha)}{(1 + \tan\alpha)}$$

$$= W\frac{(\tan\phi - \tan\alpha)}{(1 + \tan\phi\tan\alpha)}$$

$$= W\tan(\phi - \alpha) \qquad \text{...(7.26)}$$

Frictional torque required to lower the load,

$$T = \frac{Pd_m}{2} = \left(\frac{Wd_m}{2}\right)\tan(\phi - \alpha) \qquad \text{...(7.27)}$$

3. Efficiency of Screw

The efficiency of the screw is defined as the ratio of the effort required without friction P_o and with friction P.

$$\eta = \frac{P_o}{P}$$

(a) For lifting the load,
$$\eta = \frac{\tan \alpha}{\tan (\alpha + \phi)} \quad ...(7.28a)$$

(b) For lowering the load,
$$\eta = \frac{\tan (\alpha - \phi)}{\tan \alpha} \quad ...(7.28b)$$

Torque will remain positive if $\phi > \alpha$. Such a screw is called self locking. If $\phi < \alpha$, then torque will become negative. In other words, the screw will lower of its own. Such a screw is called over hauling.

For $\alpha = \phi$ the efficiency of the screw becomes,

$$\eta = \frac{\tan \phi}{\tan (2\phi)} \quad ...(7.29)$$

For a self locking screw,

$$\eta \le \frac{\tan \phi}{\tan (2\phi)} \le \left[\frac{1}{2} - \tan^2\left(\frac{\phi}{2}\right)\right] \le 50\% \quad ...(7.30)$$

7.2.3 Force Analysis of Trapezoidal Threads

For a screw having trapezoidal threads, as shown in Fig. 7.12, let 2β be the angle of the threads.

Then normal reaction on the threads,

$$R = \frac{W}{\cos \beta}$$

Frictional force,
$$F = \mu R = \left(\frac{\mu}{\cos \beta}\right) \cdot W = \mu_e W$$

where μ_e is called the virtual or equivalent coefficient of friction.

Therefore, for a screw having trapezoidal threads, the virtual coefficient of friction should be used to calculate the torque required to lift or lower the load.

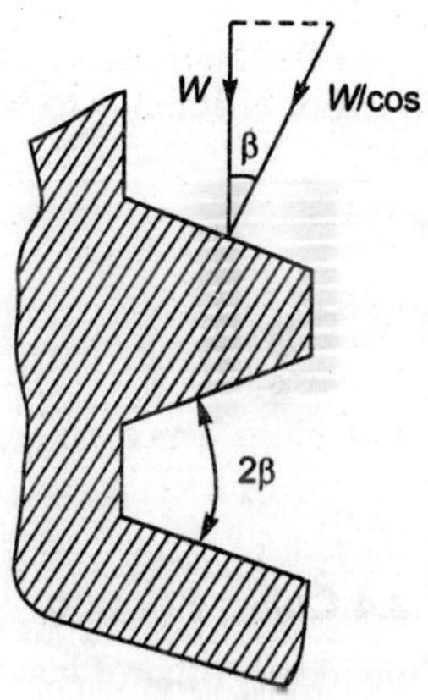

Fig. 7.12 Force analysis of trapezoidal thread.

1. Lifting Load

$$P = W\left[\frac{\tan \alpha + \mu \sec \beta}{1 - \mu \sec \beta \tan \alpha}\right] \quad ...(7.31)$$

$$T_f = \left[\frac{W d_m}{2}\right]\left[\frac{\tan \alpha + \mu \sec \beta}{1 - \mu \sec \beta \tan \alpha}\right] \quad ...(7.32)$$

$$\eta = \tan \alpha \left[\frac{1 - \mu \sec \beta \tan \alpha\ 0}{\mu \sec \beta + \tan \alpha}\right] \quad ...(7.33)$$

2. Lowering Load

$$P = W\left[\frac{\mu \sec \beta - \tan \alpha}{1 + \mu \sec \beta \tan \alpha}\right] \qquad \text{...(7.34)}$$

$$T_f = \left[\frac{Wd_m}{2}\right]\left[\frac{\mu \sec \beta - \tan \alpha}{1 + \mu \sec \beta \tan \alpha}\right] \qquad \text{...(7.35)}$$

Condition for Efficiency to be Maximum

Efficiency $\eta = \dfrac{\tan \alpha}{\tan (\alpha + \phi)}$ for lifting of load

For efficiency to be maximum, $\dfrac{d\eta}{d\alpha}$

Now
$$\eta = \tan \alpha \left[\frac{1 - \tan \alpha \tan \phi}{\tan \alpha + \tan \phi}\right]$$

$$= \left[\frac{\sin \alpha}{\cos \alpha}\right]\left[1 - \left(\frac{\sin \alpha}{\cos \alpha}\right)\left(\frac{\sin \phi}{\cos \phi}\right)\right]\left[\frac{\sin \alpha}{\cos \alpha} + \frac{\sin \phi}{\cos \phi}\right]$$

$$= \frac{\sin \alpha\ [\cos \alpha \cos \phi - \sin \alpha \sin \phi]}{\cos \alpha\,[\sin \alpha \cos \phi + \sin \phi \cos \alpha]}$$

$$= \frac{\sin \alpha \cos (\alpha + \phi)}{\cos \alpha \sin (\alpha + \phi)} = \frac{\sin (2\alpha + \phi) - \sin \phi}{\sin (2\alpha + \phi) + \sin \alpha}$$

For efficiency to be maximum,

$$\sin (2\alpha + \phi) = 0$$

or
$$2\alpha + \phi\ 90° \qquad \text{...(7.36)}$$

or
$$\alpha = 45° - \frac{\phi}{2} \qquad \text{...(7.37)}$$

$$\eta_{max} = \left[\frac{1 - \sin \phi}{1 + \sin \phi}\right] \qquad \text{...(7.38)}$$

7.2.4 Collar Friction

Consider a collar of inside diameter D_i and outside diameter D_o, as shown in Fig. 7.13. The collar of the power screw rotates with respect to the cup which remains stationary due to load W. This result in friction from diameter D_i to D_o. According to the uniform pressure theory, the torque required to overcome friction at the collar,

$$T_c = \left(\frac{\mu_c W}{3}\right)\left[\frac{D^3 - D_i^3}{D_o^2 - D_i^2}\right] \qquad \text{...(7.38a)}$$

and according to uniform wear theory

$$T_c = \left(\frac{\mu_c W}{4}\right)(D_o + D_i) \qquad \text{...(7.38b)}$$

where μ_c = collar coefficient of friction

Total torque,
$$T = T_f + T_c \qquad \text{...(7.38c)}$$

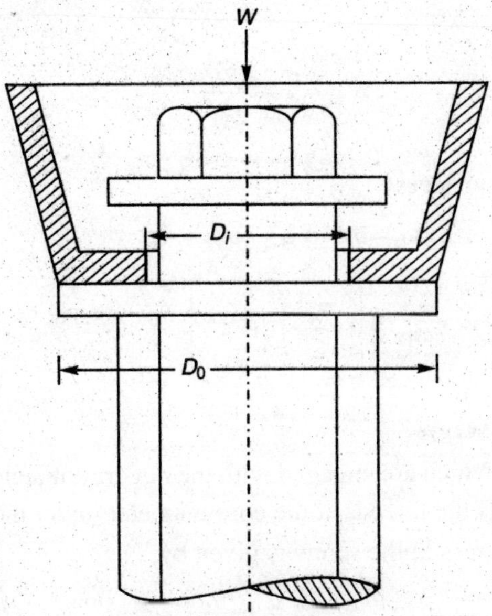

Fig. 7.13 Collar friction.

7.2.5 Stresses in Screw

Let,

d_c = core diameter of screw

W = axial load on the screw

T = total torque on the screw

The stresses acting on the screw shown in Fig. 7.14 are :

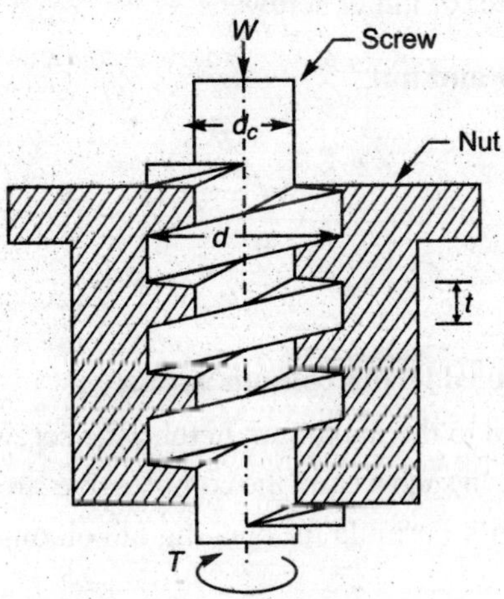

Fig. 7.14 Stresses in screw.

1. Direct compressive stress

$$\sigma_c = \frac{4W}{\pi d_c^2} \qquad \text{...(7.39)}$$

2. Torsional shear stress

$$\tau = \frac{16T}{\pi d_c^3} \quad \text{...(7.40)}$$

3. Maximum normal and shear stress

$$\sigma_1 = 0.5\,[\sigma_c + \{\sigma_c^2 = 4\tau^2\}^{1/2}] \quad \text{...(7.41a)}$$

$$\tau_{max} = \left[\left(\frac{\sigma_c}{2}\right)^2 + \tau^2\right]^{1/2} \quad \text{...(7.41b)}$$

4. Transverse shear stress in screw

The threads of the screw which are engaged with the nut are subjected to transverse shear stresses. The screw will tend to shear off the threads at the core diameter under the action of load.

$W \cdot$ the transverse shear stress in the screw is given by,

$$\tau_n = \frac{W}{\pi d_c t n} \quad \text{...(7.42)}$$

where t = thickness of threads

n = number of threads in engagement

5. Transverse shear stress in nut

$$\tau_n = \frac{W}{\pi d t n} \quad \text{...(7.43)}$$

where d = outside or major diameter of nut or screw

6. Being pressure between screw and nut

$$p_b = \frac{4W}{\pi(d^2 - d_c^2)n} \quad \text{...(7.44)}$$

$$= \frac{W}{\pi d_m t n}$$

7. Compressive stress in screw considering column action

When the load has been lifted to the maximum height, the screw may be considered as column. With suitable end conditions. In such a case, the compressive stress in screw may be determined by Rankine : Gordon or Johnson's Formula for bucking of columns.

Critical load,
$$W_{cr} = \sigma_y \times A_c \left[1 - \frac{\sigma_y \left(\frac{L}{K}\right)^2}{4C\pi^2 E}\right] \quad \text{...(7.45)}$$

where σ_y = yield stress

L = screw length

K = least radius of gyration $= \frac{d_c}{r}$

C = end condition coefficient = 0.25

E = Modulus of elasticity $\approx 20 \text{ kN/mm}^2$

A_c = core area of screw

$= \left(\frac{\pi}{4}\right) d_c^2$

■ **Example 7.3** A power screw having double-start screw threads of 26 mm nominal diameter 9 normal series and 5 mm pitch is used to lift a load of 10 kN. The outer and inner diameters of screw collar are 50 mm and 20 mm respectively. The coefficient of thread friction and collar friction may be assumed as 0.2 and 0.15 respectively. The screw rotates at 15 rpm. Assuming uniform wear condition at the collar and allowable thread bearing pressure of 6 MPs, determine (a) the stress in the screws, and (b) number of threads of nut in engagement with screws.

Solution : Given : $d = 26$ mm, $p = 5$ mm, $W = 10$ kN, $D_o = 50$ mm, $D_i = 20$ mm, $\mu = 0.2$, $\mu_c = 0.15$, $N = 15$ rpm, $p_b = 6$ MPa

Mean diameter of screw,

$$d_m = d - \frac{p}{2} = 26 - \frac{5}{2} = 23.5 \text{ mm}$$

Core diameter, $\quad d_c = d - p = 26 - 5 = 21 \text{ mm}$

Thread thickness, $\quad t = \frac{p}{2} = 2.5 \text{ mm}$

Lead $\quad l = 2p = 10 \text{ mm}$

Helix angle, $\quad \alpha = \tan^{-1}\left[\frac{l}{\pi d_m}\right] = \tan^{-1}\left[\frac{10}{\pi \times 23.5}\right] = 7.714°$

Friction angle at threads, $\quad \phi = \tan^{-1}\mu = \tan^{-1} 0.2 = 11.31°$

Friction torque, $T_f = \left(\frac{Wd_m}{2}\right)\tan(\alpha + \phi)$

$$= \left(10{,}000 \times \frac{23.5}{2}\right)\tan(7.714° + 11.31°) = 40514 \text{ N-mm}$$

Collar torque, $\quad T_c = \left(\frac{\mu_c W}{4}\right)(D_o + D_i)$

$$= \left(\frac{0.15 \times 10000}{4}\right) = (50 + 20) = 26250 \text{ N-mm}$$

Total torque, $\quad T = T_f + T_c = 40514 + 26250 = 66764 \text{ N mm}$

(a) Stresses in the screw

Direct compressive stress,

$$\sigma_c = \frac{4W}{\pi d_c^2} = \frac{4 \times 10000}{\pi \times 21^2} = 28.87 \text{ N/mm}^2$$

Torsional shear stress, $\tau = \dfrac{16T}{\pi \times d_c^3} = \dfrac{16 \times 66764}{\pi \times 21^3} = 36.71 \text{ N/mm}^2$

Maximum shear stress,

$$\tau_{max} = \left[\left(\frac{\sigma_c}{2}\right)^2 + \tau^2\right]^{1/2} = \left[\left(\frac{28.87}{2}\right)^2 + (36.71)^2\right]^{1/2} = 39.45 \text{ MPa}$$

(b) Number of threads of nut in engagement.

$$n = \frac{W}{\pi d_m t p_b} = \frac{10000}{\pi \times 23.5 \times 2.5 \times 6} = 9$$

■ **Example 7.4** The lead screw of a lathe has Acme threads of 50 mm nominal diameter and 8 mm pitch. The screw must exert an axial perrssure of 2 kN in order to drive the tool carriage. The axial thrust is taken by a collar 100 mm outside diameter and 50 mm inside diameter. The lead screw rotates at 25 rpm. The coefficient of friction for screw and collar is 0.15 and 0.10 respectively. Determine (a) the power required to drive the screw, and (b) the efficiency of the screw.

Solution : Given : $d = 50$ mm, $p = 8$ mm, $W = 2$ kN, $D_o = 100$ mm, $D_i = 50$ $\mu = 0.15$ $\mu_c = 0.10$,

$$N = 25 \text{ rpm}, 2\beta = 29°$$

Mean diameter of screw, $d_m = d - \dfrac{p}{2} = 50 - \dfrac{8}{2} = 46$ mm

Core diameter, $d_c = d - p = 50 - 8 = 42$ mm

Threads thickness, $t = \dfrac{p}{2} = 4$ mm

Helix angle, $\alpha = \tan^{-1}\left(\dfrac{p}{\pi d_m}\right) = \tan^{-1}\left[\dfrac{8}{\pi \times 46}\right] = 3.17°$

Virtual coefficient of friction,

$$\mu_e = \frac{\mu}{\cos\beta} = \frac{0.15}{\cos 14.5°} = 0.155$$

Friction angle at threads, $\phi = \tan^{-1} \mu_e \tan^{-1} 0.155 = 8.8°$

Frictional torque, $T_f = \left(\dfrac{W d_m}{2}\right) \tan(\alpha + \phi)$

$$= \left(\frac{2000 \times 6}{2}\right) \tan(3.17° + 8.8°) = 9742.43 \text{ N-m}$$

Collar torque, $T_c = \left(\dfrac{\mu_c W}{4}\right)(D_o + D_i)$

$$= \left(\frac{0.10 \times 2000}{4}\right)(100 + 50) = 7500 \text{ N-mmm}$$

Collar torque, $T = T_f + T_c = 9742.43 + 7500 = 17242.43$ N-mm

(a) Power required $P = \dfrac{2\pi NT}{60\times10^3} = \dfrac{2\pi\times25\times17242.43}{60\times10^6} = 0.045$ kW

(b) Efficiency of screw

Torque required to drive the screw without friction,

$$T_o = \left(\frac{Wd_m}{2}\right)\tan\alpha$$

$$= \frac{2000\times46}{2}\tan3.17° = 2547.64 \text{ N-mm}$$

Efficiency of screw, $\eta = \dfrac{T_o}{T} = \dfrac{2547.64}{17242.43} = 0.1477 \text{ or } 1477\%$

■ **Example 7.5** A C-clamp as shown in Fig. 7.15 has trapezoidal threads of nominal diameter 10 mm and pitch 3 mm. The coefficient of friction for screw threads is 0.15 and for the collar 0.20. The man diameter of collar is 15 mm. If the force exerted by the operator at the end of lever is 60 N, determine (a) the length of handle, (b) the maximum shear stress in screw, and (c) bearing pressure on the threads.

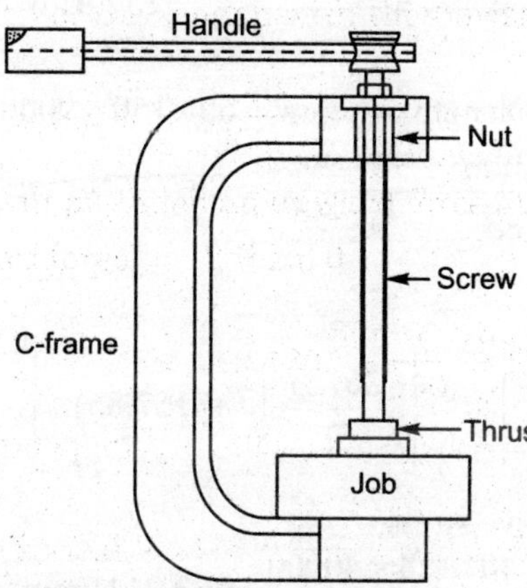

Fig. 7.15 C-clamp.

Solution : Given : $d = 10$ mm, $p = 3$ mm, $\mu = 0.15$, $\mu_c = 0.20$, $d_c = 15$ mm $F = 60$ N, $W = 4$ kN, $2\beta = 30°$, $h = 25$ mm, screw free length $= 150$ mm

Mean diameter of screw, $d_m = d - \dfrac{p}{2} = 100 - \dfrac{3}{2} = 8.5$ mm

Core diameter $d_c = d - p = 10 - 3 = 7$ mm

Helix angle, $\alpha = \tan^{-1}\left[\dfrac{p}{\pi d_m}\right] = \tan^{-1}\left[\dfrac{3}{\pi\times8.5}\right] = 6.41°$

Virtual coefficient of friction, $\mu_c = \dfrac{\mu}{\cos\beta} = \dfrac{0.15}{\cos15°} = 0.1553$

Friction angle at threads,

$$\phi = \tan^{-1}\mu_e = \tan^{-1}0.1553 = 8.827°$$

Frictional torque, $T_f = \left(\frac{Wd_m}{2}\right) \tan(\alpha + \phi)$

$$= \left(\frac{4000 \times 8.5}{2}\right) \tan (6.41° + 8.827°$$

$$= 4630.6 \text{ N-mm}$$

Collar torque, $T_c = \left(\frac{\mu_c W}{2}\right) d_c = \left(\frac{0.20 \times 4000}{2}\right) \times 15 = 6000 \text{ N-mm}$

Total torque, $T = T_f + T_c = 4630.6 + 6000 = 10630.6 \text{ N-mm}$

(a) Length of handle.

$$Fl = T$$

$$l = \frac{10630.6}{60} = 177.2 \text{ mm}$$

(b) Maximum shear stress in screw.

Section 1. 1 : Shear stress, $\tau = \frac{16T}{\pi \times d_c^3} = \frac{16 \times 10630.6}{\pi \times 7^3} = 157.85 \text{ N/mm}^2$

Bending moment, $M = F \times$ free length of screw $= 60 \times 150 = 9000$ N-m

Bending stress, $\sigma_b = \frac{32M}{\pi d_c^3} = \frac{32 \times 9000}{\pi \times 7^3} = 267.27 \text{ N/mm}^2$

Maximum shear stress,

$$\tau_{max} = \left[\left(\frac{\sigma_b}{2}\right)^2 + \tau^2\right]^{1/2} = \left[\left(\frac{267.27}{2}\right)^2 + (157.85)^2\right]^{1/2} = 206.82 \text{ MPa}$$

Section 2 : 2 : Direct compressive stress,

$$\sigma_c = \frac{4W}{\pi d_c^2} = \frac{4 \times 40000}{\pi \times 7^2} = 103.94 \text{ N/mm}^2$$

Torsional shear stress, $\tau = \frac{16T_c}{\pi \times d_c^3} = \frac{16 \times 1600}{\pi \times 7^3} = 89.1 \text{ N/mm}^2$

Maximum shear stress,

$$\tau_{max} = \left[\left(\frac{\sigma_c}{2}\right)^2 + \tau^2\right]^{1/2} = \left[\left(\frac{103.94}{2}\right)^2 + (89.1)^2\right]^{1/2} = 103.15 \text{ MPa}$$

(c) Bearing pressure on threads.

Height of nut, $h = np$

$$n = \frac{25}{3} = 8.33 \approx 9$$

Thickness of threads, $t = \frac{p}{2} = \frac{3}{2} = 1.5 \text{ mm}$

■ **Example 7.6** A machine slide weighting 25 kN is elevated by a double start Acme threaded screw at the rate of 0.96 m/min. If coefficient of friction is 0.14, calculate the power required to drive the slide. The end of the screw carries a thrust collar of 30 mm inside and 60 mm outside diameter. The coefficient of friction of collar is 0.15. Pitch of screw threads is 6 mm and outside diameter is 36 mm. If the screw is of 45C8 steel ($S_{yt} = 380$ MPa) calculate the factor of safety.

Solution : Given : $W = 25\,\text{kN}, 2\beta = 29°, v = 0.96\,\text{m/min}, \mu = 0.14, D_i = 30\,\text{mm}, D_o = 66\,\text{mm}, \mu_c = 90.15,$

$$p = 6\,\text{mm},\ d = 36\,\text{mm},\ S_{ut} = 630\,\text{MPa},\ S_{yt} = 380\,\text{MPa}$$

Mean diameter of screw, $d_m = d - \dfrac{p}{2} = 36 - \dfrac{6}{2} = 33$ mm

Core diameter, $d_c = d - p = 36 - 6 = 30$ mm

Helix angle, $\alpha = \tan^{-1}\left[\dfrac{1}{\pi d_m}\right] = \tan^{-1}\left[\dfrac{12}{\pi \times 33}\right] = 6.6°$

Virtual coefficient of friction, $\mu_e = \dfrac{\mu}{\cos\beta} + \dfrac{0.14}{\cos 14.5°} = 0.1446$

Fruition angle at threads, $\phi = \tan^{-1}\mu_e = \tan^{-1} 0.1446 = 8.23°$

Frictional torque, $T_f = \left(\dfrac{Wd_m}{2}\right)\tan(\alpha + \phi)$

$$= \left(\frac{25{,}000 \times 33}{2}\right)\tan(6.6° + 8.23°) = 109205\ \text{N-mm}$$

Collar torque,

$$T_C = \left(\frac{\mu_c W}{4}\right)(D_i + D_i) = \left(\frac{0.15 \times 25{,}000}{4}\right)(60 + 30) = 84375\ \text{N-mm}$$

Total torque, $T = T_f + T_c = 10920 + 84375$ N-mm

Speed of screw, $N = \dfrac{v}{l} = \dfrac{960}{12} = 80$ rpm

Power required to drive the slide,

$$P = \frac{2\pi NT}{60 \times 10^3} = \frac{2\pi \times 80 \times 193580 \times 10^{-3}}{60 \times 10^3} = 1.62\ \text{kW}$$

Direct compressive stress,

$$\sigma_c = \frac{4W}{\pi d_c^2} = \frac{4 \times 25000}{\pi \times 30^2} = 35.37\ \text{MPa}$$

Torsional shear stress,

$$\tau = \frac{16T}{\pi \times d_c^3} = \frac{16 \times 193580}{\pi \times 30} = 36.51\ \text{Mpa}$$

Maximum tensile stress, $\sigma_1 = \left(\dfrac{1}{2}\right)[\sigma_c + (\sigma_c^2 + 4\tau^2)^{1/2}]$

$$=0.5\,[35.37 + \{(35.37)^2 + 4(26.51)^2\}]^{1/2}$$

$$=0.5\,[35.37 + 81.13] = 58.35 \text{ MPa}$$

Maximum shear stress,

$$\tau_{max} = \left[\left(\frac{\sigma_c}{2}\right)^2 + \tau^2\right]^{1/2} = 40.5 \text{ MPa}$$

$$\text{Factor of safety in tension} = \frac{S_{yt}}{\sigma_1} = \frac{380}{58.35} = 6.51$$

$$\text{Factor of safety in tension} = \frac{S_{yt}}{2\tau_{max}} = \frac{380}{2 \times 40.5} = 4.69$$

■ **Example 7.7** The following date refer to a bench vice having single start square threads as shown in Fig. 7.16

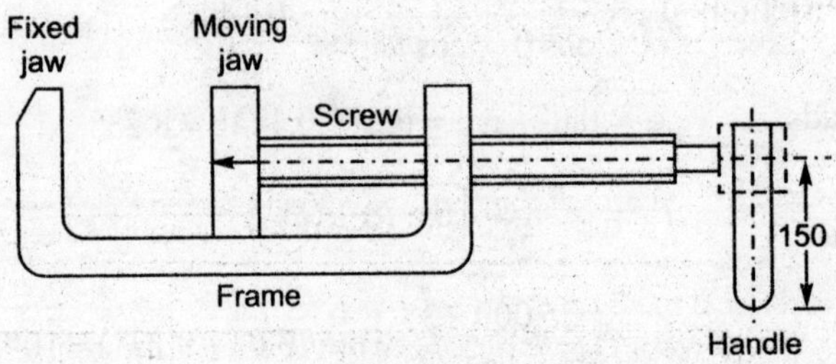

Fig. 7.16 Machine vice.

Major diameter of screw	= 12 mm
Minor diameter of screw	= 10 mm
Pitch	= 2 mm
Collar diameter	= 10 mm
Coefficient of friction for screw	= 0.15
Coefficient of friction for collar	= 0.18
Effort applied at the end of the lever	= 100 N
Length of lever	= 150 mm

Determine (a) the clamping force between the jaws, (b) the efficiency of the vice, and (c) the torque in the screw body at section *A-A* and *B-B*.

Solution : Given : $d = 12$ mm, $d_c = 10$ mm $p = 2$ mm, $d_c = 10$ mm, $\mu = 0.15\,\mu_c = 0.18$, $F = 100$ N, $l = 150$ mm

Mean diameter of screw, $d_m = d_c + \frac{p}{2} = 10 + 1 = 11$ mm

Helix angle, $\alpha = \tan^{-1}\left[\frac{p}{\pi d_m}\right] = \tan^{-1}\left[\frac{2}{\pi \times 11}\right] = 3.31°$

Friction angle at threads, $\phi = \tan^{-1}\mu = \tan^{-1} 0.15 = 8.53°$

Frictional torque,

$$T_f = \left(\frac{Wd_m}{2}\right)\tan(\alpha + \phi) = \left(\frac{W \times 11}{2}\right)\tan(3.31° + 8.53°) = 1.153\text{ W N-mm}$$

Collar torque, $T_c = \left(\frac{\mu W}{3}\right)d_c$, for uniform pressure

$$= \left(\frac{0.18 \times W}{3}\right) \times 10 = 0.6\text{ W N-mm}$$

Total torque at $B - B, T = T_f + T_c = (1.153 + 0.6)W = 1.753\, W$ N-mm

Torque applied by the operator, $T = Fl = 100 \times 150 = 15000$ N-mm

$$1.753\, W = 15000$$

(a) Clamping force, $W = 857$ N

(b) Efficiency of vice

$$= \frac{T_o}{T} = \left(\frac{Wd_m}{2}\right)\tan\alpha = \left(\frac{8557 \times 11}{2}\right)\tan\frac{3.1}{15000} = 0.1816 \text{ or } 18.15\%$$

(c) Torque at $A - A = 1.153 \times 8557 = 9866.2$ N-mm

Torque at $B - B = 15000$ N-mm

7.2.6 Design of Screw Jack

A screw jack for lifting load is shown in Fig. 7.17. The various part of the screw are designed as follows :

1. Screw diameter

Compressive stress in the stress, $\sigma_c = \frac{4W}{\pi d_c^2}$

$$d_c = \left[\frac{4W}{\pi\sigma_c}\right]^{1/2} \quad \ldots(7.46)$$

where σ_c = permissible compressive stress in screw material

W = load lifted

The screw is square threaded whose standard dimensions can be determined

Nominal or outside diameter, $d = d_c + p$

Mean diameter, $d_m = d_c + \frac{p}{2}$

2. Maximum tensile and shear stress in screw

Direct compressive stress, $\sigma_c = \frac{4W}{\pi d_c^2}$

Torsional shear stress, $\tau = \frac{16T_f}{\pi \times d_c^3}$

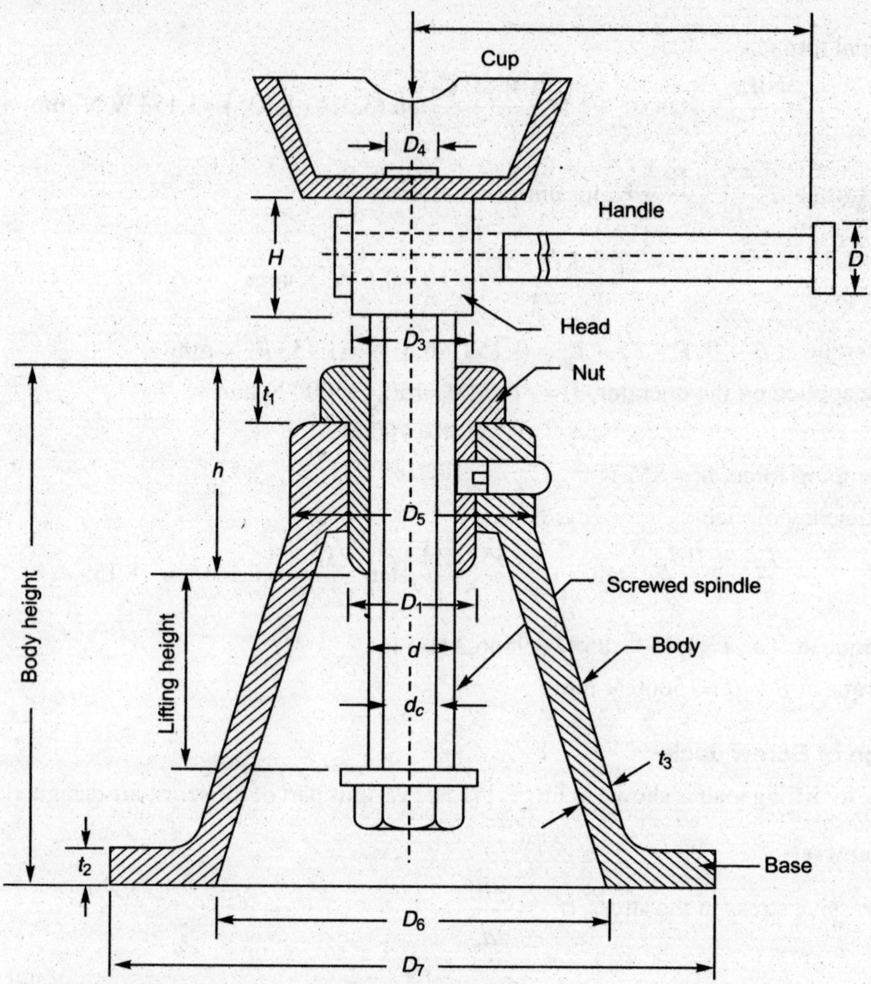

Fig. 7.17 Screw jack.

Maximum tensile stress,

$$\sigma_{1,2} = \left(\frac{1}{2}\right)[\sigma_c \pm (\sigma_c^2 + 4\tau^2)^{1/2}] \qquad \text{...(7.47)}$$

Maximum shear stress,

$$\tau_{\max} = \left[\left(\frac{\sigma_c}{2}\right)^2 + \tau^2\right]^{1/2} \qquad \text{...(7.48)}$$

where

$$T_f = \left(\frac{Wd_m}{2}\right)\tan(\alpha + \phi) \qquad \text{...(7.49)}$$

$$\alpha = \tan^{-1}\left[\frac{p}{pd_m}\right]$$

$$\phi = \tan^{-1}\mu$$

μ = thread friction coefficient

The maximum shear should be less than the permissible shear stress in screw material.

3. Height of Nut

Number of threads in contact with screw spindle,

$$n = \frac{W}{\left[\left(\frac{\pi}{4}\right)(d^2 - d_c^2) p_b\right]} \quad \text{...(7.50)}$$

where p_b = permissible bearing pressure on the nut

Height of nut, $h = np$ ...(7.51)

where p = pitch of threads.

4. Check for stresses in Screw and Nut

Shear stress in screw, $$\tau_s = \frac{W}{\pi n d_c t} \quad \text{...(7.52a)}$$

Shear stress in nut, $$\tau_n = \frac{W}{\pi n d t} \quad \text{...(7.52b)}$$

where $t = \frac{p}{2}$ = thickness of screw

5. Nut Collar size

Let D_2 = outside diameter of collar

D_1 = inside diameter of collar

The inside diameter of nut collar is found by considering the tearing strength of nut.

$$\left(\frac{\pi}{4}\right)[D_1^2 - d^2]\sigma_1 = W \quad \text{...(7.53)}$$

where σ_i = permissible tensile strength of nut material

The outside diameter of nut collar is found by considering the crushing strength of nut collar.

$$\left(\frac{\pi}{4}\right)[D_2^2 - D_1^2]\sigma_c = W \quad \text{...(7.54)}$$

where σ_c = permissible compressive strength of nut collar material.

The thickness of nut collar is found by considering the shearing strength of nut collar.

$$\pi D_1 t_c \tau_c = W \quad \text{...(7.55)}$$

where t_c = nut collar thickness

τ_c = permissible shear stress in nut collar material

6. Head dimensions

Diameter of head on the top of screw,

$$D_3 = 1.75d \quad \text{...(7.56)}$$

Pin diameter in the cup, $$D_4 = \frac{D_3}{4} \quad \text{...(7.57)}$$

7. Torque required between Cup and Head

$$T_c = \frac{\mu_c W}{3}\left[\frac{D_3^3 - D_4^3}{D_3^2 - D_4^3}\right], \text{ according to uniform pressure theory} \quad ...(7.58a)$$

$$= \left(\frac{\mu_c W}{4}\right)(D_4 + D_3), \text{ according to uniform wear theory} \quad ...(7.58b)$$

where μ_c = collar coefficient of friction

Total torque, $$T = T_f + T_c \quad ...(7.59)$$

8. Length of lever

A normal person can apply a force $F = 60$ to 100 N.

Length of lever, $$l = \frac{T}{F} \quad ...(7.60)$$

9. Diameter of lever

The lever is under bending.

Bending moment, $$M = Fl$$

Diameter of lever, $$d_1 = \left[\frac{32M}{\pi\sigma_b}\right]^{1/3} \quad ...(7.61)$$

where σ_b = permissible bending stress in lever.

10. Height of Head

$$H = 2d_1$$

11. Check of Screw in Bucking

Supported length of screw,

$$L = \text{lift of screw} + 0.5 \times \text{height of nut}$$

Bucking or critical load for screw,

$$W_{cr} = A_c\sigma_y\left[1 - \left\{\frac{\sigma_y}{4C\pi^2 E}\right\}\left(\frac{L}{R}\right)^2\right] \quad ...(7.63)$$

where σ_y = yield strength of screw material

C = end condition coefficient.

= 0.25, assuming the screw with lower end fixed and upper end free

K = least radius of gyration of screw $= \frac{d_c}{4}$

For a safe design $W_{cr} > W$

12. Efficiency of Screw

Efficiency of screw, $$\eta = \frac{T_o}{T} \quad ...(7.64)$$

where $$T_o = \left(\frac{Wd_m}{2}\right)\tan\alpha$$

13. Body dimensions

Diameter of body at top, $D_5 = 1.5\,D_2$ ...(7.65)

Thickness of base, $t_2 = 2t_1$

Thickness of body, $t_3 = 0.25d$

Inside diameter at the bottom, $D_6 = 2.25D_2$

Outer diameter a the bottom, $D_7 = 1.75D_6$

Height of body = Maximum lift + height of nut + 100 mm

■ **Example 7.8** Design a screw jack to lift a load of 100 kN through of 400 mm. The permissible stresses in screw and nut are given below :

Steel screw : Tension/compression = 100 MPa, shear = 60 MPa

Phosphor - bronze nut : Tension = 100 MPa, compression = 90 MPa, shear = 80 MPa

Bearing pressure between nut and screw = 15 MPa

Solution : Gioven : W = 100 kN, lift = 400 mm, $\sigma_{ts} = \sigma_{cs} = 100$ MPa, $\tau_s = 60$ MPa, σ_{tn} = 50 MPa, σ_{cu} = 45 MPa, τ_n = 40 MPa, p_b = 15 MPa.

Refer to Fig. 8.9.

1. Screw diameter

Compressive stress in the screw,

$$\sigma_{cs} = \frac{4W}{\pi d_c^2}$$

$$d_c = \left[\frac{4W}{\pi \sigma_{cs}}\right]^{1/2} = \left[\frac{4 \times 100000}{\pi \times 100}\right]^{1/2} = 35.68 \approx 40 \text{ mm}$$

according for shear stress.

We select normal series square threads.

$$p = 7 \text{ mm}$$

Nominal or outside diameter,

$$d = d_c + p = 40 + 7 = 47$$

Mean diameter,

$$d_m = d_c + \frac{p}{2} = 40 + 3.5 = 43.5 \text{ mm}$$

Thread thickness,

$$t = \frac{p}{2} = 3.5 \text{ mm}$$

2. Maximum tensile and shear stresses in screw.

Direct compressive stress,

$$\sigma_c = \frac{4W}{\pi d_c^2} = \frac{4 \times 100000}{\pi \times 40^2} = 79.6 \text{ MPa}$$

Helix angle

$$\alpha = \tan^{-1}\left[\frac{p}{\pi d_m}\right] = \tan^{-1}\left[\frac{7}{\pi \times 43.5}\right] = 2.93°$$

Assume coefficient of friction between screw and nut,

$$\mu = 0.2$$

Friction angle at threads,

$$\phi = \tan^{-1} \mu = \tan^{-1} 0.2 = 11.31°$$

Frictional torque, $T_1 = \left(\frac{Wd_m}{2}\right) \tan(\alpha + \phi)$

$$= \left(\frac{100000 \times 43.5}{2}\right) \tan(29.3° + 11.31°) = 552067 \text{ N-mm}$$

Torsional shear stress,

$$\tau = \frac{16T_f}{\pi \times d_c^3} = \frac{16 \times 552067}{\pi \times 40^3} = 43.93 \text{ MPa}$$

Maximum tensile stress,

$$\sigma_{1,2} = \left(\frac{1}{2}\right)[\sigma_c \pm (\sigma_c^2 + 4\tau^2)^{1/2}]$$

$$= 0.5[79.6 \pm \{(79.6)^2 + 4 \times (43.93)^2\}^{1/2}]$$

$$= 0.5[79.6 \pm 118.56]$$

$$= 99 \text{ MPa (max)} < 100 \text{ MPa. Hence safe.}$$

Maximum shear stress

$$\tau_{max} = \left[\left(\frac{\sigma_c}{2}\right)^2 + \tau^2\right]^{1/2} \quad 59.28 \text{ MPa} < 60 \text{ MPa. Hence safe.}$$

3. Height of nut

Number of threads in contact with screw spindle,

$$n = \frac{W}{\left(\frac{\pi}{4}\right)(d^2 - d_c^2)p_b} = \frac{100000}{\frac{\pi}{4}(47^2 - 40)^2\ 15} = 13.94 \approx 14$$

Height of nut, $h = np \times 14 \times 7 = 98$ mm

4. Check for stresses in Screw and Nut.

Shear stress in screw,

$$\tau_s = \frac{W}{\pi n d_c t} = \frac{100000}{\pi \times 14 \times 40\,3.5} = 16.24 \text{ MPa}$$

Shear stress in screw

$$\tau_n = \frac{W}{\pi n d_c t} = \frac{100000}{\pi \times 14 \times 47 \times 3.5} = 13.82 \text{ MPa}$$

These stresses are within permissible limits. Hence design is safe.

5. Nut Collar size

Let D_2 = outside diameter of collar

D_1 = inside diameter of collar

The inside diameter of nut collar is found by considering the tearing straight of nut.

$$\frac{\pi}{4}[D_1^2 - d^2]\sigma_{tn} = W$$

$$\frac{\pi}{4}[D_1^2 - 47^2]\times 50 = 100000$$

$$D_2 = 87.9 \approx 90 \text{ mm}$$

The outside diameter of nut collar is found by considering the crushing strength of nut collar.

$$\frac{\pi}{4}[D_2^2 - D_1^2]\sigma_{cn} = W$$

$$\frac{\pi}{4}[D_2^2 - 70_2^2]\times 45 = 100000$$

The thickness of nut collar is found by considering the shearing strength of nut collar.

$$\pi D_1 t_c \tau_{cn} = W$$

$$\pi\times 70\times t_c \times 40\times 100000$$

$$\tau_c = 11.36 \approx 12 \text{ mm}$$

6. Head dimensions

Diameter head on top of screw, $D_3 = 1.75d = 1.75\times 47 = 82.25 \approx 84$ mm

$$P_{in} \text{ diameter in the cup, } D_4 = \frac{D_3}{4} = \frac{84}{4} = 21 \text{ mm}$$

7. Torque required between Cup and Head

$$T_c = \frac{\mu_c W}{3}\left[\frac{D_3^3 - D_1^3}{D_3^2 - D_4^2}\right], \text{ according to uniform pressure theory}$$

Assuming $\mu_c = 0.15$

$$T_c = \frac{0.15\times 10000}{3}\left[\frac{84^3 - 21^3}{84^2 - 21^3}\right] - 441000 \text{ N-mm}$$

Total torque, $T = T_f + T_c = 552067 + 441000 - 993067$ N-mm

8. Length of lever

A normal person can apply a force $F = 100$ to 300 N ≈ = 300 N

Length of lever, $l = \frac{T}{F} = \frac{993067}{300} = 3310 \approx 3300$ mm

9. Diameter of lever

The lever is under bending.

Bending moment, $M = Fl = 300\times 3000 = 99000$ N-mm

Diameter of lever $d_l = \left[\frac{32M}{\pi\sigma_b}\right]^{1/3} = \left(\frac{32\times 99000}{\pi\times 100}\right)^{1/3} = 46.5 \approx 48$ mm

10. Height of Head

$$H = 2d_1 = 2\times 48 = 96 \text{ mm}$$

11. Check of Screw in Bucking

Unsupported length of screw, L = lift of screw + 0.5 × height of nut = 400 + 0.5 × 98 = 449 mm

$$W_{cr} = A_c\sigma_y\left[1 - \left(\frac{\sigma_y}{4C\pi^2 E}\right)\left(\frac{L}{K}\right)^2\right]$$

where $K = \frac{d_c}{4} = \frac{40}{4} = 10$ mm, $c = 0.25$, $\sigma_y = 200$ MPa

$$W_{cr} = \frac{\pi}{4}\times 40^2\times 200\left[1 - \left(\frac{200}{4\times 0.25\times\pi^2\times 200\times 10^3}\right)\left(\frac{499}{10}\right)^2\right]$$

$$= 199990 \approx 200 \text{ kN} > 100 \text{ kN}$$

12. Efficiency of screw

Efficiency of screw, $\eta = \frac{T_o}{T}$

where $T_o = \frac{Wd_m}{2}\tan\alpha = \frac{100000\times 43.5}{2}\tan 2.93° = 111322$ N-mm

$$\eta = \frac{111322}{993067} = 0.112 \text{ or } 11.2\%$$

13. Body dimensions

Diameter of body at top, $D_5 = 1.5D_2 = 1.5\times 90 = 135$ mm

Thickness of base, $t_2 = 2t_c = 2\times 12 = 24$ mm

Thickness of body, $t_3 = 0.25d = 0.25\times 47 = 11.75 \approx 12$ mm

Inside diameter of bottom, $D_6 = 2.25\,D_2 = 2.25\times 90 = 202.5 \approx 205$ mm

Outer diameter at the bottom, $D_7 = 1.75\,D_6 \times 1.75\times 205 = 359.75 \approx 360$ mm

Height of body = Maximum lift + height of nut + 100 mm

$= 400 + 98 + 100 = 598 \approx 600$ mm

SOLVED QUESTIONS FROM UNIVERSITY EXAMINATIONS

■ **Example 7.9** A wall bracket is attached to a wall by means of four identical bolts two at A and two at B as shown in Fig. 7.18. Assuming that the bracket is held against the wall and prevented from slipping about point C by all from bolts and using an allowable tensile stress in the bolts as 35 N/mm^2, determine the standard size of the bolts on the basis of a suitable theory of failure.

[MDU, Dec., 2008]

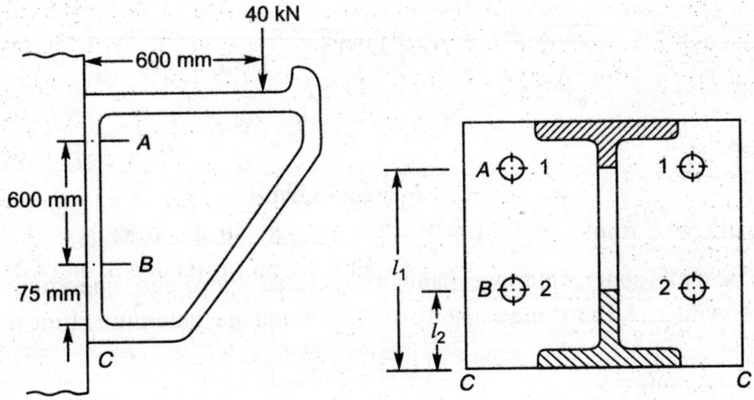

Fig. 7.18

Solution : Given : $P = 40$ kN, $e = 600$ mm, $n = 4$, $l_1 = 675$ mm, $l_2 = 75$ mm

Direct shearing load on each bolt,

$$P_s = \frac{P}{n} = \frac{40}{4} = 10 \text{ kN}$$

Let P_1, P_2 = loads on the bolts at A and B respectively.

Then
$$Pe = 2(P_1 l_1 + P_2 l_2)$$

Also the forces in the bolts are proportional to the distance from the edge cc.

$\therefore$
$$P_2 = \frac{P_1 l_2}{l_1}$$

$$Pe = 2\left(P_1 l_1 + \frac{P_1 l_2^2}{l_1}\right)$$

$$= \frac{2P_1}{l_1}(l_1^2 + l_2^2)$$

Tensile load in both A,

$$P_1 = \frac{Pe\, l_1}{2(l_1^2 + l_2^2)} = \frac{40 \times 600 \times 675}{2(675^2 + 75^2)} = 17561 \text{ kN}$$

$$P_2 = \frac{17.561 \times 75}{675} = 1.951 \text{ kN}$$

Maximum tensile load in bolt at A,

$$(P_1)_{max} = \frac{1}{2}[P_1 + \sqrt{P_1^2 + 4P_s^2}]$$

$$= \frac{1}{2}[17.561 + \sqrt{(17.561)^2 + 4 \times 10^2}]$$

$$= \frac{1}{2}[17.561 + 26.616] = 22.088 \text{ kN}$$

Let $\quad d_c$ = core diameter of bolts.

Then
$$\frac{\pi}{4} d_c^2 \sigma_t = (P_1)_{max}$$
$$\frac{\pi}{4} d_c^2 \times 35 = 22.088 \times 10^3$$
$$d_c = 28.35 \text{ mm}$$

Nearest core diameter from Table 7.1 is 31.670 mm and bolt size is M36.

■ **Example 7.10** Design a general purpose manually operated screw jack supporting the machine parts on the shop floor with a load carrying capacity of 120 kN and the maximum lifting height of 500 mm. [MDU, May, 2009]

Solution : Refer to Example 7.8.

EXERCISES

1. A bolted assembly is initially tightened by means a spanner so as to induce a preload of 3 kN in the bolt. Then it is subjected to external load of 6 kN. The bolt with coarse threads is made of plain carbon steel of tensile yield strength 450 MPa and the factor of safety is 3. The effective stiffness of the parts held together by the bolt is 3 times the stiffns of the bolt. Specify the size of the bolt.
2. Two circular plates are clamped together by means of a bolt. It is subjected to a vertical force P varying from zero to 15 kN. The bolt is made of carbon steel having $S_{ut} = 630$ Mpa, $S_{yt} = 380$ MPa, and $E = 210$ GPa. The two circular plates are made of aluminium having $E = 70$ GPa. The bolt is machined and the expected reliability is 90%. The fatigue stress-concentration factor is 2.3. The initial preload on the bolt is 5 kN. Determine the size of bolt with 3 as the factor of safety.
3. The shaft of an axial flow rotary compressor is subjected to a maximum torque of 2.5 kN-m and a maximum bending moment of 5 kN-m. The combined shock fatigue factors in torsion and bending are 1.5 and 2.0 respectively. Determine the diameter of the shaft if the permissible shear stress in the shaft is 55 MPa. What percentage saving in material shall be obtained if a hollow shaft is designed?
4. A steel plate 20 mm thick is attached to a girder by means four reverts of 20 mm diameter as shown in Fig. 7.19. Determine the distance 'e' which permits the application of a 50 kN load to limit shear stress to 45 MPa in rivets and crushing stress to 70 MPa in rivets and plate.

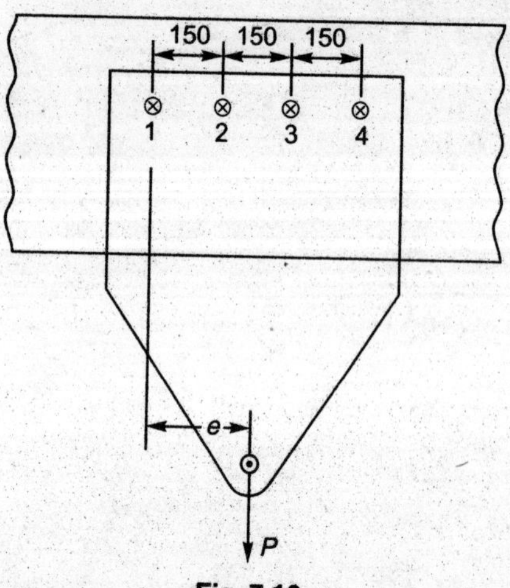

Fig. 7.19

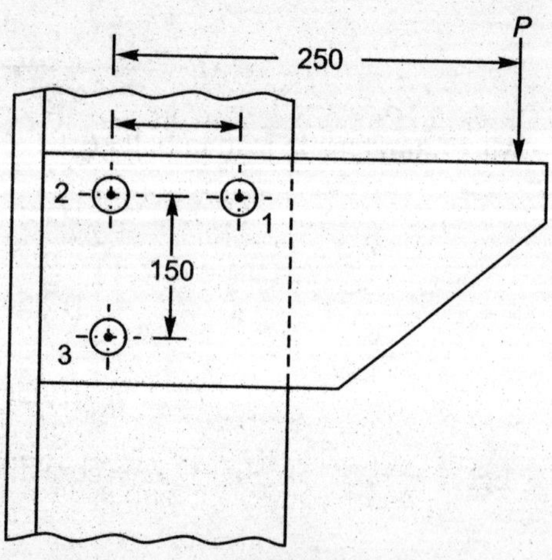

Fig. 7.20

5. An eccentric bearing plate connected by bolts is loaded as shown in Fig. 7.20. Locate the bolt with the greatest force and find its diameter if the permissible shear stress in bolts is 40 MPa.
6. A steel plate is attached to a column by four bolts as shown in Fig. 7.21. The tensile strength of bolt material is 400 MPa and the factor of safety is 2. Calculate the diameter of the bolts.

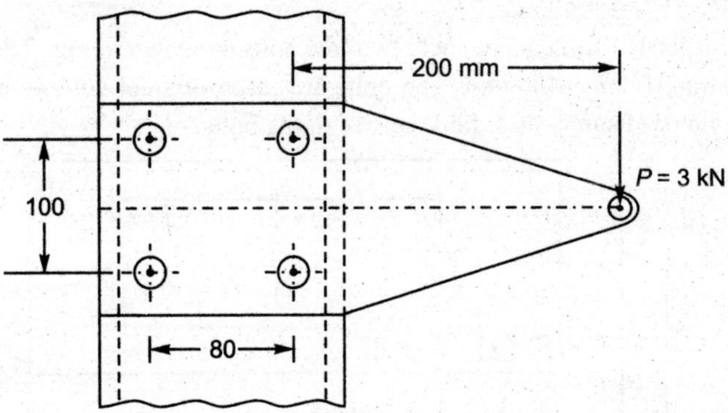

Fig. 7.21

7. The base of a pillar crane is fastened to the foundation by eight bolts spaced equally on a bolt circle of 1.6 m diameter as shown in Fig. 7.22. The diameter of the pillar base is 2 m. Determine the size of bolts when the crane carries a load of 100 kN at a distance of 5 m from the centre of the base. The allowable stress for the bolt material is 100 MPa.

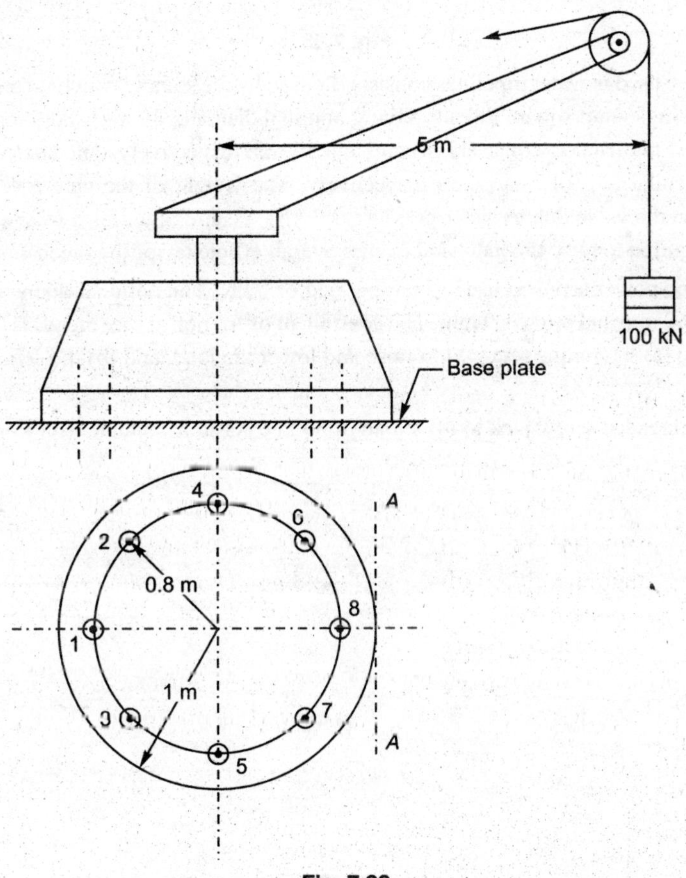

Fig. 7.22

8. A cylinder head is fastened to the cylinder of a compressor of 100 mm diameter using 8 bolts of size M16. For the bolt material $S_{yt} = 450$ and $S_{es} = 210$ MPa. The maximum pressure of fluid in cylinder is 3.5 MPa. Calculate the factor of safety in the following cases :
(a) A soft gasket ($k_g = 1$) is used which required initial compressive load of about 70 kN for sealing.
(b) An O-ring seal ($k_g = 0$) is used.

9. A load of 50 kN is suspended from a plate held by three bolts as shown in Fig. 7.23 thickness of the plate is 12 mm and $l = 250$ mm, $l_1 = l_1 = 100$ mm. The bolts are made of steel with an elastic limit in tension of 250 MPa. Taking a factor of safety of 2, find the size of the bolts.

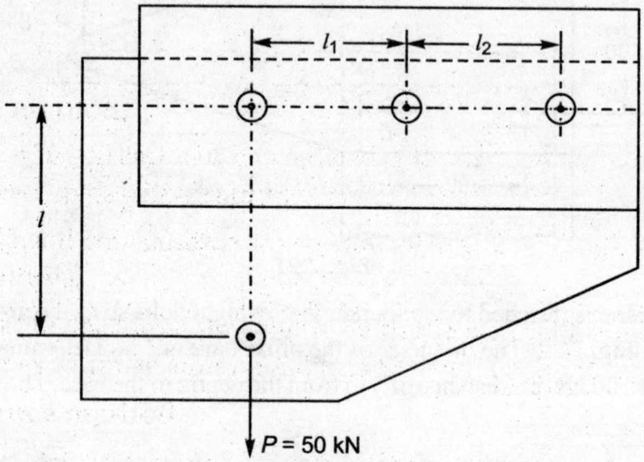

Fig. 7.23

10. A sluice gate valve used in water pipelines consists is raised by the spindle which is rotated by the handwheel. The spindle has single-start square threads having nominal diameter 36 mm and 6 mm pitch. The inner and outer diameters of the friction collar are 30 mm and 50 mm respectively. The coefficient of friction at the threads and the collar are 0.15 and 0.20 respectively. The weight of the gate is 8 kN and the frictional resistance to open the valve due to water pressure is 3 kN. Assuming uniform wear at the collar, determine (a) the torque required to raise the gate, and (b) the overall efficiency of the mechanism.

11. A double-threaded power screw is used to raise a load of 6 kN. The nominal diameter of the screw having Acme threads is 60 mm and pitch is 9 mm. The coefficient of friction at the threads is 0.15. Neglecting collar friction, calculate (a) the torque required to raise and lower the load, and (b) the efficiency of the screw for lifting the load.

12. Design a square threaded screw jack to lift a load of 50 kN. Assume suitable data from the Data Book.

❑❑❑

8

Chapter

Shafts

Inside This Chapter

8.1 Introduction
8.2 Materials for Shafts
8.3 Theories of Strength for Shafts
8.4 Design of Shaft against Static Loading

8.1 INTRODUCTION

A shaft is a machine members which is used to support pulleys, gears, flywheel, etc. The shaft is attached to these elements by means of a key or splines. It is a rotating member used for transmitting power and is subjected to torsional, bending and thrust stresses. An axle is a machine member which supports transverse load and may or may not rotate. The axle does not transmit power. A spindle is a short shaft or axle. A shaft may be right or flexible. It may be solid or hollow.

A line shaft is a long and conditions shaft used in workshops for the distribution of power. A counter shaft is used in between a line shaft and machine. A shaft directly coupled to the power source is called a jack shaft. Flexible shafts are used where the axis of rotation of the driving and driven members are not coaxial.

8.2 MATERIALS FOR SHAFTS

The material to be selected for shafts should have the following properties :

1. High strength
2. Low roted sensitivity factor
3. Good machinability
4. Less prone to environmental effects
5. Wear resistant and good heat treatment properties.

The mechanical properties of material used for shaft are given in Table 8.1.

Table 8.1 : Properties of steels used for shafts.

Steel Designation	*Tensile strength, S_{ut}, MPa*	*Yield strength, S_{yt}, MPa*
1. 40C8	569–667	324
2. 45C8	618–696	353
3. 50C8	747–765	373
4. 50C12	706 mm	392

8.3 THEORIES OF STRENGTH FOR SHAFTS

The theories of failure used for shaft are :

1. Maximum shear stress theory
2. Maximum distortion energy theory.

According to maximum shear stress theory,

Maximum shear stress,
$$\tau_{max} = \frac{\sigma_1 - \sigma_2}{2} \quad ...(8.1)$$

where σ_1, σ_2 = maximum and minimum principal stresses

$$\tau_{max} = \frac{\text{yield shear strength}}{\text{Factor of safely}} = \frac{S_{ys}}{n} \quad ...(8.2)$$

$$S_{ys} = \frac{S_{yt}}{2} \quad ...(8.3)$$

$$\sigma_{1,2} = \frac{1}{2}\left[(\sigma_x + \sigma_y) \pm \sqrt{(\sigma_x - \sigma_y)^2 + 4\tau_{xy}^2}\right] \quad ...(8.4)$$

$$\tau_{max} = \frac{1}{2}\sqrt{(\sigma_x - \sigma_y)^2 + 4\tau_{xy}^2} \quad ...(8.5)$$

$$\sigma_1 - \sigma_2 = \frac{S_{yt}}{n} \quad ...(8.6)$$

According to maximum distortion energy theory,

$$[(\sigma_1 - \sigma_2)^2 + (\sigma_2 - \sigma_3)^2 + (\sigma_3 - \sigma_1)^2]^{1/2} = \frac{\sqrt{2}\, S_{yt}}{n} \quad ...(8.7)$$

For two-dimensional stress system,

$$\sqrt{\sigma_x^2 + \sigma_y^2 - \sigma_x\sigma_y + 3\tau_{xy}^2} = \frac{S_{yt}}{n} \quad ...(8.8)$$

8.4 DESIGN OF SHAFT AGAINST STATIC LOADING

1. Maximum Shear Stress Theory

(a) Shaft subjected to pure torsion

(i) Solid shaft :

Maximum shear stress under pure torsion,

$$\tau_{max} = \frac{16T}{\pi d^3} \quad ...(8.9)$$

where T = torque applied, N.m

$$= \frac{10^3 \times 60 \times P}{2\pi N}$$

P = power transmitted, kW

N = speed, rpm

Allowable (or working) shear strength for shaft material,

$$\tau_w = \frac{S_{yt}}{2n}$$

∴ Shaft diameter,
$$d = \left[\frac{16T}{\pi \tau_w}\right]^{1/3} \quad \text{...(8.10)}$$

(ii) Hollow Shaft

Let
$$k = \frac{d_i}{d_0}$$

where d_i, d_0 = inner and outer diameters of shaft

$$\tau_{max} = \frac{16T}{\pi d_0^3 (1 - k^4)} \quad \text{...(8.11)}$$

$$d_0 = \left[\frac{16T}{\pi(1 - k^4)\tau_w}\right]^{1/3} \quad \text{...(8.12)}$$

(b) Shaft subjected to combined bending and torsion.

(i) Solid Shaft :

Bending stress,
$$\sigma_b = \frac{32M}{\pi d^3}$$

Shear stress,
$$\tau = \frac{16T}{\pi d^3}$$

$$\tau_{max} = \frac{1}{2}\sqrt{\sigma_b^2 + 4\tau^2} = \frac{1}{2}\sqrt{\left(\frac{32M}{\pi d^3}\right)^2 + 4\left(\frac{16T}{\pi d^3}\right)^2}$$

$$= \frac{16}{\pi d^3}\sqrt{M^2 + T^2} = \frac{16}{\pi d^3} T_e \quad \text{...(8.13)}$$

where
$$T_e = \sqrt{M^2 + T^2} = \text{Equivalent torque}$$

∴
$$d = \left[\frac{16T_e}{\pi \tau_w}\right]^{1/3} \quad \text{...(8.14)}$$

(ii) Hollow Shaft :

$$d_0 = \left[\frac{16T_e}{\pi(1-k^4)\tau_w}\right]^{1/3} \qquad \text{...(8.15)}$$

(c) Shaft subjected to combined axial thrust, bending and torsion

Stress due to axial thrust load F, $\sigma_x = \dfrac{4F}{\pi d_0^2 (1-k^2)}$ ...(8.16)

To account for buckling of shaft, the RHS of above equation is multiplied by a constant α, which accounts for the magnitude of column action of unsupported shaft length between bearings.

$$\alpha = \frac{1}{1-0.0044\left(\dfrac{L}{K}\right)} \text{ for } \frac{L}{k} < 115 \qquad \text{...(8.17a)}$$

$$= \frac{S_{yc}\left(\dfrac{L}{K}\right)^2}{\pi^2 n_0 E} \qquad \text{...(8.17b)}$$

where S_{yc} = yield strength in compression

n_0 = 1 for hinged ends

= 2.25 for fixed ends

E = modulus of elasticity

$K = \dfrac{d_0^2}{16}(1+k^2)$ = radius of gyration of shaft

L = Length of shaft between the supporting bearings

$$\frac{S_{yt}}{2n} = \frac{16}{\pi d_0^3 (1-k^4)}\left[\left\{\frac{\alpha d_0 (1+k^2)F}{8} + M\right\}^2 + T^2\right]^{1/2} \qquad \text{...(8.18)}$$

2. Maximum Distortion Energy Theory

$$\frac{S_{yt}}{n} = \frac{32}{\pi d_0^3 (1-k^4)}\left[\left\{\frac{\alpha d_0 (1+k^2)F}{8} + M\right\}^2 + 0.75\,T^2\right]^{1/2} \qquad \text{...(8.19)}$$

The preferred shaft diameters are given in Table 8.2.

Table 8.2 : Preferred shaft dameters, mm

6, 8, 9, 10, 14, 16, 18, 19, 20, 25, 28, 30, 40, 45, 50, 56, 60, 63, 71, 75, 80, 85, 90, 95, 100, 125, 140, 150, 160, 170, 180, 190, 200, 250, 280, 300, 400, 450, 500, 530, 560, 600 and 630 mm

8.4.1 ASME Code for the Design of Transmission Shafting

The ASME code for the design of transmission shafting is based on the maximum shear stress theory and takes into account the effect of fatigue and shock by introducing two constants into the equation of maximum shear stress. These are the numerical combined shock and fatigue factores K_m and K_t to be apolied to bending and taristing moments respectively.

$$d_0^3 = \frac{16n}{\pi S_{ys}(1-k^4)}\left[\left\{\frac{\alpha d_0(1+k^2)F}{8} + K_m\,M\right\}^2 + (K_t T)^2\right]^{1/2} \qquad \ldots(8.20)$$

The values of K_m and K_t are given in Table 8.3.

Table 8.3 : Values of combined fatigue and shock factors.

Nature of load	K_m	K_t
Gradually applied or steady load	1.5	1.0
Suddently applied loads with minor shocks only	1.5–2.0	1.0–1.5
Suddenly applied loads with heavy shocks	2.0–3.0	1.5–3.0

Permissible shear stress,

$$\tau_w = 0.30\,S_{yt} \text{ or } 0.18\,S_{ut}, \text{ whichever is less} \qquad \ldots(8.21)$$

■ **Example 8.1** A line shaft carrying two pulleys A of diameter 200 mm and B of 400 mm receives power at pulley A by means of a vertical belt, which is then transmitted to pulley B carrying horizontal belt. The pulley A is located at the middle of bearings 1 m apart and pulley B overhangs on the right of right bearing by 200 mm. The ratio of tensions is 2.5 : 1 and the maximum tension in either belt is limited to 2.5 kN. For the shaft material, $S_{ut} = 650$ MPa and $S_{yt} = 380$ MPa. Determine the shaft diameter using ASME code and taking into account the keyway effect.

Solution : Given : $S_{ut} = 650$ MPa and $S_{yt} = 380$ MPa,

$$\frac{F_1}{F_2} = 2.5,\ F_{max} = 2.5 \text{ kN},\ d_\alpha = 200 \text{ mm},\ d_b = 400 \text{ mm}$$

Design shear stress,

$$\tau_d = 0.30 S_{yt} = 0.30 \times 380 = 114 \text{ MPa}$$

$$= 0.18 S_{ut} = 0.18 \times 650 = 117 \text{ MPa}$$

$$= 114 \text{ MPa (minimum value)}$$

Accounting for keyway effect,

$$\tau_d = 0.75 \times 114 = 85.5 \text{ MPa}$$

The maximum belt tension shall occur on pulley A, being of smaller diameter.

The arrangement of pulleys is shown in Fig. 8.1.

Pulley A :

$$F_1 = 2500 \text{ N},\ F_2 = 1000 \text{ N}$$

Torque supplied to shaft, $T = (F_1 - F_2)\dfrac{d_a}{2}\ (2500 - 1000) \times \dfrac{200}{2} = 150{,}000$ N.mm

Resultant pull on pulley A, $F_a = F_1 + F_2 = 2500 + 1000 = 3500$ N↓

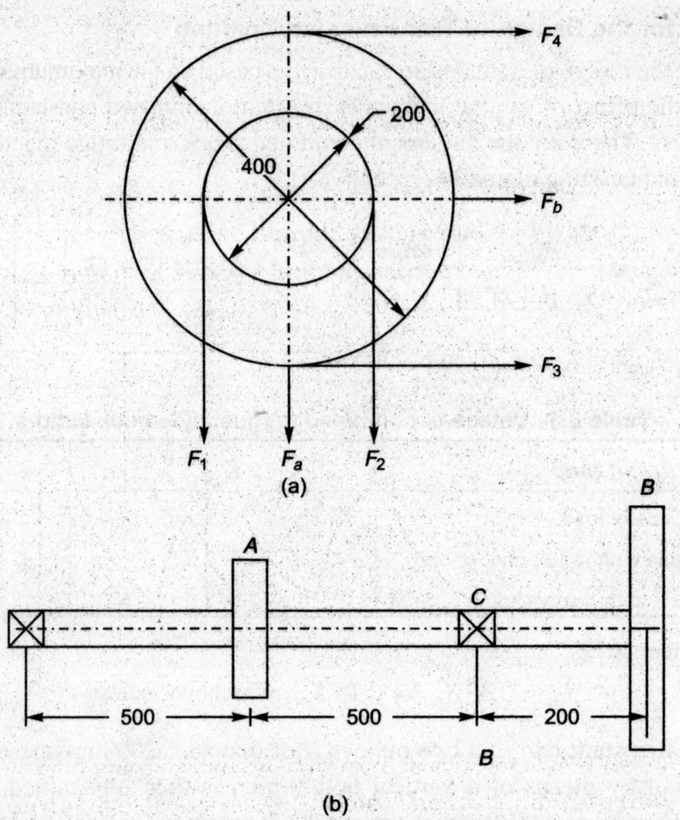

Fig. 8.1

Pulley B :

Let F_3, F_4 = tensions in the belt.

$$T = \frac{(F_3 - F_4)\, d_b}{2}$$

$$F_3 - F_4 = 2 \times \frac{150{,}000}{400} = 750 \text{ N}$$

$$\frac{F_3}{F_4} = 2.5$$

$$F_4 = \frac{750}{1.5} = 500 \text{ N}$$

$$F_3 = 1250 \text{ N}$$

Resultant pull on pulley B, $F_b = F_3 + F_4 = 1250 + 500 = 1750 \text{ N} \rightarrow$

Bending Moments :

(a) Vertical force

$$M_{av} = \frac{F_a L}{4} = \frac{3500 \times 1000}{4} = 875000 \text{ N.mm}$$

(b) Horizontal force

$$M_c = F_b \times 200 = 1750 \times 200 = 350{,}000 \text{ N.mm}$$

$$M_{ah} = \frac{350{,}00}{2} = 175{,}000 \text{ N.mm}$$

Resultant bending moment at A,

$$M = [M_{av}^2 + M_{ah}^2]^{1/2}$$

$$= 1000 \times [875^2 + 175^2]^{1/2} = 892328 \text{ N.mm}$$

Shaft Diameter :

$$d^3 = \left[\frac{16}{(\pi\tau_d)}\right][(K_m M)^2 + (K_t T^2)]^{1/2}$$

$$= \left[\frac{16}{(\pi \times 85.5)}\right][(1.5 \times 892328)^2 + (1.0 \times 150{,}000)^2]^{1/2}$$

$$= 80228$$

$$d = 43.13 \text{ mm} \approx 45 \text{ mm}$$

■ **Example 8.2** A shaft transmits 5 kW at 1000 rpm. Using maximum shear strain energy theory, find the diameter of the shaft. Take yield point stress in tension as 300 N/mm^2, factor of safety as 2.0, and Poisson's ratio 0.25.

Solution : $P = 5$ kW, $N = 1000$ rpm, $S_{yt} = 300$ N / mm^2, $n = 2$, $v = 0.25$

$$P = \frac{2\pi NT}{60 \times 10^3}$$

$$T = 5 \times 60 \times \frac{10^3}{2 \times \pi \times 1000} = 47.746 \text{ N.m}$$

Shear stress,

$$\tau = \frac{16T}{\pi d^3}$$

The shear stress induces tensile and compressive stresses in the shaft on the diagonals of equal magnitude and equal to the shear stress. Thus,

$$\sigma_1 = \tau, \sigma_2 = 0, \sigma_3 = -\tau$$

According to maximum shear strain energy theory,

$$\sigma_1^2 + \sigma_3^2 + (\sigma_3 - \sigma_1)^2 = 2\left(\frac{S_{yt}}{n}\right)^2$$

$$\tau^2 + (-\tau)^2 + (-2\tau) = 2\left(\frac{S_{yt}}{n}\right)^2$$

$$6\tau^2 = 2\left(\frac{S_{yt}}{n}\right)^2$$

$$\tau = \left(\frac{1}{3}\right)^{1/2} \frac{S_{yt}}{n} 16 \times \frac{47746}{\pi d^3} = \left(\frac{1}{3}\right)^{1/2} \frac{300}{2}$$

$$d = 14.1 \approx 15 \text{ mm}$$

■ **Example 8.3** A transmission shaft supports a spur gear B and a pulley D as shown in Fig. 8.2. The diameters of pulley and gear are 400 mm and 300 mm respectively. The shaft transmits 25 kW power at 600 rpm from the pulley to the gear. The ratio of the belt tensions is 3. The pressure angle α of gear is 20° and the relationship between radial force F_r and tangential force F_t on gear tooth is : $F_r = F_t \tan \alpha$. The pulley and gear are keyed to the shaft. The material of the shaft is steel 50 C4 having $S_{ut} = 700$ MPa and $S_{yt} = 460$ MPa.

Determine the shaft diameter using the ASME code taking, $K_m = K_t = 1.5$.

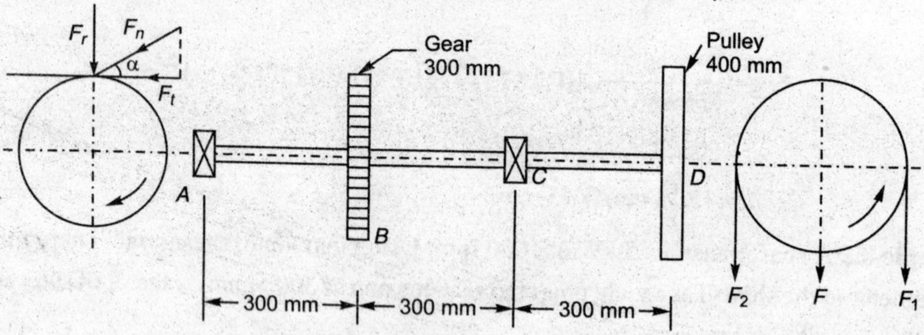

Fig. 8.2

Solution : Given : $S_{ut} = 700$ MPa, $S_{yt} = 460$ MPa, $\frac{F_1}{F_2} = 3$, $d_g = 300$ mm, $d_p = 400$ mm,

$$P = 25 \text{ kW}, N = 600 \text{ rpm}, \alpha = 20°$$

Design shear stress,

$$\tau_d = 0.30 S_{yt} = 0.30 \times 460 = 138 \text{ MPa}$$

$$= 0.18\, S_{ut} = 0.18 \times 700 = 126 \text{ MPa}$$

$$= 126 \text{ MPa (minimum value)}$$

Accounting for keyway effect,

$$\tau_d = 126 \times 0.75 = 94.5 \text{ MPa}$$

Pulley D :

$$P = \frac{2\pi NT}{60 \times 10^3}$$

$$T = \frac{25 \times 60 \times 10^3}{2\pi \times 600} = 398 \text{ N.m}$$

$$(F_1 - F_2)\frac{d_p}{2} = T$$

$$F_1 - F_2 = \frac{398 \times 10^3 \times 2}{400} = 1990 \text{ N}$$

$$\frac{F_1}{F_2} = 3$$

$$F_2 = \frac{1990}{2} = 995 \text{ N}$$

$$F_1 = 2985 \text{ N}$$

$$F = F_1 + F_2 = 2985 + 995 = 3980 \text{ N}$$

Gear B :

$$F_t \times \frac{d_g}{2} = T$$

$$F_t = \frac{398 \times 10^3 \times 2}{300} = 2653.2 \text{ N}$$

$$F_r = F_t \tan\alpha = 2653.2 \tan 20° = 965.7 \text{ N}$$

Bearing Reactions :

Vertical force [Fig. 8.3(a)]

$$R_A \times 600 + 965.7 \times 300 = 3980 \times 300$$

$$R_A = 1507.15 \text{ N}\downarrow$$

$$R_C = 1507.15 + 965.7 + 3980 = 6452.85 \text{ N}\uparrow$$

$$M_A = 0,\ M_B = -1507.15 \times 300 = 4521.45 \text{ N.mm}$$

$$M_C = -3980 \times 300 = -1194000 \text{ N.mm}$$

Horizontal forces [Fig.8.3(c)]

$$M_B = \frac{2653.2 \times 600}{4} = 397980 \text{ N.mm}$$

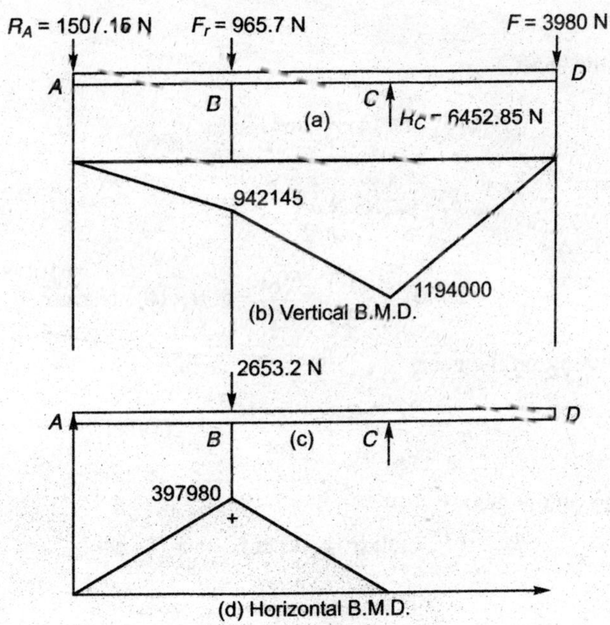

Fig. 8.3

Resultant B.M. at B,

$$M_B = [(942145)^2 + (397980)^2]^{1/2}$$

$$= 1022754 \text{ N.mm}$$

$$\therefore \quad M_{max} = M_C = 1194000 \text{ N.mm}$$

$$T = 398000 \text{ N.mm}$$

According to ASME code,

$$d^3 = \frac{16}{\pi \tau_d} \sqrt{(K_m\ M)^2 + (K_t\ T)^2}$$

$$= \frac{16}{\pi \times 94.5} \sqrt{(1.5 \times 1194000)^2 + (1.5 \times 398000)^2}$$

$$d = 46.7 \simeq 47 \text{ mm}$$

■ **Example 8.4** A shaft is supported on bearings *A* and *B*, 1 m between centres. A 20° pressure angle spur gear having 500 mm pitch circle diameter is located 250 mm to the right of left hand bearing *A* and a 600 mm diameter pulley is mounted 300 mm towards the left of bearing *B*. The gear is driven by a pinion with a downward tangential force while the pulley drives a horizontal belt having 180° angle of arc of contact. The pulley also serves as a flywheel and weighs 2 kN. The maximum belt tension is 2.5 kN and the ratio of tensions is 3 : 1. Determine the shaft diameter if the allowable shear stress of its material is 42 MPa.

Solution : Given : $L = 1$ m, $\alpha = 20°$, $d_g = 500$ mm, $L_1 = 250$ mm,

$$d_p = 600 \text{ mm}, L_2 = 300 \text{ mm},$$

$$W_p = 2 \text{ kN}, F_1 = 2.5 \text{ kN}, \frac{F_1}{F_2} = 3, \tau_d = 42 \text{ MPa}$$

Refer to Fig. 8.4

Torque acting on the shaft,

$$T = (F_1 - F_2)\frac{d_p}{2}$$

$$= F_1\left(1 - \frac{F_2}{F_1}\right)\frac{d_p}{2}$$

$$= 2500\left(1 - \frac{1}{3}\right) \times \frac{600}{2} = 500 \times 10^3 \text{ N.mm}$$

Tangential force acting on the gear,

$$F_{tg} = \frac{2T}{d_g} = \frac{2 \times 500 \times 10^3}{500} = 2000 \text{ N}$$

Radial force acting on the gear,

$$F_{rg} = F_{tg} \tan \alpha = 200 \tan 20° = 727.94 \text{ N}$$

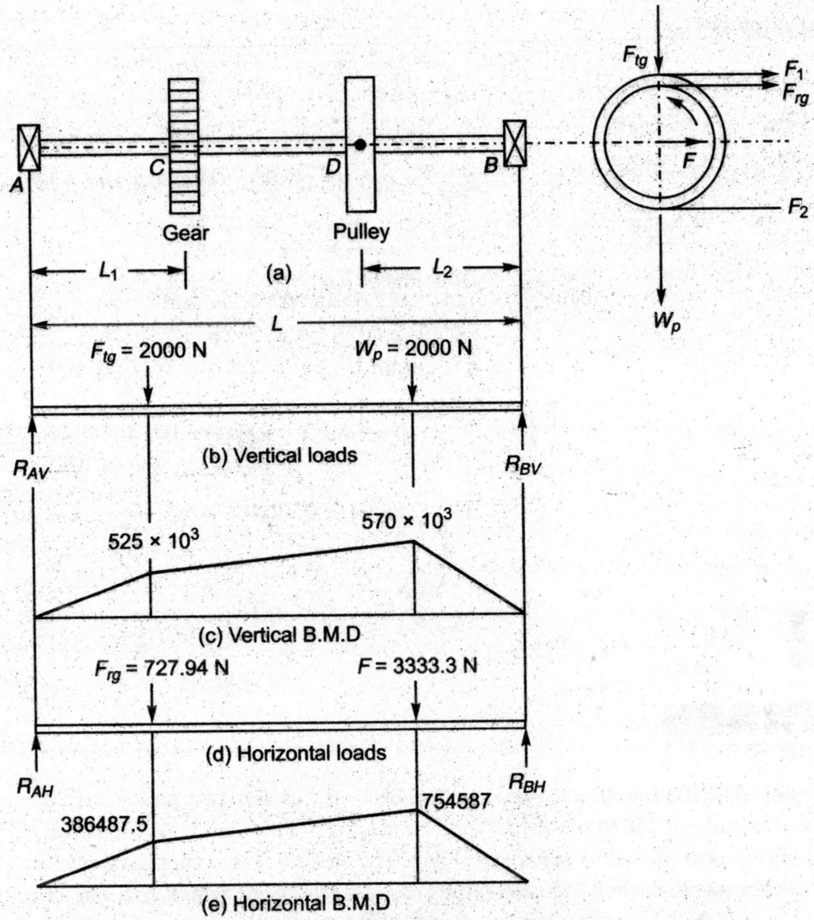

Fig. 8.4

Pulley :

$$F_2 = \frac{F_1}{3} = \frac{2500}{3} \text{ N}$$

$$F = F_1 + F_2 = 2500 + \frac{2500}{3} = 3333.3 \text{ N}$$

$$W_p = 2000 \text{ N}$$

Vertical Loads :

$$R_{AV} \times 1000 = 2000 \times 750 + 2000 \times 300$$

$$R_{AV} = \frac{2000 \times 1050}{1000} = 2100 \text{ N}$$

$$R_{BV} = 4000 - 2100 = 1900 \text{ N}$$

$$M_{CV} = R_{AV} + L_1 = 2100 \times 250 = 52500 \text{ N.mm}$$

$$M_{DV} = R_{BV} \times L_2 = 1900 \times 300 = 570000 \text{ N.mm}$$

Horizontal loads :

$$R_{AH} \times 1000 = 727.94 \times 750 + 3333.3 \times 300$$

$$R_{AH} = 1545.95 \text{ N}$$

$$R_{BH} = (727.94 + 3333.3) - 1545.95 = 2515.29 \text{ N}$$

$$M_{CH} = R_{AH} \times L_1 = 1545.95 \times 250 = 386487.5 \text{ N.mm}$$

$$M_{DH} = R_{BH} \times L_2 = 2515.29 \times 300 = 754587 \text{ N.mm}$$

$$M_{max} = M_D = \sqrt{M_{DV}^2 + M_{DH}^2}$$

$$= \sqrt{(570 \times 10^3)^2 + (754587)^2}$$

$$= 955675 \text{ N.mm}$$

Equivalent torque, $T_e = \sqrt{(M_{max})^2 + T^2} = \sqrt{(955675)^2 + (500 \times 10^3)^2} = 1069720 \text{ N.mm}$

Shaft diameter,

$$d^3 = \frac{16T_e}{\pi\tau_d} = \frac{16 \times 1069720}{\pi \times 42}$$

$$d = 50.6 \text{ mm} \simeq 51 \text{ mm}$$

EXERCISES

1. A transmission shaft 1 m long and supported on bearings, carries two pulleys of 200 mm and 500 mm diameters at a distance of 250 mm from the right hand and left hand bearings respectively. The bigger pulley shaft is supplied power by means of a vertical belt. The smaller pulley supplies power by a horizontal belt. The maximum tension in the belt of bigger pulley is 2 kN. The coefficient of frictioin on both pulleys is 0.25. The shaft is made of a material having $S_{yt} = 420 \text{ N/mm}^2$ and factor of safety desired is 3. Determine the shaft diameter.
2. A transmission shaft $AB = 1.2$ m long is supported on bearings and carries two pulleys at C and D of weight 300 N and 100 N respectively. $AC = BD = 300$ mm. The diameter of pulleys at C and D are 300 mm and 500 mm respectively. The shaft transmits 7.5 kW at 360 rpm from pulley at C to pulley at D. Both the belts are vertically down wards. The ratio of tensions is 3. The yield strength in tension of shaft material is 390 MPa and factor of safely is 3. Determine the shaft diameter.
3. A line shaft supports two pulleys A and B of diameters 200 mm and 500 mm respectively. Pulley A is at a distance of 500 mm from left end bearing and pulley B is at a distance of 300 mm on the overhanging shaft from right hand bearing. The belt on pulley A is vertical and on pulley B is horizontal. The ratio of belt tensions is 3 and value of limiting tension in the belts is 2.5 kN. $S_{ut} = 650$ MPa and $S_{yt} = 380$ MPa; $K_m = 1.5$ and $K_t = 1.25$. Determine the diameter of shaft.
4. A transmission shaft is supporting a spur gear of pitch circle diameter 500 mm between two bearings 750 mm apart and placed symmetrical. The shaft is overhanging from right hand bearing by 300 mm and carries a pulley of diameter 450 mm at the right end of shaft. The pulley transmits 20 kW at 600 rpm to gear by means of a vertical belt. The ratio of belt tensions is 3. For the gear, $F_r = F_t \tan 20°$. $S_{ut} = 700$ MPa, $S_{yt} = 460$ MPa, $K_m = K_t = 1.5$. Determine the shaft diameter using ASME code.

□□□

9

Chapter

Belt, Chain and Rope Drives

Inside This Chapter

9.1 Introduction
9.2 Flat Belt Drive
9.3 V-Belts Drive
9.4 Chain Drive
9.5 Ropes

9.1 INTRODUCTION

Power can be transmitted either by flexible or non-flexible type of machine elements. In **flexible** elements, the centre distance can be changed conveniently and is **non-flexible** type of machine elements, the centre distance cannot be changed. The flexible type of elements are : belts, chains, and ropes. The non-flexible type of elements are : gears, clutches, couplings, and power screws. Flexible elements are used for power transmission over long distances and non-flexible type over short distances.

9.2 FLAT BELT DRIVE

9.2.1 Factors for Selection of Belt Drive

The following factors must be considered while selecting a belt drive.

1. Speed of driving and driven shaft.
2. Speed reduction ratio.
3. Power to be transmitted.
4. Centre distance of shafts.
5. Space available.
6. Layout of shaft.

9.2.2 Factors on which Coefficient of Friction Depends

The coefficient of friction between belt and pulley depends upon the following factors :

1. Belt material
2 Pulley material
3. Belt slip
4. Belt speed

9.2.3 Design of Flat Belts

Flat belts can be used for horizontal, vertical, quarter-turn, right-angled, crossed and reversed drive. A horizontal open type flat belt drive is shown in Fig. 9.1. Flat belt drives are classified as :

Light : $v \le 12$ m/s and upto 7.5 kW

Medium : $v = 12 - 24$ m/s and 7.5 – 15 kW

Heavy : $v > 24$ m/s and above 15 kW

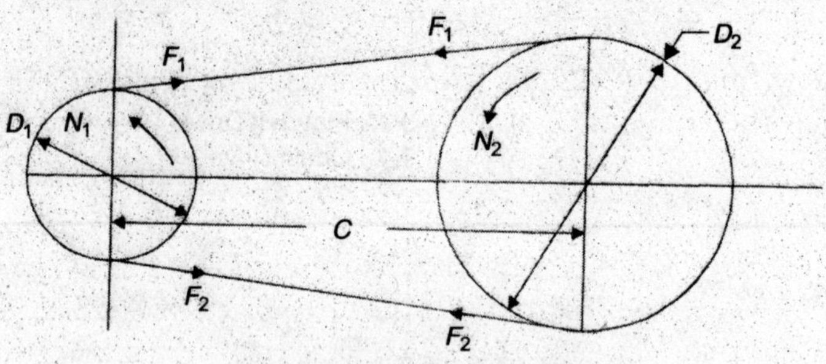

Fig. 9.1 Open flat belt drive.

1. Speed Ratio

Neglecting slip and thickness of belt, the speed ratio.

$$i = \frac{N_2}{N_1} = \frac{D_1}{D_2} \qquad \text{...(9.1a)}$$

where N_1, N_2 = rpm of smaller and bigger pulley respectively.

D_1, D_2 = diameters of smaller and bigger pulley respectively, mm.

Considering slip and thickness of belt.

Speed ratio,
$$i = \left(1 - \frac{s}{100}\right)\left[\frac{D_1 + t}{D_2 + t}\right] \qquad \text{...(9.1b)}$$

where t = thickness of belt

s = total percentage slip on the pulleys

$= s_1 + s_2 - 0.01\, s_1 s_2$

s_1, s_2 = % slip on smaller and bigger pulley respectively.

2. Angle of Arc of Contact

For open belt drive,

Angle of arc of contact,

$$\theta = \pi \pm 2\sin^{-1}\left[\frac{D_2 - D_1}{2C}\right] \text{rad} \qquad \text{...(9.2a)}$$

where C = centre distance

Use +ve sign for the bigger pulley and –ve sign for the smaller pulley.

For crossed belt drive (Fig. 9.2),

$$\theta = \pi \pm 2\sin^{-1}\left[\frac{D_2 + D_1}{2C}\right] \text{rad} \quad \text{...(9.2b)}$$

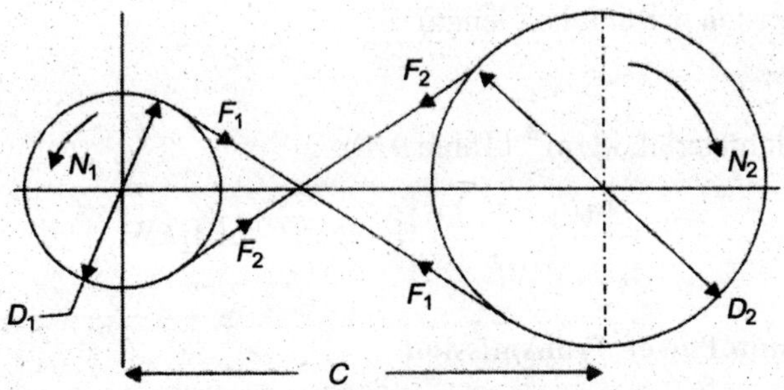

Fig. 9.2 Crossed flat belt drive.

3. Belt Tension and Power Transmitted Neglecting Centrifugal Tension

At low speeds ($v < 12$ m/s), the effect of centrifugal tension may be neglected. The ratio of belt tensions is,

$$\frac{F_1}{F_2} = \exp(\mu\theta) \quad \text{...(9.3)}$$

where F_1, F_2 = belt tension on the tight and slack side respectively.

μ = coefficient of friction between belt and pulley.

For tanned leather belts on C.I. pulleys,

$$\mu = 0.54 - \frac{0.712}{v + 2.542} \; v \text{ in m/s} \quad \text{...(9.4)}$$

Peripheral speed of belt, $$v = \frac{\pi D_1 N_1}{60 \times 10^3} \text{ m/s}$$

Corrected power, $$P_c = \frac{P \times K_s}{K_j} \quad \text{...(9.5)}$$

where K_s = service correction factor

K_j = correction factor for belt joint

Power transmitted, $$P_t = \frac{Fv}{10^3} \text{ kW} \quad \text{...(9.6)}$$

$F = F_1 - F_2$, tangential force in N

Power transmitted per square mm of belt cross-section,

$$P_c = \left[\frac{\sigma_1 v}{10^3}\right][1 - \exp(-\mu\theta)] \quad \text{...(9.7)}$$

where $\sigma_1 = \frac{F_1}{bt}$ is the allowable stress in the belt material

b = belt width, mm

4. Belt Tensions and Power Transmitted Considering Centrifugal Tension

At high speeds ($v > 12$ m/s), the effect of centrifugal tension must be considered. The ratio of belt tension is

$$\frac{F_1 - F_c}{F_2 - F_c} = \exp(\mu\theta) \qquad \text{...(9.8)}$$

where F_c = centrifugal tension per unit belt length

$= \rho b t v^2$

ρ = density of belt material, kg/m^3 (Table 9.4)

Power transmitted, $$P_c = \left[\frac{(F_1 - F_c)\, v}{10^3}\right][1 - \exp(-\mu\theta)] \text{ kW} \qquad \text{...(9.9)}$$

5. Condition for Maximum Power Transmission

$$v = \left[\frac{\sigma_d}{3\rho}\right]^{1/2} \qquad \text{...(9.10)}$$

where σ_d = design stress for the belt material per unit width of belt.

6. Initial Tension

$$F_i = \frac{F_1 + F_2}{2} \qquad \text{...(9.11)}$$

7. Width of Belt

Neglecting centrifugal tension,

Width of belt, $$b = \frac{F_1}{\sigma_d} \qquad \text{...(9.12)}$$

Considering centrifugal tension,

$$bt = \left[\frac{10^3\, P_c}{v(\sigma_d - \rho v^2)}\right]\left[\frac{1}{1 - \exp(-\mu\theta)}\right] \qquad \text{...(9.13)}$$

8. Maximum Stress in the Belt

σ_{max} = stress due to maximum belt tension + centrifugal stress + bending stress while turning over the pulleys

$$= \frac{F_1}{bt} + \rho v^2 + \frac{Et}{D_1} \qquad \text{...(9.14)}$$

where E = modulus of elasticity of belt material.

9. Length of Belt

The length of open belt is,

$$L_o = \left(\frac{\pi}{2}\right)(D_1 + D_2) + 2C + \frac{(D_2 - D_1)^2}{4C} \qquad \text{...(9.15)}$$

The length of crossed-belt is,

$$L_c = \left(\frac{\pi}{2}\right)(D_1 + D_2) + 2C + \frac{(D_2 + D_1)^2}{4C} \qquad \text{...(9.16)}$$

10. Centre Distance

$$C = \frac{L_o}{4} - 0.3925\,(D_1 + D_2) + \left[\left\{\frac{L_o}{4} - 0.3925\,(D_2 - D_1)\right\}^2 - \frac{(D_2 - D_1)^2}{8}\right]^{1/2} \qquad \text{...(9.17)}$$

9.2.4 Pulleys for Flat Belts

(a) Cast Iron Pulley

A cast iron pulley is shown in Fig. 9.3.

Material : Cast iron of Grade FG 200 of IS : 210 – 1978

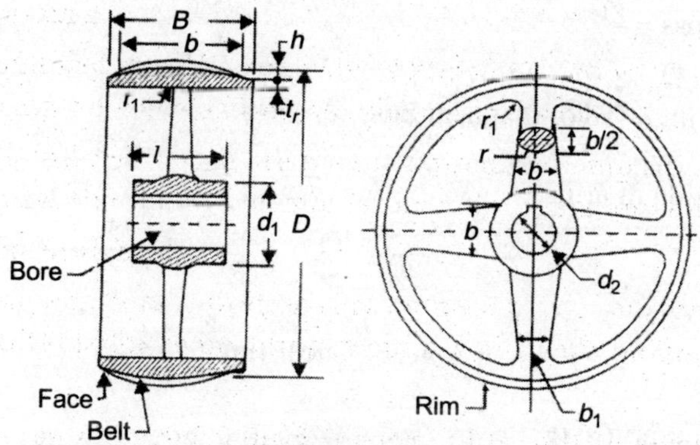

b = Width of belt

Fig. 9.3 Cast iron flat belt pulley.

1. Rim Size

Bartho's formula for width of rim

$$B = 1.19b + 10 \text{ mm of single belt} \qquad \text{...(9.18a)}$$

$$= 1.10b + 5 \text{ mm for double belt} \qquad \text{...(9.18b)}$$

Bartho's formula for thickness of pulley rim.

$$t_r = 0.25\sqrt{D} + 1.5 \text{ mm for light duty} \qquad \text{...(9.19a)}$$

$$= 0.375\sqrt{D} + 3.2 \text{ mm for heavy duty} \qquad \text{...(9.19b)}$$

For cast iron flat pulleys, rim thickness,

$$t_r = \frac{D}{200} + 3 \text{ mm for single belt} \qquad \text{...(9.20a)}$$

$$= \frac{D}{200} + 6 \text{ mm for double belt} \qquad \text{...(9.20b)}$$

2. Crowning of Pulley Rim

A convex curvature (called crowning) is given to the pulley rim to make the belt run in the middle of the pulley width. The crowning depends upon the diameter and width of the pulley rim.

3. Number of Arms

Provide web for pulleys upto 200 mm diameter. The number of arms for other diameters are as follows :

For $D = 200 - 450$ mm, number of arms, $n = 4$ ...(9.21a)

For $D > 450$ mm, $n = 6$ ...(9.21b)

4. Cross-section of Arms

The cross-section of arms is elliptical.

$$\text{Major axis of arm near boss, } b_1 = 2.94\left[\frac{BD}{4n}\right]^{1/3} \text{ for single belt} \quad \text{...(9.22a)}$$

$$= 2.94\left[\frac{BD}{2n}\right]^{1/2} \text{ for double belt} \quad \text{...(9.22b)}$$

$$\text{Minor axis near boss} = \frac{b_1}{2} \quad \text{...(9.23)}$$

$$\text{Cross-section near rim} = \frac{2}{3}\text{rd of that near boss} \quad \text{...(9.24)}$$

Radius of cross-section of arms,

$$r = \frac{3b_1}{4} \quad \text{...(9.25)}$$

$$\text{Radius,} \quad r_1 = r_2 = \frac{b_1}{2} \text{ near rim} \quad \text{...(9.26)}$$

Analytical method for arm size :

Maximum bending moment on arms,

$$M_{max} = (F_1 - F_2)\frac{D}{2} \quad \text{...(9.27)}$$

Bending moment in each arm, assuming that only half the number of arms take the bending moment,

$$M = \frac{2M_{max}}{n} = \left(\frac{\pi}{64}\right) b_1^3 \, \sigma_t \quad \text{...(9.28)}$$

where σ_t = permissible tensile strength for material of arms.

5. Size of Boss

$$\text{Minimum length of bore,} \quad l = \frac{2B}{3} \quad \text{...(9.29)}$$

The diameter d_1 of boss should be such that,

$$\frac{d_1 - d_2}{2} = 0.412\,(BD)^{1/3} + 6 \text{ mm for single belt} \quad \text{...(9.30a)}$$

$$= 0.529\,(BD)^{1/3} + 6 \text{ mm for double belt} \quad \text{...(9.30b)}$$

(b) Steel Pulley

A steel pulley is shown in Fig. 9.4.

Material : Mild steel conforming to IS : 226 – 1975

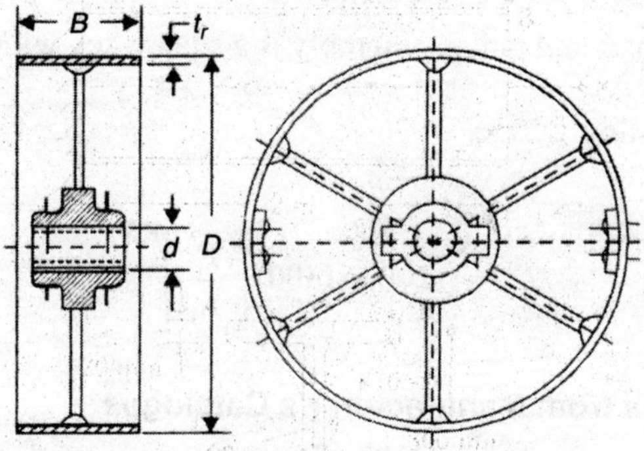

Fig. 9.4 Mild steel flat belt pulley.

Thickness of rim, $t_r = 5$ mm ...(9.31)

Length of boss $= \frac{1}{2} \times$ width of face

$= 100$ mm minimum for pulleys with 19 mm diameter spokes

$= 138$ mm minimum for pulleys with 22 mm diameter spokes

Cross-section of spokes = circular

Rows of arms = single for pulleys upto 300 mm width ...(9.33a)

= two for width more than 300 mm ...(9.33b)

9.2.5 Shaft and Key Size

Resultant pull on pulley due to belt tensions,

$$P = \left[\frac{F}{K-1}\right][K^2 + 1 + 2K \cos(180° - \theta)]^{1/2} \quad \text{...(9.34)}$$

where $K = \dfrac{F_1 - F_c}{F_2 - F_c}$, $F = F_1 - F_2$

Let W = weight of the pulley, and α = angle between P and W

Resultant force on the pulley,

$$S = [P^2 + W^2 + 2PW \cos\alpha]^{1/2} \quad \text{...(9.35)}$$

Let x = overhang of pulley

$M = Sx$, bending moment due to S

Torque in the shaft, $T = (F_1 - F_2)\dfrac{D}{2}$ ...(9.36)

Equivalent torque, $T_e = [(K_m M)^2 + (K_t T)^2]^{1/2}$ ...(9.37)

where K_m and K_t are shock and fatigue factors for bending and torsion respectively.

Let d_2 = diameter of shaft

$$T_e = \left(\frac{\pi}{16}\right) d_2^3 \, \tau \qquad \ldots(9.38)$$

where τ = permissible shear stress for the shaft material.

A suitable key for the shaft and pulley assembly is a square key with a gib head.

9.2.6 Diameter of Motor Pulley

Diameter of motor pulley,

$$D_1 = (1200 \text{ to } 1400)\left[\frac{P}{N_1}\right]^{1/3} \qquad \ldots(9.39)$$

9.2.7 Selection of Flat-Belts from Manufacturer's Catalogue

The following information is required to select the flat-belt from manufacturer's catalogue :

1. Power to be transmitted
2. Input and output speeds
3. Centre distance available

The basic procedure for belt selection from the catalogue of the Dunlop Rubber Co. (India) Ltd. is as follows :

1. Calculate the maximum power transmitted by the belt from the rated power.

$$(\text{kW})_{max} = F_a \, (\text{kW}) \qquad \ldots(9.40)$$

where F_a = load correction factor.

2. The rated power is based on 180° angle of arc of contact. For this angle less than 180°, the belt is reuqired to carry additional tension. For this purpose a factor called 'arc of contact factor' F_d is used to calculate the corrected power to be transmitted by the belt.

$$(\text{kW})_{corrected} = F_a \, F_d \, (\text{kW}) \qquad \ldots(9.41)$$

where F_d = arc of contact factor.

3. Calculate the corrected power rating for the belt.

The power transmitting capacities of belts are given in Table 9.1.

Table 9.1 : Power transmission capacities of belts.

Type of belt	*Power rating in kW per mm per ply*	*Corrected kW rating*	*Optimum belt speed, m/s*
HI-SPEED	0.0118	$\frac{0.0118 \times v}{5.08}$	17.8 – 22.0
FORT	0.0147	$\frac{0.0147 \times v}{5.08}$	17.8 – 22.0

$$\text{Corrected kW rating} = \frac{0.0118 \times v}{5.08} \text{ for HI-SPEED belt} \qquad \ldots(9.42a)$$

$$= \frac{0.0147 \times v}{5.08} \text{ for FORT belt} \qquad \ldots(9.42b)$$

where v = belt velocity, m/s

4. Calculate the product of (width × number of plies),

$$\text{Width} \times \text{number of plies} = \frac{\text{corrected power}}{\text{corrected kW rating}} \quad \ldots(9.43)$$

5. Select the standard belt width and number of plies.

9.2.8 Creep in Belts

The tension on the tight side of a belt drive is more than the tension on the slack side. As a result of this, the belt is stretched more on the tight side as compared to the slack side. Therefore, the driver pulley receives more length of the belt and delivers less. Hence, the belt creeps (jumps) forward. The reverse occurs on the driven pulley, which receives less length of the belt and delivers more. Thus the belt jumps backward. This phenomenon is called creeping of belt.

$$\text{Creep} = \frac{F_1 - F_2}{bt\,E} \quad \ldots(9.44)$$

where F_1, F_2 = tensions on the tight and slack sides respectively.

b, t = width and thickness of belt respectively

E = Young's modulus of elasticity of belt material.

■ **Example 9.1** It is required to select a flat-belt drive for a compressor running at 720 rpm, which is driven by a 25 kW, 1440 rpm motor. Space is available for a centre distance of 3 m. The belt is open type.

Solution : Given : $N_1 = 1440$ rpm, $N_2 = 720$ rpm, $P = 25$ kW, $C = 3$ m

We take, $D_1 = 280$ mm

$$D_2 = \frac{280 \times 1440}{720} = 560 \text{ mm}$$

Angle of arc of contact,

$$\theta = \pi - 2\sin^{-1}\left[\frac{D_2 - D_1}{2C}\right]$$

$$= \pi - 2\sin^{-1}\left[\frac{560 - 280}{2 \times 3000}\right]$$

$$= 174.65° \text{ or } 3.048 \text{ rad}$$

Length of open belt,

$$L_o = 2C + \left(\frac{\pi}{2}\right)(D_1 + D_2) + \frac{(D_2 - D_1)^2}{4C}$$

$$= 2 \times 3 + \left(\frac{\pi}{2}\right)(0.280 + 0.560) + \frac{(0.560 + 0.280)^2}{12}$$

$$= 7.326 \text{ m}$$

Pitch line velocity,

$$v = \pi D_1\, N_2 / (60 \times 10^3)$$

$$= \frac{\pi \times 280 \times 1440}{60 \times 10^3} = 21.11 \text{ m/s}$$

Load correction factor for compressor,

$$F_a = 1.3$$

Arc of contact factor, $F_d = 1.04 - \dfrac{(1.04-1.00)\,(174.65-170)}{180-720}$

$$= 1.0214$$

Corrected power, $P_c = P \times F_a \times F_d = 25 \times 1.3 \times 1.0214 = 33.1955$ kW

We select of HI-SPEED Dunlop transmission duck belting of power rating 0.012 kW per mm width per ply. The optimum range of belt velocity for this belt is 17.8 – 22.9 m/s.

$$\text{Corrected belt rating} = \frac{0.0118v}{5.08} = \frac{0.012 \times 21.11}{5.08} = 0.049 \text{ kW}$$

$$\text{Width} \times \text{No. of plies} = \frac{\text{Corrected power}}{\text{Corrected belt rating}}$$

$$= \frac{33.1955}{0.049} = 677.459$$

Belt widths :

$$4 \text{ plies, } b = \frac{677.459}{4} = 169.36 \text{ mm}$$

$$5 \text{ plies, } b = \frac{677.459}{5} = 135.49 \text{ mm}$$

$$6 \text{ plies, } b = \frac{677.459}{6} = 112.91 \text{ mm}$$

We select a belt of 152 mm preferred width and 5 plies.

■ **Example 9.2** A leather belt is used in a flat-belt drive to transmit 15 kW from a motor pulley running at 1440 rpm to another pulley running at 480 rpm. The centre distance between the pulleys is twice the diameter of the bigger pulley. The belt should operate at a velocity of 20 m/s approximately. The maximum stress in the belt should not exceed 2.25 MPa. The density of leather is 950 kg/m^3 and the coefficient of friction is 0.35. The belt thickness is 5 mm. Calculate the length and width of belt.

Solution : Given : $P = 15$ kW, $N_1 = 1440$ rpm, $N_2 = 480$ rpm, $C = 2D_2$, $v \approx 20$ m/s, $\sigma_w = 2.25$ MPa, $\rho = 950$ kg/m^3, $\mu = 0.35$, $t = 5$ mm

1. Diameter of motor pulley,

$$D_1 = \frac{600 \times 10^3 \times 20}{\pi \times 1440} = 265.26 \approx 250 \text{ mm}$$

$$v = \frac{\pi \times 250 \times 1440}{60 \times 10^3} = 18.65 \text{ m/s}$$

$$D_2 = \frac{250 \times 1440}{480} = 750 \text{ mm}$$

2. Centre distance, $C = 2D_2 = 2 \times 750 = 1500$ mm

3. Length of open belt, $L_o = 2C + \dfrac{\pi(D_1 + D_2)}{2} + \dfrac{(D_2 - D_1)^2}{4C}$

$$= 2 \times 15.5 + \frac{\pi(0.25 + 0.75)}{2} + \frac{(0.75 - 0.25)^2}{4 \times 155}$$

$$= 4.61246 \text{ m}$$

4. Mass of belt per metre length,

$$m = \rho bt = 950 \times b \times 5 \times 10^{-6} \times 1$$

$$= 4.75 \times 10^{-3} \times b \text{ kg/m}$$

5. Centrifugal tension, $F_c = mv^2 = 4.75 \times 10^{-3} \times 18.85^2 \times b = 1.6878\, b$ N

6. Angle of arc of contact, $\theta = 180° - 2\sin^{-1}\left[\dfrac{D_2 - D_1}{2C}\right]$

$$= 180° - 2\sin^{-1}\left[\frac{750 - 250}{2 \times 1500}\right]$$

$$= 160.81° \text{ or } 2.8067 \text{ rad}$$

7. Exp $(\mu\,\theta) = \exp(0.35 \times 2.8067) = 2.67$

8. $$\frac{F_1 - F_c}{F_2 - F_c} = \exp(\mu\theta)$$

$$\frac{F_1 - 1.6878\, b}{F_2 - 1.6878\, b} = 2.67 \qquad \ldots(1)$$

9. Maximum belt tension, $F_1 = \sigma_w\, bt = 2.25 \times b \times 5 = 11.25\, b$ ...(2)

10. Tangential force, $F_1 - F_2 = \dfrac{10^3 P}{v} = \dfrac{10^3 \times 15}{18.85} = 795.756$ N ...(3)

11. Solving Eqs. (1) to (3), we have

$$\frac{11.25\, b - 1.6878\, b}{11.25\, b - 795.756 - 1.6878\, b} = 2.67$$

$$\frac{9.5622\, b}{9.5622\, b - 795.756} = 2.67$$

$$3.5183\, b = 9.5622\, b - 795.756$$

$$b = 133 \text{ mm or 4 ply 140 mm}$$

We select belt width, $b = 140$ mm

9.3 V-BELTS DRIVE

V-belts give more positive drive and are used for short centre distance between the driving and driven pulleys.

9.3.1 Characteristics of V-Belt Drive

1. It requires a grooved pulley.
2. The belt comes into contact with the sides of the pulley groove only.

3. The belts are endless.
4. It gives more positive drive than flat belts.
5. The slip of the belt is less as compared to flat belts.
6. The wedging action of the belt in the pulley groove offset the decrease in the arc of contact.
7. V-belts are used as a matched set.
8. The pulley groove angle is 40°.
9. They are suitable for speed ratio upto 7 and belt speed upto 25 m/s.
10. The size of a V-belt is specified by the cross-section followed by inside length.

9.3.2 Design of V-Belt Drive

1. Equivalent Coefficient of Friction

Fig. 9.5 shows a V-belt in a grooved pulley. The normal force on the belt face is,

$$F_n = \frac{F}{2\sin\beta}$$

where 2β = angle of the groove = 40°

Tractive force $= 2\mu F_n$

$$= \left(\frac{\mu}{\sin\beta}\right) F = \mu_e F \qquad \text{...(9.45)}$$

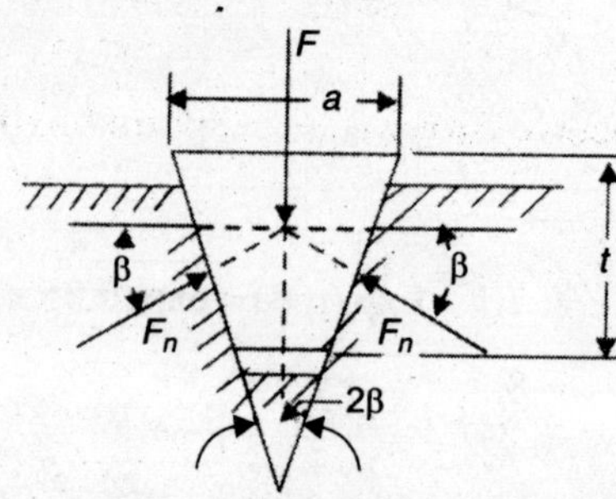

Fig. 9.5

where $\mu_e = \dfrac{\mu}{\sin\beta}$ is the equivalent of effective coefficient of friction

2. Selection of Belt Section

The cross-section suitable for a given power to be transmitted can be selected from standard.

3. Number of Belts

The number of belts n, may be determined from the following formula :

$$n(F_1 - F_c)\left[\frac{1}{1-\exp(\mu_e\theta)}\right] = \frac{10^3 \times P \times K_s}{v} \qquad \text{...(9.46)}$$

where $v = \dfrac{\pi D_1 N_1}{60 \times 10^3}$ m/s

D_1 = diameter of smaller pulley, mm

N_1 = rpm of smaller pulley

K_s = correction factor according to service

$$\theta = 2\cos^{-1}\left[\frac{D_1 - D_1}{2C}\right] \qquad \text{...(9.47)}$$

According to IS : 2494 – 1974, the number of belts are given by,

$$n = \frac{P \times K_a}{\text{kW} \times K_c \times K_d} \qquad \text{...(9.48)}$$

where P = drive power, kW

kW = rating of V-belts at 180° arc of contact

K_a = correction factor for industrial service

K_c = correction factor for belt length

K_d = correction factor for arc of contact

Equivalent pitch diameter, $d_e = d_p \times K_b$

K_b = small diameter factor

4. Power Rating of V-Belts

$$\text{kW} = \left(\frac{\text{nav}}{10^3}\right)\left[\sigma_d - \frac{\rho v^2}{10^6}\right]\left[1 - \frac{1}{\exp(\mu_e \theta)}\right] \quad \text{...(9.49)}$$

where σ_d = design stress = 2.26 N/mm^2

ρ = mass density of belt material = 1080 kg/m^3

a = area of cross-section of belt, mm^2

5. Initial Tension

$$F_i = \frac{F_1 + F_2}{2} \quad \text{...(9.50)}$$

6. Optimum Velocity of Belt for Maximum Power Transmission

$$v = \left(\frac{F_i}{3m}\right)^{1/2} \quad \text{...(9.51)}$$

where m = mass per meter length of belt

7. Ratio of Belt Tensions

$$\frac{F_1 - mv^2}{F_2 - mv^2} = \exp(\mu_e \theta) \quad \text{...(9.52)}$$

8. Pitch Length of Belt

$$L_p = 2C + 15.57(D_1 + D_2) + \frac{(D_2 - D_1)^2}{4C} \quad \text{...(9.53)}$$

9. Centre Distance

$$C = A + (A^2 - B)^{1/2} \quad \text{...(9.54)}$$

where $A = \frac{L_p}{4} - \frac{\pi(D_1 + D_2)}{8}$

$B = \frac{(D_2 - D_1)^2}{8}$

$$C_{max} = 2\,(D_1 + D_2) \qquad \text{...(9.55)}$$

$$C_{min} = 0.55\,(D_1 + D_2) + t$$

Refer to Table 9.24 for selection of centre distance.

10. Pulley Size

Refer to standard tables for dimensions of standard V-grooved pulleys and standard pulley pitch diameters.

■ **Example 9.3** Design a V-belt drive to transmit 7.5 kW from an induction motor running at 1440 rpm to a fan running approximately at 480 rpm. The service is for 24 hours per day. Centre distance is 1.25 m.

Solution : Given : $P = 7.5$ kW, $N_1 = 1440$ rpm, $N_2 \approx 480$ rpm, service = 24 hours, $C = 1.25$ m.

1. Select cross-section of belt.

   For $P = 0.8 - 13$ kW, cross-section C of V-belt is suitable.

   Top width of belt, $w = 22$ mm, Nominal thickness,

   $$t = 14 \text{ mm}$$

   Weight per metre = 2.962 N/m,

   Minimum pulley diameter, $D_1 = 229$ mm, $v_{max} = 25$ m/s

   Let $\qquad D_1 = 240$ mm

2. Diameter of fan pulley, $D_2 = \dfrac{240 \times 1440}{480} = 920$ mm

3. Belt speed, $v = \dfrac{\pi \times 0.24 \times 1440}{60} = 18$ m/s

4. Pitch length of belt, $L = 2C + \dfrac{\pi\,(D_1 + D_2)}{2} + \dfrac{(D_2 - D_1)^2}{4C}$

   $$= 2 \times 1.25 + \frac{\pi\,(0.25 + 0.72)}{2} + \frac{(0.72 - 0.24)^2}{(4 \times 15.25)}$$

   $$= 4.013 \text{ m or } 4013 \text{ mm}$$

   The standard nominal pitch length of C-belt nearest to above length is 4069 mm.

5. Exact centre distance $C = A + (A^2 - B)^{1/2}$

   where $\qquad A = \dfrac{L_p}{4} - \dfrac{\pi\,(D_1 + D_2)}{8}$

   $$= \frac{4069}{4} - \frac{\pi\,(240 + 720)}{8} = 640.26 \text{ mm}$$

   $$B = \frac{(D_2 - D_1)^2}{8} = \frac{(720 - 240)^2}{8} = 28800$$

   $$C = 640.26 + [(640.26)^2 - 28800]^{1/2}$$

   $$= 1257.6 \text{ mm or } 1.2576 \text{ m}$$

6. Inside length of belt, $L_i = 4013$ mm
   The belt size is C-4013; V-belt.
7. Angle of arc of contact, $\theta = 180° - 2\sin^{-1}\left[\dfrac{D_2 - D_1}{2C}\right]$

$$= 180° - 2\sin^{-1}\left[\frac{0.72 - 0.24}{2 \times 1.2576}\right]$$

$$= 158°$$

8. Small diameter factor, $K_b = 1.14$ or speed ratio = 3
   Correction factor for industrial service,
   $$K_a = 1.3$$
   Correction factor for belt length,
   $$K_c = 1.02$$
   Correction factor for arc of contact,
   $$K_d = 0.80$$
9. Equivalent pitch diameter, $D_e = D_1 \times K_b = 240 \times 1.14 = 273.6$ mm
   Maximum value of d_e in the C-belt power rating formula,
   $$d_e = 300 \text{ mm}$$
   $$\text{Power rating for C-belt} = \left(1.47\, v^{-0.09} - \frac{142.7}{d_e} - 2.34 \times 10^{-4}\, v^2\right) v$$
   $$= \left(1.47 \times 18^{-0.09} - \frac{142.7}{273.6} - 2.34 \times 10^{-4} \times 18^2\right) \times 18$$
   $$= 9.646 \text{ kW}$$
10. $$\text{Number of belts} = \frac{(\text{Transmitted power}) \times K_a}{(\text{kW rating} \times K_c \times K_d)}$$
    $$= \frac{7.5 \times 1.3}{(9.464 \times 1.02 \times 0.8)}$$
    $$= 1.239 \approx 2 \text{ belts.}$$

9.4 CHAIN DRIVE

9.4.1 Introduotion

In order to avoid slipping, steel chains are used. The chains are made up of number of rigid links which are hinged together by pin joints in order to provide the necessity flexibility for wrapping round the

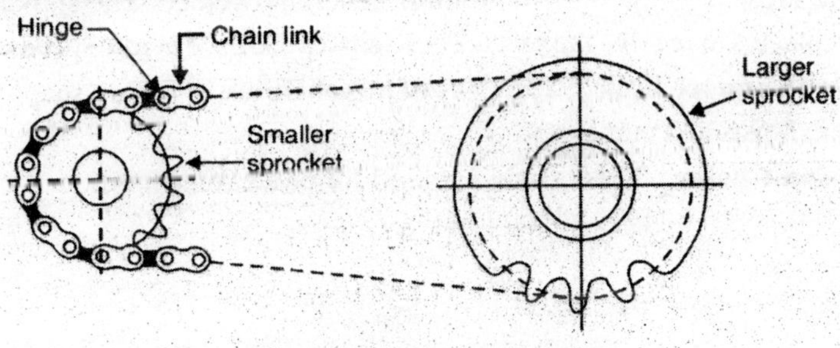

Fig. 9.6 Sprockets and chain.

driving and driven sprockets. The sprockets have projecting teeth of special profile and fit into the corresponding recesses in the links of the chain as shown in Fig. 9.6. The sprockets and chain are constrained to move together without slipping and ensure perfect velocity ratio.

9.4.2 Advantages and Disadvantages of Chain Drive Over Belt and Rope Drives

The advantages and disadvantages of chain drive over belts or ropes are given below :

Advantages :

1. Can be used for short as well as long centre distances.
2. A number of shafts can be driven in the same or opposite directions from a single driving sprocket.
3. More compact drive as compared to belt and rope drives.
4. High efficiency.
5. More positive drive, *i.e.,* less slip.
6. No fire hazards and temperature effects.
7. Can be easily replaced.

Disadvantages :

1. Unsuitable for precise motion due to polygonal effect and wear of joints.
2. Require precise alignment of shafts.
3. Require proper maintenance, lubrication and slack adjustment.
4. Generates noise.

9.4.3 Types of Chains

The chains are classified as :

(a) Hoisting and bauling chains,

(b) Transmission chains.

(a) Hoisting and Hauling Chains : The hoisting and hauling chains are used for lifting of load. These chains are of the following two types as shown in Fig. 9.7.

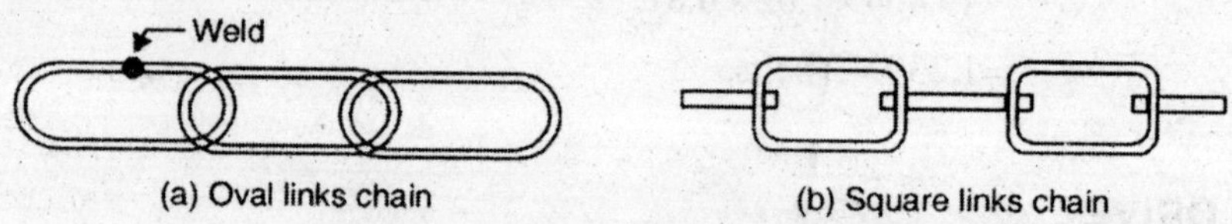

Fig. 9.7 Hoisting and hauling chains.

1. **Chain with Oval Links :** The links of this chain are oval in shape. The cross-section of the link is generally circular—The joint should be perpendicular to the line of action of the load where the stress is minimum. The joint is generally welded. The sprockets for these chains have receptacles to receive the links. These chains are used at low speeds such as chain hoists.
2. **Chains with Square Links :** The links of this chain are of square shape. Such type of chains are used in hoists and cranes, etc.

(b) Transmission Chains : These chain are used for power transmission.

9.4.4 Design of Transmission Roller Chain

The chain on the sprocket of a transmission chain is shown in Fig. 9.8. The pitch (p) of the chain is the distance between the hinge centre of a link and the corresponding hinge centre of the adjacent link. The pitch circle diameter (d) of chain sprocket is the diameter of the circle on which the hinge centres of the chain lie, when the chain is wraped round a sprocket. The points A, B, C and D are the hinge centres of the chain and the circle drawn through these centres is called the pitch circle and its diameter is called the pitch circle diameter.

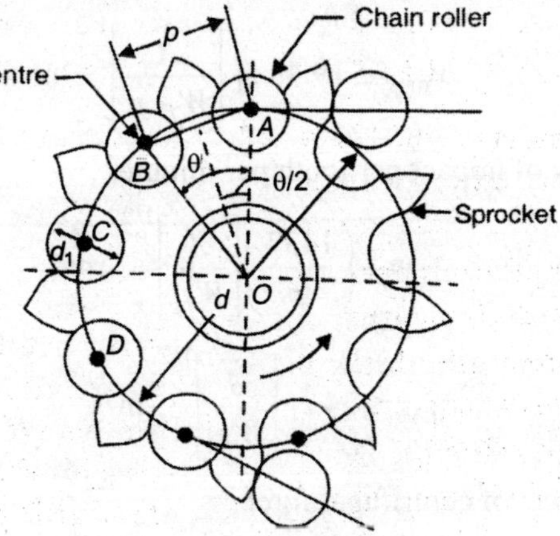

Fig. 9.8 Terms used in chain drive.

1. Transmission Ratio

Transmission ratio,
$$i = \frac{n_1}{n_2} = \frac{d_2}{d_1} = \frac{z_2}{z_2} \quad \ldots(9.56)$$

where $n = rpm$, d = pitch diameter of sprocket, z = number of teeth on sprocket.

2. Average Chain Speed

$$v = \frac{p z_1 n_1}{60 \times 10^3} \text{ m/s} \quad \ldots(9.57)$$

where p = pitch of chain, mm

3. Minimum Number of Teeth on Sprocket

The minimum number of teeth on smaller sprocket are 17 and the maximum number of teeth on the bigger sprocket are 114.

$$z_{\min} = \frac{4d_r}{p} + 5 \text{ for pitches of 16.54 mm} \quad \ldots(9.58a)$$

$$= 4d_r + 4 \text{ for pitches of 3.175 to 5.785 mm} \quad \ldots(9.58b)$$

where d_r = roller pin diameter, mm

4. Maximum Sprocket Speed

(a) Based on allowable amount of impact between a roller and a sprocket, the speed of sprocket, according to Bartlett formula, is given by :

$$n \leq 1170 \left[\frac{A}{(W_f \, p)} \right]^{1/2} \text{ rpm} \qquad \ldots(9.59a)$$

where W_f = weight of chain, N/m

$A = b_1 \times d_r$ = projected area of roller, m^2

b_1 = length of roller, m

p = pitch, m

$$v_{max} \leq 19.48 \, z_1 \left[\frac{A}{W_f \, p} \right]^{1/2} \text{ m/s} \qquad \ldots(9.59b)$$

(b) Based on energy of impact per tooth per minute,

$$n \leq \left(\frac{1437}{p} \right) \left[\frac{A}{W_f} \right]^{1/2} \text{ rpm} \qquad \ldots(9.60a)$$

$$v_{max} \leq 24 \, z_1 \left[\frac{A}{W_f} \right]^{1/2} \text{ m/s} \qquad \ldots(9.60b)$$

(c) Based on the effect of centrifugal force,

$$n \leq \left[\left(\frac{36350}{p} \right) W_f \, z_1 \right]^{1/2} \text{ rpm} \qquad \ldots(9.61a)$$

$$v_{max} \leq 600 \left[\frac{A z_1}{W_f} \right]^{1/2} \text{ m/s} \qquad \ldots(9.61b)$$

5. Chain Pitch

$$p \leq 250 \left(\frac{90}{n_1} \right)^{2/3} \text{ mm} \qquad \ldots(9.62)$$

6. Sprocket Pitch Diameter

$$d = \frac{p}{\sin\left(\frac{180°}{z} \right)} \qquad \ldots(9.63a)$$

Root diameter, $d_f = d - d_r$ ...(9.63b)

7. Centre Distance

The centre distance in a transmission chain is shown in Fig. 9.9

$$C_{min} = K_{min} \, C' \qquad \ldots(9.64)$$

where $C' = \dfrac{d_{a1} + d_{a2}}{2}$

d_a = tip diameter of sprocket

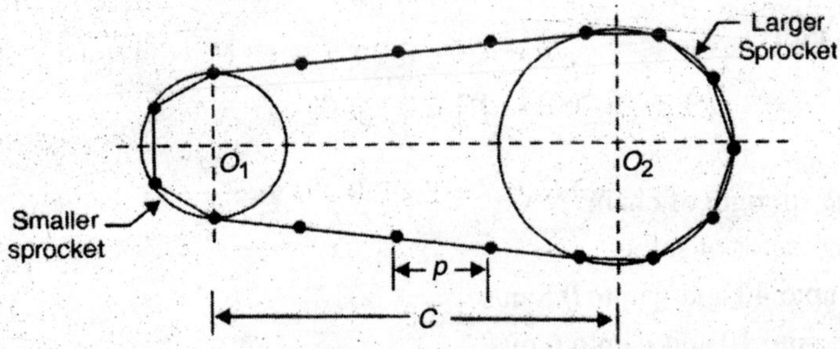

Fig. 9.9 Length of chain.

$$= \frac{p}{\tan\left(\frac{180°}{z}\right)} + 0.6p$$

K_{min} = a constant (Table 16.6)

$$C_{max} = 80p \qquad ...(9.65)$$

Centre distance in pitches,

$$C_p = (20 - 30)\ p \text{ for highly pulsating loads} \qquad ...(9.66a)$$

$$= (30 - 50)\ p \text{ for uniform loads} \qquad ...(9.66b)$$

8. Chain Length

Chain length in pitches,

$$L_p = 2C_p \cos\alpha + \frac{(z_1 + z_2)}{2} + \frac{\alpha\,(z_1 + z_2)}{180°} \qquad ...(9.67a)$$

$$= 2C_p + \frac{(z_1 + z_2)}{2} + \frac{k\,(z_1 - z_2)^2}{C_p} \qquad ...(9.67b)$$

where $\alpha = \sin^{-1}\left[\frac{d_2 - d_1}{C_p}\right]$

k = a variable dependent on $\frac{(z_1 - z_2)}{C_p}$

Chain length,

$$L = 2C\cos\alpha + z_1\,p\left[\frac{180° + 2\alpha}{360°}\right] + z_2\,p\left[\frac{180° - 2\alpha}{360°}\right] \qquad ...(9.68)$$

$$L = mp \qquad ...(9.69)$$

where m = number of chain links

$$= \frac{2C}{p} + \frac{(z_1 + z_2)}{2} + \frac{p\,(z_2 - z_1)^2}{4\pi^2 C}$$

Exact centre distance,

$$C = \left(\frac{P}{4}\right)\left[m - \frac{(z_1 + z_2)}{2}\right] + \left[\left\{m - \frac{(z_1 + z_2)}{2}\right\}^2 - 8\left\{\frac{z_2 - z_1}{(2\pi)}\right\}^2\right]^{1/2} \qquad ...(9.70)$$

9. Allowable Chain Pull

$$F_a = \frac{F_u}{K_f} \quad \text{...(9.71)}$$

where F_u = ultimate strength of chain, N

K_f = working factor of safety

= 5 for z upto 40 and v upto 0.5 m/s

= 18 for z upto 10 and v upto 6 m/s

10. Power Transmitted

$$P = \frac{Fv}{10^3 \times K_s} \text{ kW} \quad \text{...(9.72)}$$

where K_s = service factor.

11. Design Power

$$\text{Design power} = \text{rated power} \times K_s$$

Rated power per strand,

$$\text{kW} = p^2 \left[0.98v - (0.1988v) \left\{ 1 + 50 \sin^2 \left(\frac{90°}{z_1} \right) \right\} \right] \quad \text{...(9.73)}$$

12. Actual Factor of Safety

$$K_{fa} = \frac{F_u}{F + F_c + F_s} \quad \text{...(9.74)}$$

where $F = \dfrac{10^3 \times P \times K_s}{v}$, is the tangential force in N

$F_c = \dfrac{W_f\ v^2}{g}$, is the centrifugal force in N

$F_s = K_{sg}\ W_f C$, is the sag force in N

W_f = weight of chain per meter, N/m

K_{sg} = coefficient of sag

For a safe design, $K_{fa} > K_f$

13. Number of Strands

If $F > F_a$, ten number of strands,

$$i = \frac{F}{F_a} \quad \text{...(9.75)}$$

14. Bearing Pressure on Roller

$$p_b = \frac{F}{A} \quad \text{...(9.76)}$$

where $A = b_1 d_r$ = area of roller.

Power transmitted on the basis of allowable stress,

$$P = \frac{p_b\, Av}{K_s} \qquad \ldots(9.77)$$

where p_b = allowable bearing pressure on rollers, N/mm^2

15. Load on Shaft

$$F_{\text{shaft}} = K_L F \qquad \ldots(9.78)$$

where K_L = load factor.

16. Diametral Dimensions of Wheel Rims and Profile

The diametral dimensions of wheel rims are shown in Fig. 9.10 and wheel rim profile in Fig. 9.11.

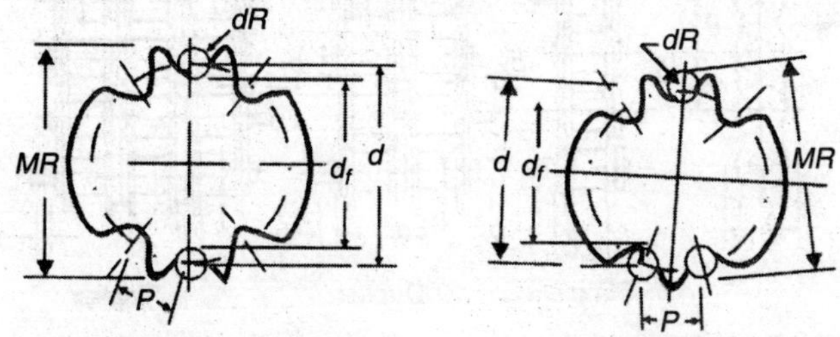

Even number of teeth | Odd number of teeth

p = chordal pitch = chain pitch | d = pitch circle diameter = p (sin 180°/z)

dR = measuring pin diameter | dt = root diameter

z = number of teeth | MR = Measurement over pins

Fig. 9.10 Diametral dimensions of wheel rims.

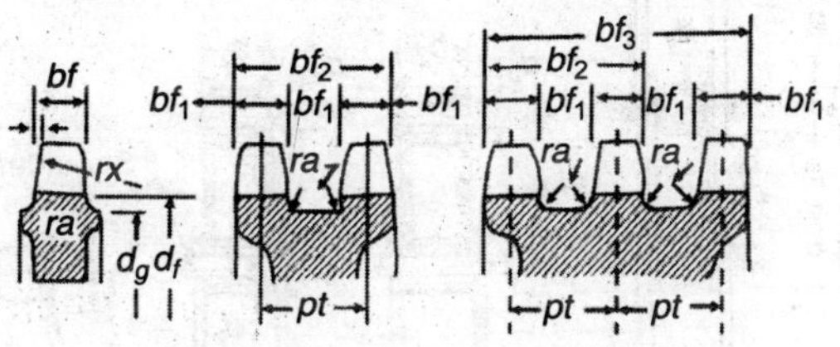

Fig. 9.11 Wheel of rim profile.

17. Wheel Tooth Gap Forms

The wheel tooth gap forms are shown in Fig. 9.12.

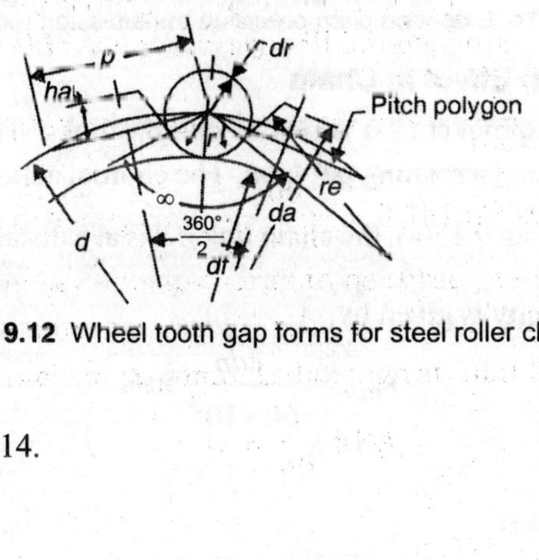

Fig. 9.12 Wheel tooth gap forms for steel roller chain.

18. Sprocket Proportions

Refer to Figs. 9.13 and 9.14.

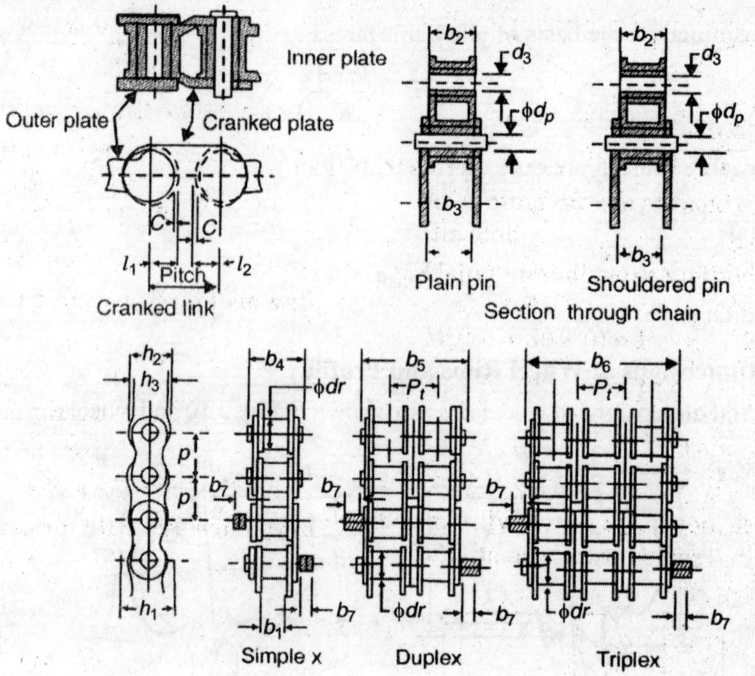

Fig. 9.13 Dimensions for transmission steel roller chains.

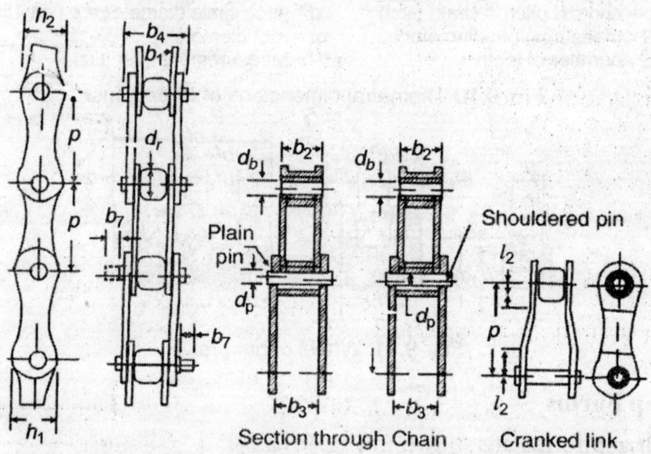

Fig. 9.14 Extended pitch precision transmission roller chain.

9.4.5 Chordal (or Polygonal) Effect in Chain

The chain passes around the sprocket as a series of chordal links. This action is similar to that of a non-slipping belt wrapped around a rotating polygon. The chordal action on a sprocket having only four teeth is shown in Fig. 9.15. In Fig. 9.15(a), the chain link AB is at a distance of $\left(\frac{d}{2}\right)$ from the centre of the sprocket wheel. Its linear velocity is given by,

$$v_{max} = \frac{\pi d n}{60 \times 10^3} \text{ m/s} \quad \ldots(1)$$

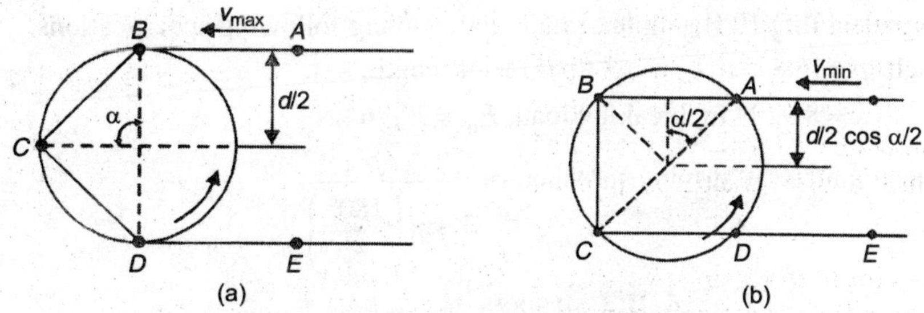

Fig. 9.15 Chordal action on a chain sprocket.

As the sprocket rotates through an angle $\left(\frac{\alpha}{2}\right)$, the position of the chain link AB is shown in Fig. 9.15(b). In this case, the link is at a distance of $\left(\frac{d}{2}\right)\cos\left(\frac{\alpha}{2}\right)$ from the centre of the sprocket and its linear velocity is given by,

$$v_{min} = \frac{\pi dn \cos\left(\frac{\alpha}{2}\right)}{60 \times 10^3} \text{ m/s} \qquad \text{...(2)}$$

Therefore, the linear speed of the chain is not uniform but varies from v_{max} to v_{min} during every cycle of tooth engagement. This results in a pulsating and jerky motion. The variation in velocity is given by,

$$(v_{max} - v_{min}) \propto \left[1 - \cos\left(\frac{\alpha}{2}\right)\right]$$

or

$$\propto \left[1 - \cos\left(\frac{180°}{z}\right)\right]$$

In order to decrease the velocity variation, the number of teeth on the sprocket should be increased. For smooth operation at moderate and high speeds, it is considered a good practice to use a driving sprocket with at least 17 teeth. From durability and noise considerations, the minimum number of teeth on the driving sprocket is 19 or 21. It is better to use odd number of teeth on the driving sprocket with an even number of chain links for more uniform wear. This also ensures that the same link does not come into contact with the same sprocket teeth in every rotation. The maximum number of teeth recommended on the sprocket wheel are 100 to 120.

■ **Example 9.4** Design a chain drive to connect a 7.5 kW, 1440 rpm electric motor to a centrifugal pump running at 720 rpm. The service conditions involve moderate shocks.

Solution : Given : $P = 7.5$ kW, $n_1 = 1440$ rpm, $n_2 = 720$ rpm, $i = 2$

1. Preliminary decisons : In order to reduce polygonal effect, the number of teeth on the driving sprocket for transmission ratio 2 are,

$$z_1 = 27$$

It is further assumed that the chain is a simplex roller chain with a single strand.

2. Minimum factor of safety, $K_f = 11.7 + \frac{(13.2 - 11.7)(1440 - 1200)}{400} = 12.6$

3. We select ISO 10 B simplex roller chain having following specifications.
   Pitch, $p = 15.875$ mm, weight per meter length,

   $w = 8.92$ N/m, breaking load, $F_u = 22.26$ kN

4. Pitch diameter of driving sprocket, $d_1 = \dfrac{p}{\sin\left(\dfrac{180^\circ}{z_1}\right)}$

$$= \frac{15.875}{\sin\left(\dfrac{180^\circ}{27}\right)} = 136.74 \text{ mm}$$

5. Number of teeth on driven sprocket,

$$z_2 = i \times z_1 = 2 \times 27 = 54$$

6. Pitch line velocity,

$$v = \frac{\pi d_1 n_1}{60 \times 10^3} = \frac{\pi \times 136.74 \times 1440}{60 \times 10^3} = 10.31 \text{ m/s}$$

7. Center distance, $C = 40p = 40 \times 15.875 = 635$ mm

   Length of chain in pitches,

$$L_p = \frac{2C}{p} + \frac{z_1 + z_2}{2} + \frac{p(z_2 - z_1)^2}{4C\pi^2}$$

$$= 2 \times 40 + \frac{(27 + 54)}{2} + \frac{15.875\,(54 - 27)^2}{4 \times 635 \times \pi^2}$$

$$= 120.96 \text{ or } 121 \text{ links}$$

8. Service factor :

$$K_s = K_1 K_2 K_3 K_4 K_5 K_6$$

   $K_1 = 1.25$ for load with minor shocks

   $K_2 = 1.0$ for adjustable supports

   $K_3 = 1.0$ for $C = (30 - 50)\,p$

   $K_4 = 1.0$ for horizontal drive

   $K_5 = 1.0$ for drop lubrication

   $K_6 = 1.0$ for single shift of 8 hours per day

   $K_s = 1.25 \times 1 \times 1 \times 1 \times 1 \times 1 = 1.25$

9. Power transmitted on the basis of F_u

$$P = \frac{F_u v}{n K_s} = \frac{22260 \times 10.31}{12.6 \times 1.25 \times 10^3} = 14.571 \text{ kW}$$

10. Tangential force, $F = \dfrac{10^3 P}{v} = \dfrac{10^3 \times 7.5}{10.31} = 727.45$ N

    Centrifugal force, $F_c = \dfrac{wv^2}{g} = \dfrac{8.92 \times (10.31)^2}{9.81} = 96.65$ N

Sag tension, $F_s = K_{sag}\, wC$

$K_{sag} = 6$ for horizontal position of drive

$$F_s = 6 \times 8.92 \times 0.635 = 33.985 \text{ N}$$

11. Actual factor of safety, $K_{fa} = \dfrac{F_u}{F + F_c + F_s}$

$$= \frac{22260}{727.45 + 96.65 + 33.985}$$

$$= 25.94 > 12.6$$

Hence the design is safe.

9.5 ROPES

Ropes are used for long distance transmission and lifting of loads in mines, drilling rigs, multistorey building construction, etc. Basically there are two types of ropes : fibre ropes and metallic ropes. Fibre ropes are made of hemp, cotton, sisal, manila, jute, nylon, poly-propylene and coir. Metallic ropes are made of steel, aluminium alloys, copper, bronze and stainless steel. Fibre ropes are generally used where flexibility is of utmost importance, low power transmission, as in instrument devices. Metallic ropes are generally made of steel wire with either fibre core or metallic core.

9.5.1 Selecting Fibre Ropes

Fibre ropes are selected based on the load to be lifted. A factor of safety of about 35 is acceptable. The velocity should be from 15 – 30 m/s. The diameter of rope sheave should be (36 – 40) d, where d = rope diameter. The number of grooves on the pulley should not be more than 24. Fibre ropes in pulley groove are shown in Fig. 9.16.

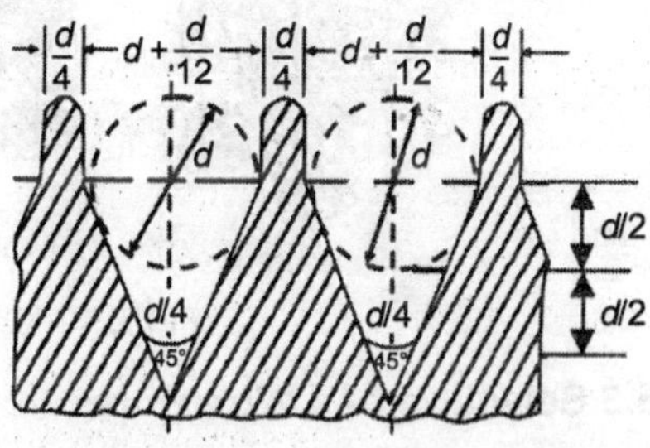

Fig. 9.16 Fibre rope pulley grooves.

Length of rope,

$$L = \frac{\pi (D_1 + D_2)}{2} + (D_2 - D_1) \sin^{-1} + \left[\frac{D_2 - D_1}{2C}\right] + 2C \cos \beta \qquad \ldots(9.79)$$

where D_1, D_2 = diameters of smaller and bigger pulley respectively

C = centre distance

β = semi-groove angle of pulley

Minimum centre distance,

$$C_{min} = (D_2 - D_1) + 1.5\, D_2 \qquad \ldots(9.80)$$

Considering centrifugal tension,

$$\frac{F_1 - F_c}{F_2 - F_c} = \exp(\mu_e \theta) \qquad \ldots(9.81)$$

where $\mu_e = \dfrac{\mu}{\sin \beta}$

Total power transmitted, $$P = \frac{n(F_1 - F_2)v}{10^3} \text{ kW} \quad ...(9.82)$$

where n = number of ropes

v = rope velocity, m/s

Fibre ropes are joined together either by splicing or with the help of eyelet thimbles.

9.5.2 Steel Wire Ropes

A wire rope is made up of strands and a strand is made up of one or more layers of wires. The number of strands in a rope denotes the number of groups of wires that are laid over the central core. For example a 6×19 wire rope construction means that the rope has 6 strands and each strand is composed of 19 (12/6/1) wires. It means that there are 114 wires in the rope. The tensile designation of the wire rope corresponds to the minimum tensile strength of the wire in N/mm^2.

Some of the steel wire ropes are shown in Fig. 9.17.

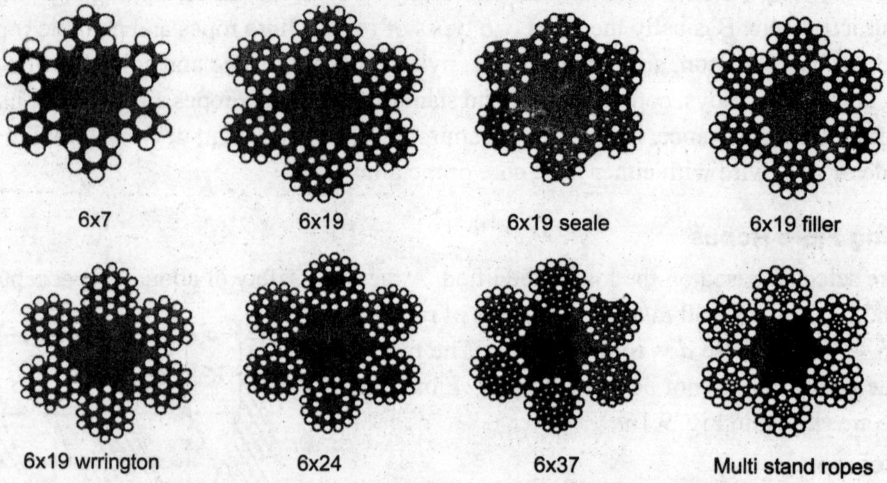

Fig. 9.17 Examples of rope construction, designation.

9.5.3 Selection of a Wire Rope

The following points must be considered while making a choice for a wire rope :

1. Strength
2. Abrasion resistance
3. Flexibility
4. Resistance to crushing
5. Fatigue strength, and
6. Corrosion resistance.

1. Number of Bends

When a wire rope passes over drums and pulleys, it bends. It has been found that rope life is approximately inversely proportional to the number of bends. One bend means the transition of the rope either from its straight position to a bent one or from a bent position into a staight one. Reverse bending is equal to two single bends towards the same side.

2. Diameter of Wire Rope

The diameter of wire rope is approximately given by,

$$d = 1.5\, d_w \sqrt{i} \qquad \text{...(9.83)}$$

where i = number of wires inthe rope

d_w = wire diameter

3. Stress in the Wire Rope

Individual wires in a loaded rope experience a complex state of stress consisting of tension, bending and twisting combined with mutual compression and rubbing of the wires and strands. The stress in the wire rope is given by,

$$\sigma = \frac{\sigma_{br}}{K_f} = \frac{W}{A} + \frac{d_w E_r}{D_{min}} \qquad \text{...(9.84)}$$

where σ_{br} = ultimate breaking strength of the rope wire material, N/mm^2

K_f = rope factor of safety

A = area of useful cross-section of the rope, mm^2

W = tension in the rope, N

$E_r = \frac{3E}{8}$, corrected elastic modulus of rope accounting for torsion effects in the wires, N/mm^2

4. Maximum Allowable Tension in the Rope

$$W_a = \frac{F_{br}}{K_f} \text{ N} \qquad \text{...(9.85)}$$

where F_{br} = breaking strength of the rope, N.

5. Steel Rope Weight and Breaking Load

The approximate steel rope weight is given by,

$$W_f = Kd^2 \text{ N/m length} \qquad \text{...(9.86)}$$

where K = empirical factor for weight of steel cored ropes

d = diameter of rope, mm

Minimum breaking load,

$$F_{br} = \frac{K' d^2 R_0}{10^3} \text{ N}$$

where R_0 = tensile grade of wire, N/mm^2 ...(9.87)

K' = empirical factor for the minimum breaking load for a given rope construction.

9.5.4 Stresses in Hoisting Ropes

The following stresses are induced in hoisting ropes :

1. Direct tensile stress due to load hoisted and the weight of the rope.
2. Stresses due to bending of the rope over the sheave.

3. Starting stresses, and
4. Stresses due to change of rope speed including stops.

1. Direct Tensile Stress

The direct tensile stress is given by,

Effective load, $W_e = W + W_r$

$$\sigma_d = \frac{W + W_r}{A} \quad \text{...(9.88)}$$

where W = load lifted

W_r = self weight of the rope

A = area of cross-section of the rope

$= \frac{i \pi d_w^2}{4}$

2. Bending Stresses

The bending stresses are induced in the rope when the rope passes over the sheaves and drums. The bending stress is given by,

$$\sigma_b = \frac{d_w E_r}{D} \quad \text{...(9.89)}$$

where D = sheave diameter

Bending load, $W_b = \sigma_b \times A$

Effective load, $W_e = W + W_r + W_b$

3. Starting Stresses

If the rope starts from rest without any slackness and acquires a velocity T m/s in t seconds, then acceleration of rope is,

$$a = \frac{v}{t}$$

Acceleration load, $W_a = \frac{(W + W_r)\, a}{g}$

Effective load, $W_e = W + W_r + W_b + W_a$ ...(9.90)

If there is a slack in the rope of amount h before starting, then impact load in the rope during starting is given by,

$$W_{st} = (W + W_r)\left[1 + \left\{1 + \frac{v_{st}^2 E_r}{\sigma L_g}\right\}^{1/2}\right] \quad \text{...(9.91)}$$

where v_{st} = velocity of rope at the instant when the rope is taut = $(2ah)^{1/2}$

$\sigma = \frac{W + W_r}{A}$, is the static stress in the rope

L = length of the rope

Effective load during starting,

$$W_e = W_{st} + W_b \qquad ...(9.92)$$

4. Stresses due to Change in Speed or Sudden Stopping of Hoisting Drum

(a) Let the speed of rope change from v_1 to v_2 in time t. Then acceleration of rope.

$$a = \frac{v_2 - v_1}{t}$$

Acceleration load, $$W_a = \frac{(W + W_r)\,a}{g} \qquad ...(9.93)$$

Effective load, $$W_e = W + W_r + W_a + W_b \qquad ...(9.94)$$

(b) Let x = distance in which the load being lifted is stopped.

L = length of the rope

Then $$x_s = \frac{(W + W_r)\,L}{AE_r}$$

$$\frac{x}{x_s} = \frac{\sigma_{sud}}{\sigma}$$

where σ_{sud} = sudden stress in the rope

Sudden load, $$W_{sud} = \sigma_{sud} \times A$$

Effective load, $$W_e = W + W_r + W_b + W_{sud} \qquad ...(9.95)$$

5. Factor of Safety

$$\text{Factor of safety} = \frac{\text{Breaking load}}{\text{Maximum effective load on the rope}} \qquad ...(9.96)$$

■ **Example 9.5** Select a suitable steel wire rope for a mine hoist carrying a load of 2 tonnes to be lifted from a depth of 100 m. A rope speed of 10 m/s must be attained in 10 seconds.

Solution :

1. Preliminary decisions :

   Minimum factor of safety for mine ropes = 10

   The rope is also subjected to sudden and bending loads; each of which double the load. Therefore, a nominal factor of safety of about 40 may be assumed.
2. Breaking load for selection of rope, $W_{br} = 2 \times 40 = 80$ tonnes ≈ 785 kN
3. We find that 6 × 19 (9/9/1) steel wire rope with swtcel core and 1570 tensile designation of wire gives minimum breaking load of 898 kN with nominal diameter of 40 mm.

   Approximate mass = 656 kg/100 m
4. Average diameter of sheave, $D = 45d = 45 \times 40 = 1800$ mm.
5. Bending load, $$W_b = \frac{Ae_r\, d_w}{D}$$

$$A = 0.38\, d^2$$

$$E_r = 64700\ \text{N/mm}^2$$

$$d_w = 0.063\, d$$

$$W_b = \frac{(0.38 \times 40^2) \times 64700 \times 0.063 \times 40}{1800}$$

$$= 55073 \text{ N}$$

6. Acceleration, $a = \dfrac{v}{t} = \dfrac{10}{10} = 1 \text{ m/s}^2$

7. Weight of the rope, $W_r = \dfrac{656 \times 9.81 \times 100}{100} = 6.435 \text{ kN}$

   Weight to be hoisted, $W = 2 \times 9.81 = 19.62$ kN

8. Acceleration load, $W_a = \dfrac{(W + W_r)\, a}{g} = \dfrac{(19.62 + 6.435) \times 1}{9.81} = 2.656 \text{ kN}$

9. Effective load, W_e

   (a) Starting load without slack, $W_{st} = 2\,(W + W_r) = 2\,(19.62 + 6.435) = 52.11$ kN

   $$W_e = W_{st} + W_b = 52.11 + 55.073 = 107.183 \text{ kN}$$

   (b) During the first 10 seconds, after starting

   $$W_e = W + W_r + W_b + W_a$$
   $$= 19.62 + 6.435 + 55.073 + 2.656 = 83.784 \text{ kN}$$

   (c) Maximum effective load, $(W_e)_{max} = 107.183$ kN

10. Operating factor of safety $= \dfrac{\text{Breaking strength}}{(W_e)_{max}}$

   $$= \frac{898}{107.183} = 8.378 < 10$$

   We observe that the operating factor of safety is less than the minimum desired. If we increase the sheave diameter to 3000 mm, then bending load,

   $$W_b = 33.044 \text{ kN}$$
   $$W_e = 75.755 \text{ kN giving FOS} = 11.85 > 10$$

9.5.5 Wire Rope Sheaves

Small sheaves are usually cast in one piece without ribs. Large sheaves are provided with ribs and holes or with cross-shaped spokes. The arms of the sheave may be elliptical in cross-section. Sheaves are freely mounted on fixed axles on anti-friction bearings or bronze bushings. The ratio of pulley hub length to axle diameter should be 1.5 to 1.8. The unit bearing pressure is :

$$p = \frac{W}{l_a\, d_a} \qquad \text{...(9.97)}$$

where l_a = length of axle bushing

d_a = diameter of sheave axle

9.5.6 Steel Wire Rope Drums

With a power drive the drum should always be provided with helical grooves so that the rope winds up uniformly and is less subject to wear. The radius of the helical grooves should be selected so as to prevent jamming of the rope. Annular or concentric grooves should not be greater in depth than 10 per cent of the rope diameter.

Drums with one coiling rope have only one helix—right hand; drums designed for two rope members are provided with two helices—right hand and left hand (Fig. 9.18). The number of turns on the drum for one rope member is :

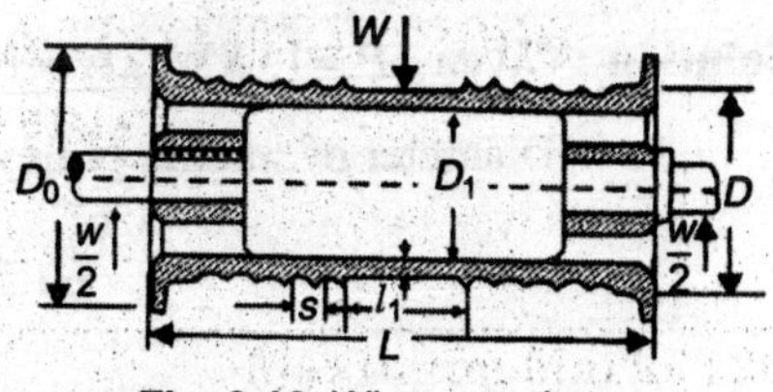

Fig. 9.18 Wire rope drum.

$$z = \frac{Hi}{\pi D} + 2 \qquad \text{...(9.98)}$$

where i = ratio of pulley system

D = drum diameter

H = height to which the load is raised

The factor 2 accounts fot the holding turns.

Length of drum, $$L = \left(\frac{Hi}{\pi D} + 7\right) s \text{ for one rope} \qquad \text{...(9.99a)}$$

$$= \left(\frac{2Hi}{\pi D} + 12\right) s + l_1 \text{ for two ropes} \qquad \text{...(9.99b)}$$

where s = pitch of the helix

l_1 = space in the middle of the drum between the right hand and left hand helices, dictated by design.

Wall thickness of a cast iron drum,

$$w = 0.02D + 6 \text{ to } 10 \text{ mm} \qquad \text{...(9.100)}$$

Outside diameter of drum, $$D_0 = D + 6d \qquad \text{...(9.101)}$$

9.5.7 Stresses in Rope Drums

Compressive stress in drum, $$\sigma_c = \frac{W}{w \times s} \qquad \text{...(9.102)}$$

Maximum bending stress, $$\sigma_b = \frac{8WLD}{\pi(D^4 - D_i^4)} \qquad \text{...(9.103)}$$

Maximum torque on drum, $$T_{max} = \frac{W(D + d)}{2} \qquad \text{...(9.104)}$$

Maximum shear stress, $$\tau_{max} = \frac{16T_{max} \times D}{\pi(D^4 - D_i^4)} \qquad \text{...(9.105)}$$

Rope drums are generally made of grey cast iron for which maximum principal stress theory is applicable. Thus maximum stress in the drum becomes :

$$\sigma = \sqrt{\sigma_b^2 + 4\tau_{max}^2 + \sigma_c^2} \le \sigma_{allowable} \qquad \text{...(9.106)}$$

■ **Example 9.6** A belt is required to transmit 15 kW from a pulley 1.2 m diameter at 300 rpm to another pulley which runs at 500 rpm. The centre distance is 3 m. The centre line of drive makes an angle of 50° with horizontal. Design a suitable flat leather belt for the drive. Also design the larger pulley.

Solution : Given : $P = 15$ kW, $D_2 = 1.2$ m, $N_2 = 300$ rpm, $N_1 = 500$ rpm, $C = 3$ m

1. **Diameter of smaller pulley,**

$$D_1 = D_2 \times \frac{N_2}{N_1} = \frac{1.2 \times 300}{500} = 0.72 \text{ m}$$

2. **Angle of ace of contact,** $\theta = \pi - 2\sin^{-1}\frac{(D_2 - D_1)}{2C}$

$$= \pi - 2\sin^{-1}\frac{(1.2 - 0.72)}{2 \times 3}$$

$$= 170.8° \text{ or } 2.981 \text{ red}$$

3. **Length of open belt,** $L_0 = 2C + \frac{\pi}{2}(D_1 + D_2) + \frac{(D_2 - D_1)^2}{4C}$

$$= 2 \times 3 + \frac{\pi}{2}(0.72 + 1.2) + \frac{(1.2 - 0.72)^2}{4 \times 3}$$

$$= 9.035 \text{ m}$$

4. **Pitch line velocity,** $v = \frac{\pi D_2 N_2}{60} = \frac{\pi \times 1.2 \times 300}{60} = 18.85$ m/s

5. **For a leather belt, working stress,**

$$\sigma_w = 2.25 \text{ MPa; density } \rho = 950 \text{ hg/m}^3$$

Coefficient of friction of leather on cast iron pulley = 0.35.

Let belt thickness, $t = 5$ mm

6. **Mass of belt per metre length,** $m = \rho bt = 950 \times b \times 5 \times 10^{-6} \times 1$

$$= 4.75 \times 10^{-3} \times b \text{ kg/m}$$

7. **Centrifugal tension,**

$$F_c = mv^2 = 4.75 \times 10^{-3} \times b \times (18.85)^2 = 1.6878 \times b \text{ N}$$

8. $\frac{F_1 - F_c}{F_2 - F_c} = e_{xp}(\mu\,\theta) = e_{xp}(0.35 \times 2.981) = 2.8387$

9. $F_1 = \sigma_w \times bt = 2.25 \times b \times 5 = 11.25 \times b$ N

10. **Tangential force,** $F_1 - F_2 = \frac{P \times 10^3}{v} = \frac{15 \times 10^3}{18.85} = 795.756$ N

11. $\frac{11.25\,b - 1.6878\,b}{(11.25\,b - 795.756) - 1.6878\,b} = 2.8387$

$$\frac{9.5622\,b}{9.5622\,b - 795.756} = 2.8387$$

$$9.5622\,b = 27.144\,b - 2258.91$$

$$b = 128.5 \text{ mm or 4 ply 140 mm}$$

$\therefore$ **Belt width,** $b = 140$ mm

Belt thickness, $t = 5$ mm

Belt length, $L_0 = 9035$ mm

12. Design of cast iron pulley

Pulley dia, $D = 1.2$ m

Rim size :

Width, $B = 1.196 + 10 \text{ mm} = 1.19 \times 140 + 10 = 176.6 \simeq 180$ mm

Thickness, $t_r = 0.25\sqrt{D} + 1.5$ mm for light duty

$$= 0.25\sqrt{1200} + 1.5 \simeq 10 \text{ mm}$$

or $t_r = \dfrac{D}{200} + 3 \text{ mm} = \dfrac{1200}{200} + 3 = 9$ mm

Adopt $t_r = 10$ mm

Number of arms, $n = 6$ for $D > 450$ mm.

Cross-section of arms :

These cross-section of arms is elliptical.

Major axis of arms near boss,

$$b_1 = 2.94\left[\frac{BD}{4n}\right]^{1/3}$$

$$= 2.94\left[\frac{180 \times 1200}{4 \times 6}\right]^{1/3} = 61 \simeq 66 \text{ mm}$$

Minor axis of arms near boss,

$$a_1 = \frac{b_1}{2} = 32.5 \simeq 33 \text{ mm}$$

Cross-section of arms near rim :

$$b_2 = \frac{2}{3}b_1 = 44 \text{ mm}$$

$$a_2 = \frac{2}{3}a_1 = 22 \text{ mm}$$

Radius near rim, $r_1 = r_2 = \dfrac{b_1}{2} = 33$ mm

Size of boss :

Minimum length of bore, $l = \dfrac{2B}{3} = \dfrac{2}{3} \times 180 = 120$ mm

Shaft size :

Resultant pull on pulley due to belt tensions,

$$P = \left(\frac{F}{K-1}\right)[K^2 + 1 + 2K\cos(180° - \theta)]^{1/2}$$

where $K = \dfrac{F_1 - F_c}{F_2 - F_c} = 2.8387$

$$F = F_1 - F_2 = 795.756 \text{ N}$$

$$\theta = 170.8°$$

$$P = \left[\frac{795.756}{2.8387-1}\right][(2.8387)^2 + 1 + 2 \times 2.8387 \times \cos(180° - 170.8°)]^{1/2}$$

$$= 1657 \text{ N}$$

Weight of pulley rim, $W_r = \pi D_2 \times B \times t_r \times P_{CI}$

$$= \pi \times 1.2 \times 180 \times 10 \times 10^{-6} \times 7800 = 53 \text{ kg}$$

Weight of 6 arms, $W_a = \pi \, abl_a \times P_{CI}$

$$= \pi\left(\frac{33+22}{2}\right)\left(\frac{44+66}{2}\right) \times 10^{-6} \times 0.5 \times 7800$$

$$= 18.5 \text{ kg}$$

Let weight of *hub* $\simeq 30$ kg

Weight of pulley, $W = 53 + 18.5 = 71.5 \text{ kg} \simeq 100 \text{ kg}$

Angle between P and W, $\alpha = 90° + 50° = 140°$

Resultant force on pulley, $S = [P^2 + W^2 + 2PW \cos 140°]^{1/2}$

$$= [(1657)^2 + (100)^2 + 2 \times 1657 \times 100 \times \cos 140°]^{1/2}$$

$$= 1582 \text{ N}$$

Let overhang of pulley, $x = 150$ mm

Bending moment on pulley shaft,

$$M = S_x = 1582 \times 0.15 = 237.3 \text{ N.m}$$

Torque on shaft,

$$T = F \times \frac{D_2}{2} = 795.756 \times 0.6 = 477.5 \text{ N.m}$$

Equivalent torque, $T_e = [(K_m \ M)^2 + (K_t \ T)^2]^{1/2}$

$$= [(1.5 \times 237.3)^2 + (1.25 \times 477.5)^2]^{1/2}$$

$$= 695 \text{ N.m}$$

$$= \frac{\pi}{16} d_2^3 \ \tau_s = \frac{\pi}{16} d_2^3 \times 42 \times 10^6$$

$$d_2 = 44 \text{ mm} \simeq 50 \text{ mm}$$

Diameter of pulley boss; $d_1 \simeq 2d_2 = 100$ mm

■ **Example 9.7** A 40 kW, 1000 rpm A.C. motor is to drive a pump at a speed of 300 rpm. The pump is for 12 hr per day service. The centre distance is to be 1.25 m. Determine the details of a multiple V-belt drive, *i.e.,* pulleys diameter, Number of belts. Also draw a neat sketch.

Solution : Given : $P = 40$ kW, $N_1 = 1000$ rpm, $N_2 = 300$ rpm, $C = 1.25$ m, service = 12 h/day.

1. Cross-section of belt :

   For $P = 3.48$ to 51.8 kW, E cross-section of V-belt is recommended.

   Top width of belt, $w = 38$ mm (nominal)

   Thickness of belt, $t = 23$ mm (nominal)

Minimum pulley diameter, $D_1 = 533 \text{ mm} \simeq 540 \text{ mm}$ (say)
Weight per metre of belt = 8.885 N/m

2. Diameter of pump pulley,

$$D_2 = 540 \times \frac{1000}{300} = 1800 \text{ mm}$$

3. Belt speed, $v = \dfrac{\pi D_1 N_1}{60 \times 10^3} = \dfrac{\pi \times 540 \times 100}{60 \times 10^3} = 28.3 \text{ m/s}$

4. Pitch length of belt, $L = 2C + \dfrac{\pi}{2}(D_1 + D_2) + \dfrac{D_2 - D_1}{4C}$

$$= 2 \times 1.25 + \frac{\pi}{2}(0.54 + 1.8) + \frac{(1.8 - 0.54)^2}{4 \times 1.25}$$

$$= 6.493 \text{ m} = 6493 \text{ mm}$$

The standard nominal pitch length of E-belt nearest to 6493 mm is 6889 mm and nominal inside length = 6807 mm.

5. Exact centre distance, $C = A + \sqrt{A^2 - B}$

where $A = \dfrac{L_p}{4} - \dfrac{\pi(D_1 + D_2)}{8}$

$$= \frac{6889}{4} - \frac{\pi(540 + 1800)}{8} = 803.3 \text{ mm}$$

$$B = \frac{(D_2 - D_1)^2}{8} = \frac{(1800 - 540)^2}{8} = 198450$$

$$C = 803.3 + \sqrt{(803.3)^2 - 198450} = 1471.7 \text{ mm or } 1.4717 \text{ m}$$

6. Belt size : E-6807; V-belt

7. Angle of arc of contact, $\theta = 180° - 2\sin^{-1}\left[\dfrac{D_2 - D_1}{2C}\right]$

$$= 180° - 2\sin^{-1}\left[\frac{1800 - 540}{2 \times 1471.7}\right] = 129.3°$$

8. Speed ratio $= \dfrac{100}{300} = 3.33$

Small diameter factor, $K_b = 1.14$
Correction factor for industrial service, $K_a = 1.1$
Correction factor for belt length, $K_c = 0.99$
Correction factor for arc of contact, $K_d = 0.86$

9. Equivalent pitch diameter, $d_e = D_1 \times K_b = 540 \times 1.14 = 615.6 \text{ mm}$
Maximum value of d_e in power rating formula for E-belt = 700 mm.

Power rating for E-belt $= \left(4.58\, v^{-0.09} - \dfrac{952}{d_e} - 7.05 \times 10^{-4}\, v^2\right) v$

$$= \left[4.58\,(28.3)^{-0.09} - \frac{952}{615.6} - 7.05 \times 10^{-4}\,(28.3)^2 \right] \times 28.3$$

$$= 36.2 \text{ kW}$$

10. Number of belts $= \dfrac{P \times K_a}{\text{kW rating} \times K_c \times K_d}$

$$= \frac{40 \times 1.1}{36.2 \times 0.99 \times 0.86} = 1.43 \simeq 2$$

11. Pulley groove size : (See Fig. 9.19)

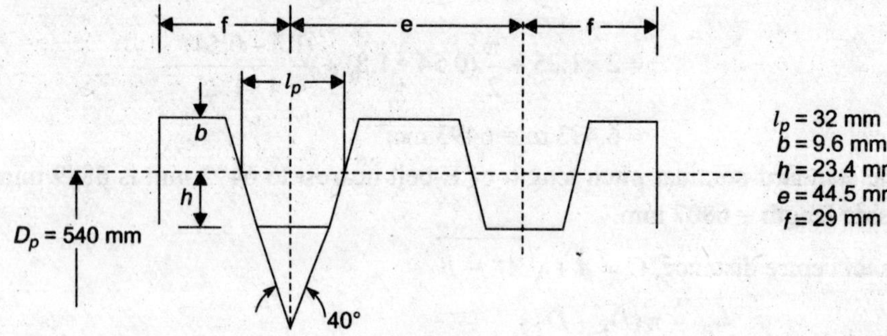

Fig. 9.19

■ **Example 9.8** Design a chain drive for the following duty :

Power to be transmitted	= 6 kW
Speed of motor shaft	= 1000 rpm
Speed of driven shaft	= 500 rpm
Centre distance	= 600 mm

Solution : Given : $P = 6$ kW, $n_1 = 1000$ rpm, $n_2 = 500$ rpm, $C = 600$ mm

1. Transmission ratio, $i = \dfrac{n_1}{n_2} = \dfrac{1000}{500} = 2$

   Minimum number of teeth on driving sprochet for $i = 2$ are, $z_1 = 27$

   Assume that the chain is a simplex roller chain with a single strand.
2. Factor of safety for $n_1 = 1000$ rpm for bush roller chain $K_f = 11$ to $16.3 \simeq 11$
3. We select ISO-10 B transmission steel roller chain having following specifications :

   Pitch, $p = 15.875$ mm

   Weight per metre length, $W_f = 8.92$ N/m,

   Breaking load, $F_u = 22.26$ kN
4. Pitch diameter of driving sprochet,

$$d_1 = \frac{p}{\sin\left(\dfrac{180°}{z_1}\right)} = \frac{15.875}{\sin\left(\dfrac{180°}{27}\right)} = 136.74 \text{ mm}$$

5. Number of teeth on driven sprochet,

$$z_2 = i\, z_1 = 2 \times 27 = 54$$

6. Pitch line velocity, $v = \dfrac{\pi d_1 n_1}{60 \times 10^3} = \dfrac{\pi \times 136.74 \times 1000}{60 \times 10^3} = 7.16 \text{ m/s}$

7. Length of chain in pitches, $L_p = \dfrac{2C}{p} + \dfrac{z_1 + z_2}{2} + \dfrac{p(z_2 - z_1)^2}{4C\pi^2}$

$$= \frac{2 \times 600}{15.875} + \frac{27 + 54}{2} + \frac{15.875\,(54 - 27)^2}{4 \times 600 \times \pi^2}$$

$$= 120.97 \simeq 121 \text{ links}$$

8. Service factor : $K_s = K_1 K_1 K_3 K_4 K_5 K_6$

$K_1 = 1.25$ for load with minor shocks
$K_2 = 1.0$ for adjustable supports
$K_3 = 1.0$ for $C = (30 - 50)\,p$
$K_4 = 1.0$ for horizontal drive
$K_5 = 1.0$ for drop lubrication
$K_6 = 1.0$ for single shift of 8 hours per day
$K_s = 1.25 \times 1.0 \times 1.0 \times 1.0 \times 1.0 \times 1.0 = 1.25$

9. Power transmitted on the basis of F_u

$$P = \frac{F_u\, v}{K_f\, K_s} = \frac{22.26 \times 7.16}{11 \times 1.25} = 11.59 \text{ kW} > 6 \text{ kW}$$

10. Tangential force, $F_t = \dfrac{P \times 10^3}{v} = \dfrac{6 \times 10^3}{9.81} = 838 \text{ N}$

Centrifugal force, $F_c = \dfrac{w_f\, v^2}{g} = \dfrac{8.92 \times (7.16)^2}{9.81} = 46.6 \text{ N}$

Tension due to sag, $F_s = K_{sag}\, w_f C$

$K_{sag} = 6$ for horizontal drive

$$F_s = 6 \times 8.92 \times 0.6 = 32.11 \text{ N}$$

11. Actual factor of safety,

$$K_{fa} = \frac{F_u}{F_t + F_c + F_s} = \frac{22260}{838 + 46.6 + 32.11} = \frac{22260}{916.71} = 24.38$$

Now, $K_{fu} > K_f$

Hence design is safe.

SOLVED QUESTIONS FROM UNIVERSITY EXAMINATIONS

■ **Example 9.9** Select a wire rope for a vertical mine hoist to lift a load of 55 kN from a depth of 300 m. A rope speed of 500 m/min is to be attained in 10 seconds. [MDU, Dec., 2007]

Solution : Given : $W = 55$ kN, depth, $h = 300$ m, $v = 500$ m/min, $t = 10$ s.

1. Preliminary decisions : The minimum factor of safety for mine ropes is 10. The rope is also subjected to sudden and bending loads. Therefore, a nominal factor of safety of about 40 may be assumed.

2. Breaking load for selection of rope,

$$W_{br} = 55 \times 40 = 2200 \text{ kN}$$

3. We find that $6 \times 19 (9/9/1)$ steel wire rope with steel core and 1960 tensile designation of wire gives minimum breaking load of 1121 kN with nominal diameter of 40 mm.
   Approximate mass of wire rope,

$$m = 656 \text{ kg/(100 m)}$$

   To take up 2200 kN load, 2 steel wire ropes of 6×19 size has to be used.
4. Average diameter of sheave, $D = 45\, d = 45 \times 40 = 1800$ mm
5. Bending load, $W_b = \dfrac{A\, E_r\, d_w}{D}$

$$A = 0.38\, d^2,\ E_r = 64700 \text{ N/mm}^2,\ d_w = 0.063\, d$$

$$W_b = \frac{0.38\,(40)^2 \times 64700 \times 0.063 \times 40}{1800} = 55073 \text{ N}$$

6. Acceleration of rope, $a = \dfrac{v}{t} = \dfrac{500}{60 \times 10} = 0.833 \text{ m/s}^2$
7. Weight of rope, $W_r = \dfrac{mgh}{100} = \dfrac{656 \times 9.81 \times 300}{100} = 19306$ N
8. Weight to be lifted, $W = 55$ kN
   Load per rope = 27.5 kN
9. Acceleration load,

$$W_a = \left(\frac{W}{2} + W_r\right) \times \frac{a}{g} = \frac{(27500 + 19306) \times 0.833}{9.81} = 3974.5 \text{ N}$$

10. Effective load, W_e

   (i) Starting load without slack, $W_{st} = 2\left(\dfrac{W}{2} + W_r\right) = 2\,(27.5 + 19.306) = 93.612$ kN

$$W_e = W_{st} + W_b = 93.612 + 55.073 = 148.685 \text{ kN}$$

   (ii) During the first 10 s after starting,

$$W_e = \frac{W}{2} + W_r + W_b + W_a$$

$$= 27.5 + 19.306 + 55.306 + 3.9745$$

$$= 106.0865 \text{ kN}$$

   Maximum effective load, $(W_e)_{max} = 148.685$ kN
11. Operating factor of safety $= \dfrac{\text{Breaking strength}}{(W_e)_{max}} = \dfrac{898}{148.685} = 6 < 10$

   We observe that the component of bending load is maximum.
   If $D = 2500$ mm, then

$$W_b = \frac{55073 \times 1800}{2500} = 39652 \text{ N}$$

$$W_e = 90.432 \text{ kN}$$

$$\text{FOS} = \frac{898}{90.432} = 9.93 \simeq 10$$

Hence OK.

The results are :

Rope size : 6×19 (9/9/1) having 1960 MPa tensile strength.

Number of ropes	= 2
Steel wire rope diameter	= 40 mm
Sheave diameter	= 2.5 m
Mass of rope	= 656 hr/(100 m) length

■ **Example 9.10** Design a chain drive to run a blower at 600 rpm. The power to the blower is available from 8 kW motor at 1500 rpm. The centre distance is to be kept at 800 mm. [MDU, May, 2008]

Solution : Given : $n_2 = 600$ rpm, $P = 8$ kW, $n_1 = 1500$ rpm, $C = 800$ mm

1. Preliminary decisions :

   Transmission ratio, $i = \frac{n_1}{n_2} = \frac{1500}{600} = 2.5$

   For $i = 2.5$, number of teeth on roller chain in order to reduce polygonal effect is 25 – 27. We take $z_1 = 26$.

   It is further assumed that the chain is a simplex roller chain with a single strand of specification ISO 10 B of pitch 15.875 mm and breaking load 22.26 kN.

   Weight of chain, $W = 8.92$ N/m

2. Minimum factor of safety for smaller sprocket speed,

   $$n_1 = 1500 \text{ rpm is,}$$

   $$K_f = \left(\frac{13.2 - 11.7}{1600 - 1200}\right)(1500 - 1200) + 11.7 = 12.8$$

3. Pitch diameter of driving sprocket,

   $$d_1 = \frac{p}{\sin\left(\frac{180°}{z_1}\right)} = \frac{15.875}{\sin\left(\frac{180°}{26}\right)} = 131.7 \text{ mm}$$

4. Number of teeth on driven sprocket,

   $$z_2 = i\, z_1 = 2.5 \times 26 = 65$$

5. Pitch line velocity,

   $$v = \frac{\pi d_1 n_1}{10^3 \times 60} = \frac{\pi \times 131.7 \times 1500}{10^3 \times 60} = 10.34 \text{ m/s}$$

6. Length of chain in pitches,

   $$L_p = \frac{2C}{p} + \frac{z_1 + z_2}{2} + \frac{p(z_2 - z_1)^2}{4\pi^2 C}$$

$$= \frac{2 \times 800}{15.875} + \frac{26+65}{2} + \frac{15.875\,(65-26)^2}{4\pi^2 \times 800}$$

$$= 100.787 + 45.5 + 0.765 = 147 \text{ links}$$

$$\frac{C}{p} = \frac{800}{15.875} = 50.4$$

7. Service factor, $K_s = K_1 K_2 K_3 K_4 K_5 K_6$

   Load factor, $K_1 = 1.25$ for variable load with mild shocks

   Distance regulation factor, $K_2 = 1.0$ for adjustable supports

   Centre distance factor, $K_3 = 1.0$ for $\frac{C}{P} = 30 - 50$

   Position of shaft factor, $K_4 = 1.0$ for horizontal drive

   Lubrication factor, $K_5 = 1.5$ for periodic lubrication

   Rating factor, $K_6 = 1.25$ for 16 hours working/day

$$K_s = 1.25 \times 1.0 \times 1.0 \times 1.0 \times 1.5 \times 1.25 = 2.34$$

8. Power transmitted on the basis of F_u

$$P = \frac{F_u\, v}{n\, K_s} = \frac{22260 \times 10.34}{12.8 \times 2.34 \times 10^3} = 7.685 \text{ kW} \simeq 8 \text{ kW}$$

9. Tangential force, $F = \frac{10^3 \times P}{v} = \frac{10^3 \times 8}{10.34} = 773.7 \text{ N}$

   Centrifugal force,

$$F_c = \frac{w v^2}{g} = \frac{8.92 \times (10.34)^2}{9.81} = 97.21 \text{ N}$$

   Sag tension, $F_s = K_{sag}\; w C$

   $K_{sag} = 6$ for horizontal position

$$F_s = 6 \times 8.92 \times 0.8 = 42.81 \text{ N}$$

10. Actual factor of safety,

$$K_{fa} = \frac{F_u}{F + F_c + F_s} = \frac{22260}{773.7 + 97.21 + 42.81} = 24.36$$

$$n_{act} > n$$

   Hence safe.

■ **Example 9.11** A V-belt drive is to transmit 15 kW to a compressor. The motor speed is 1440 rpm and compressor pulley runs at 480 rpm. Give complete design of the drive. [MDU, Dec., 2008]

Solution : Given $P = 15$ kW, $N_1 = 1400$ rpm, $N_2 = 480$ rpm

1. Preliminary decisions :

   Let centre distance, $C = 1.5$ m

   For power range of 1.65 to 32.2 kW, D cross-section of belt is recommended.

   Minimum groove pulley diameter, $D_1 = 330$ mm

Area of cross-section of belt, $A = 476.6\ \text{mm}^2$ (32 mm × 19 mm)

Weight per metre length of belt, $w = 5.923$ N/m

Pulley groove angle, $2\beta = 40°$

2. Diameter of compressor pulley, $D_2 = \dfrac{D_1 N_1}{N_2} = \dfrac{330 \times 1440}{480} = 990$ mm

Speed ratio, $i = \dfrac{N_1}{N_2} = \dfrac{1400}{480} = 3$

3. Pitch line belt speed, $v = \dfrac{\pi D_1 N_1}{10^3 \times 60} = \dfrac{\pi \times 330 \times 1440}{10^3 \times 60} = 24.88$ m/s

4. Pitch length of belt, $L_p = 2C + \dfrac{\pi}{2}(D_1 + D_2) + \dfrac{(D_2 + D_1)^2}{4C}$

$$= 2 \times 1.5 + \frac{\pi}{2}(0.33 + 0.99) + \frac{(0.99 - 0.33)^2}{4 \times 1.5}$$

$$= 5.146 \text{ m or } 5146 \text{ mm}$$

The standard nominal pitch length of D-belt nearest to 5146 mm is 5032 mm and nominal inside length of belt is 4953 mm.

5. Exact centre distance

$$5032 \times 10^{-3} = 2C + \frac{\pi}{2}(0.33 + 0.99) + \frac{(0.99 - 0.33)^2}{4C}$$

$$= 2C + 2.073 + \frac{0.1089}{C}$$

or $$2C + \frac{0.1089}{C} = 2.959$$

$$2C^2 - 2.959C + 0.1089 = 0$$

$$C^2 - 1.4795C + 0.05445 = 0$$

$$C = \frac{1}{2}[1.4795 \pm \sqrt{(1.4795)^2 - 4 \times 0.05445}]$$

$$= 1.442 \text{ m or } 1442 \text{ mm}$$

6. Angle of arc of contact,

$$\theta = 180° - 2\sin^{-1}\left[\frac{D_2 - D_1}{2C}\right]$$

$$= 180° - 2\sin^{-1}\left[\frac{990 - 330}{2 \times 1442}\right] = 153.5°$$

7. For $i = 1.275 - 1.340$, small diameter factor, $K_b = 1.09$

Correction factor for industrial service, $K_a = 1.1$

Correction factor for bolt length, $K_c = 0.96$ (for nominal inside length)

Correction factor for arc of contact, $K_d = 0.81$ (for $\theta = 153.5°$)

Equivalent pitch diameter, $d_e = D_1 \times K_b = 330 \times 1.09 = 359.7$ mm

8. Power rating of belt :

Maximum value of d_e in power rating formula for D-belt = 425 mm

$$\text{Power rating for D-belt}\left[3.22\,v^{-0.09} - \frac{506.7}{d_e} - 4.78 \times 10^{-4}\,v^2\right]v$$

$$= \left[3.22 \times (24.88)^{-0.09} - \frac{506.7}{359.7} - 4.78 \times 10^{-4}\,(24.88)^2\right] \times 24.88$$

$$= 17.58 \text{ kW}$$

9. Number of belts $= \dfrac{P \times K_a}{\text{kW rating } \times K_c \times K_d}$

$$= \frac{15 \times 1.1}{17.58 \times 0.96 \times 0.81} = 1.2 \simeq 2$$

10. Pulley groove size : Refer to Fig. 9.20.

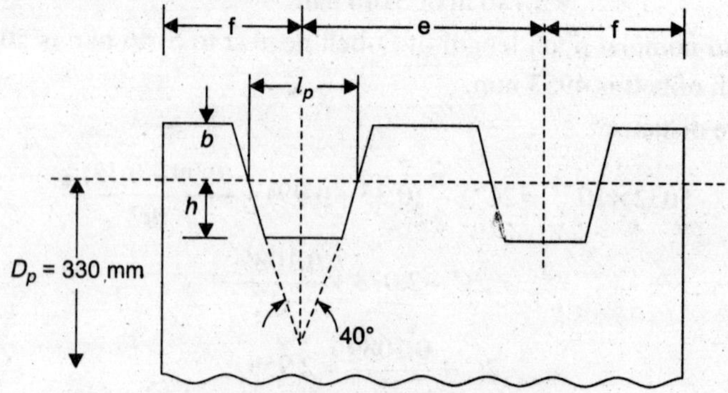

Fig. 9.20 Pulley groove dimensions.

$$l_p = 27 \text{ mm}$$
$$b = 8.1 \text{ mm}$$
$$h = 19.9 \text{ mm}$$
$$e = 37 \pm 0.6 \text{ mm}$$
$$f = 24^{+3}_{-1} \text{ mm}$$

■ **Example 9.12** Design completely a flat belt drive to drive a winch crab from an electric motor of 12 kW power. Speed of motor shaft is 750 rpm. Speed ratio is 3. Belt position is horizontal and there is considerable variation of load. [MDU, May, 2009]

Solution : Given : $P = 12$ kW, $N_1 = 750$ rpm, $i = 3$

1. Speed of winch crab shaft,

$$N_2 = \frac{N_1}{i} = \frac{750}{3} = 250 \text{ rpm}$$

2. Let us assume the following data :

Centre distance, $C = 3$ m

For $P = 7.5 - 15$ kW, $v = 12 - 24$ m/s

Let $v = 15$ m/s

Then $$\frac{\pi D_1 N_1}{10^3 \times 60} = 15$$

$$D_1 = \frac{15 \times 10^3 \times 60}{\pi \times 750} = 381.9 \simeq 400 \text{ mm}$$

$$\therefore \quad v = \frac{\pi \times 0.4 \times 750}{60} = 15.7 \text{ m/s}$$

$$D_2 = \frac{D_1 N_1}{N_2} = \frac{400 \times 750}{250} = 1200 \text{ mm}$$

Coefficient of friction between belt and pulley = 0.3.

3. Angle of arc of contact, $\theta = 180° - 2\sin^{-1}\left[\frac{D_2 - D_1}{2C}\right]$

$$= 180° - 2\sin^{-1}\left[\frac{1200 - 400}{2 \times 3000}\right] = 164.7°$$

Ratio of belt tensions,

$$\frac{F_1}{F_2} = \exp(\mu\theta) = \exp\left(0.3 \times \frac{\pi}{180} \times 164.7\right) = 2.369$$

4. Further assume the following data :

For solid woven impregnated cotton belting, permissible working stress,

$$\sigma_w = 2.25 \text{ MPa}$$

Density of belt material, $\rho = 950 \text{ kg/m}^2$

Belt thickness, $t = 6$ mm

Let b = width of belt = b mm

5. Mass of belt per metre length,

$$m = \rho bt = 950 \times b \times 6 \times 10^{-6} = 5.7 \times 10^{-3}\ b \text{ kg/m}$$

Centrifugal tension,

$$F_c = mv^2 = 5.7b \times 10^{-3}\ (15.7)^2 = 1.405b \text{ N}$$

6. Maximum tension in belt,

$$F_{max} = \sigma_w \times bt = 2.25 \times b \times 6 = 13.5b \text{ N}$$

Tight side tension,

$$F_1 = F_{max} - F_c = (13.5 - 1.405)\ b = 12.095b \text{ N}$$

$$F_2 = \frac{12.095b}{2.369} = 5.106b \text{ N}$$

$$\rho = \frac{(F_1 - F_2)\, v}{10^3}$$

$$12 \times 10^3 = (12.095 - 5.106)\, b \times 15.7$$

$$b = 109.36 \simeq 110 \text{ mm}$$

7. Length of belt, $L_0 = 2C + \frac{\pi}{2}(D_1 + D_2) + \frac{(D_2 - D_1)^2}{4C}$

$$= 2 \times 3000 + \frac{\pi}{2}(400 + 1200) + \frac{(1200 - 400)^2}{4 \times 3000} = 8566.6 \text{ mm} \simeq 8.57 \text{ m}$$

The results are :

$D_1 = 400$ mm, $D_2 = 1200$ mm, $C = 3$ m, $L_0 = 8.57$ m, $b = 110$ mm, $t = 6$ mm

■ **Example 9.13** A V-belt drive is to transmit 15 kW to a compressor. The motor speed is 750 rpm and the compressor pulley runs at 250 rpm. Give complete design of the drive. [MDU, Dec., 2009]

Solution : Refer to Example 9.11.

EXERCISES

1. A 50 kW, 1200 revolutions per minute high-torque squirrel-cage motor is used to drive a punch press. The motor pulley is cast iron 350 mm in diameter. The driven pulley is cast iron 1050 mm in diameter. The centre distance is 2.5 m and is inclined at 55° with the horizontal. Select a suitable leather belt and calculate its length required.
2. An exhaust fan fitted with a 900 mm diameter pulley is driven by a flat belt from a 30 kW, 950 rpm squirrel-cage motor. The pulley on the motor shaft is 250 mm in diameter and the centre distance is 2.25 m. The belt is 100 mm wide with a coefficient of friction of 0.25. If the allowable stress in the belt material is not to exceed 2 N/mm^2, determine the necessary thickness of the belt and its total length. Take centrifugal force effect due to belt weight into consideration when the belt weights 970 kg/m^3.
3. The main spindle of a machine is required to be driven at a speed of 550 rpm by a separately mounted A.C. motor, 5 kW, 1500 rpm. The spindle is 35 mm in diameter and extends by about 200 mm from the body of the machine. Make a complete design of the drive using a flat belt drive.
4. The following particulars refer to a belt drive from a cast iron pulley mounted on a shaft :

Power transmitted	= 25 kW
RPM of shaft	= 250
Ratio of belt tensions	= 3
Velocity of leather belt	= 800 m/min
Allowable belt tension	= 10 N/mm width
Number of arms with elliptical cross-section of the pulley	= 6

   Select a suitable belt and design (a) the pulley, (b) the shaft, and (c) the key.

   Safe stresses are as follows :

Shear stress for the shaft and the key	= 56 N/mm^2
Tensile stress for cast iron	= 17 N/mm^2
Shear stress for cast iron	= 14 N/mm^2

5. A 900 mm driving pulley and a 1200 mm driven pulley are arranged on 3 m centres. The output of the driven shaft is 150 kW. Assume a belt speed of 1400 m/min, a coefficient of friction of 0.30, a slip of 1.5 per cent at each pulley and 5 per cent friction loss at each shaft. Determine
(a) the rpm of shaft, (b) difference in belt tensions, (c) size of machine-wire-laced leather belt required and (d) shaft sizes, assuming pure torsion and an allowable stress of 56 N/mm^2.

6. A machine is driven by a 1440 rpm motor by means of a flat belt of thickness 20 mm and width 200 mm. The pulleys on the motor and compressor are 200 mm and 1200 mm diameters respectively. The centre distance is 1000 mm and an idler is used to make the angle of contact on the smaller pulley 200° and on the larger pulley 250°. Assuming suitable different values of coefficient of friction on the two pulleys, determine (i) Power of the drive (ii) To eliminate idler pulley, the smaller pulley is changed with V-groove pulley having the groove angle as 40° and if the bigger pulley and other data remain same, what shall be the power transmitted now?

7. A line shaft 75 mm in diameter requires 10 kW and runs at 300 rpm. Power is supplied by an electric motor running at 1400 rpm through an open belt drive. Design a suitable pulley for the line shaft. The motor shaft pulley is 200 mm in diameter. Also determine the size of the belt; assume ratio of belt tensions as 2. Give a neat sketch of the pulley with dimensions.

8. A 90 kW steam engine running at 90 rpm and having a 2.4 m diameter flywheel is connected to a 1.05 m diameter pulley by means of an open leather belt. The distance between the shafts is 6 m. Assume that cemented joints are used. Determine (a) the minimum angle of contact, (b) the cross-sectional area of the belt required, and (c) the thickness, width and length of belt.

9. Design a chain drive for a compressor driven by a 20 kW motor at 950 rpm. The compressor rpm being 300. The compressor is required to work 8 hours per day. The centre distance should be approximately 600 mm.

10. Design a chain drive for an electric motor to a compressor to transmit 100 kW at 750 rpm of the motor and 150 rpm of the compressor sprocket. The centre distance should not be more than 1 m. Assume 16 hours/day working.

11. A 740 rpm, 20 kW squirrel-cage motor is used to drive a reciprocating pump at 150 rpm. The pump is to run at full load for 24 hours a day. Select a suitable chain and sprocket sizes for a roller chain drive.

12. Select a roller chain drive to transmit 40 kW from a 1200 rpm motor to a line shaft at 250 rpm. The centre distance is adjustable from 600 mm. Service will be 10 hours per day, 6 days per week and good lubrication will be provided.

13. Select a silent chain for the drive between a 45 kW, 740 rpm motor and a compressor running at 135 rpm. The distance between centres is to be made as short as practical and the drive sprocket is not to exceed 1200 mm in diameter.

14. A roller chain operating under steady load conditions, transmits 3 kW from a shaft rotating at 600 rpm to one rotating at 750 rpm. Determine (a) the chain required using at least 15 teeth in the sprockets, (b) sprocket pitch diameters, (c) the shortest advisable center distance, and (d) the number of links of chain required.

15. A 10 kW, 1200 rpm motor drives a line shaft at 250 rpm. The shaft centre distance is to be approximately 500 mm. The motor shaft has a diameter of 300 mm. The starting torque of the motor is from 1.75 to 2.0 times the running torque. The load is applied with moderate shocks.
(a) Select a roller chain for this drive,
(b) Determine the sprocket pitch diameters,
(c) Number of chain links required, and
(d) Exact centre distance.

16. A 40 mm, 6 × 19 steel wire rope is used to lift the cage of a mine hoist. The mine is 200 m deep, the rope sheave is 4 m in diameter and the cage acceleration is 2 m/s^2. Determine (a) equivalent bending load, (b) load due to the rope, (c) acceleration load, and (d) useful load with a factor of safety of 5.

17. **Select a wire rope for a vertical mine hoist to lift 15000 kN of ore in each 8 hour shift from a depth of 1000 m. Assume a two compartment shaft with the hoisting skips in balance. Use a maximum velocity of 800 m/min with acceleration and deceleration periods of 15 seconds each and a rest period of 10 seconds for discharging and loading skips. A hoisting skip weighs approximately 0.6 of its load capacity. The factor of safety should be approximately 5.**

18. **An 8 × 19 steel wire rope is used to hoist 50 kN of load from a depth of 1000 m. Determine the rope diameter if the maximum speed is 2.5 m/s and the acceleration is 1.25 m/s^2, when starting under no slack conditions.**

19. **Select the steel wire rope, pulleys and drum size reuqired for an electric overhead travelling crane for the following data :**

Lifting	**= 50 kN**
Lifting height	**= 8 m**
Weight of lifting tackle	**= 1200 N**
Weight of lifting magnet	**= 20 kN**
Number of rope parts	**= 4**

Also determine the dimensions of the fastener for the rope to the drum.

❑❑❑

10

Chapter

Clutches and Brakes

Inside This Chapter

10.1 Clutches 10.2 Brakes

10.1 CLUTCHES

A clutch is a device which is used to connect a driving shaft to a driven shaft when desired without stopping the driving shaft.

10.1.1 Types of Clutches

There are two types of cluthes :

1. Position Action Cluthces

These are jaw type of clutches in which power is transmitted as long as the jaws are in contact. The part of the jaw on the driving shaft is rigidly fixed to the shaft whereas the other parts of the clutch may be made to slide on the driven shaft having splines.

2. Friction Clutches

These cluthces work on the basis of frictional forces developed between the surfaces of contact. Friction clutches may be of the axial engagement type or radial type. In the axial type of clutches, the engagement or disengagement of the clutch is achieved by an axial movement of the clutch parts attached to the driven shaft. Whereas in the radial type of clutches the contact pressure is applied in a direction perpendicular to the shaft axis. Plate or disc and cone type of clutches are of the axial type and type-pneumatic, centrifugal, over-running, block, expanding ring and band clutches are of the radial type. The plate type of clutches may be either of the single plate or multi-place friction clutches. Multi-plate friction clutches are used in small vehicles two and three wheelers whereas single plate clutches are used in heavy vehicles like trucks, buses, etc. Multi-plate clutches are wet clutches and operate in oil which acts as cushion and takes away the heat energy released due to rubbing action of friction plates.

10.1.2 Design Considerations

A clutch of good design should have the following qualities :

1. Ability to withstand and dissipate heat.
2. Adequate torque capacity.
3. Long life.
4. Positive release.
5. Smooth engagement.
6. Low operating force.
7. Ease of repair.

Clutches should be designed for overload capacities of 75-100%. If the load is variable or subjected to shock, additional service factors should be used.

10.1.3 Properties of Material for the Lining of a Clutch

The lining material for clutches should have the following properties :

1. High coefficient of friction.
2. Unaffected by moisture and oil.
3. Ability to withstand high tempeature caused by slippage.
4. High heat conductivity.
5. High resistance to wear and scoring.

Coefficient of friction and unit pressure for clutch facings.

10.1.4 Disc or Plate Clutch

Consider two friction surfaces maintained in contact by an axial force F_a, as shown in Fig. 10.1.

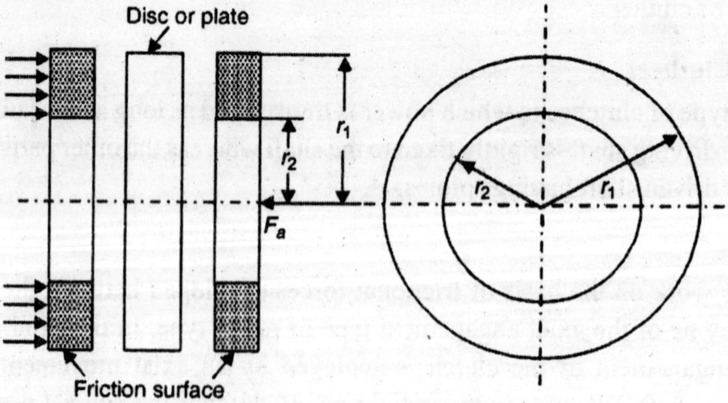

Fig. 10.1 Single plate friction clutch.

Let, T = torque transmitted by the clutch

p = intensity of axial pressure

r_1, r_2 = external and internal radii of friction surfaces respectively

r_m = mean radius of friction surface

μ = coefficient of friction

The two conditions prevailing on the clutch plate may be,

1. Uniform pressure, $p \propto r$ or $\frac{p}{r}$ = constant.

2. Uniform wear, $p \propto \frac{1}{r}$ or pr = constant.

The first condition generally prevails in a new clutch and the second condition in a worm out clutch.

(a) For uniform pressure (p = const.), we have

Axial force, $$F_a = \pi(r_1^2 - r_2^2)\,p \quad \text{...(10.1)}$$

Frictional torque, $$T_f = \mu F_a r_m \quad \text{...(10.2)}$$

where $$r_m = \left(\frac{2}{3}\right)\left[\frac{r_1^3 - r_2^3}{r_1^2 - r_2^2}\right]$$

(b) For uniform wear (pr = const.), we have

Axial force, $$F_a = 2\pi\, p_a r_2\,(r_2 - r_1) \quad \text{...(10.3)}$$

$$T_f = \mu F_a r_m \quad \text{...(10.4)}$$

where $$r_m = \frac{r_1 + r_2}{2} \quad \text{...(10.5a)}$$

$$\frac{r_2}{r_1} = 0.6 - 0.7 \text{ for general machinery} \quad \text{...(10.5b)}$$

$$= 0.748 \text{ for tractors.}$$

10.1.5 Multiple Disc Clutch

A multiple disc clutch is shown in Fig. 10.2.

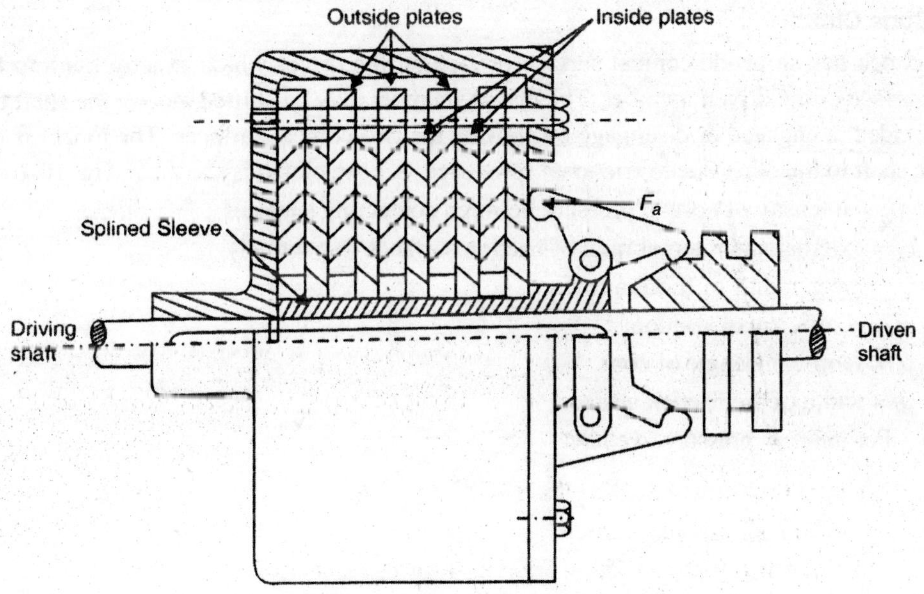

Fig. 10.2 Multi-plate friction clutch.

Let n_1, n_2 = number of discs on the driving end driven shafts respectively

$$n = n_1 + n_2 - 1 = \text{number of pairs of contact surfaces} \quad ...(10.6)$$

Total friction torque acting on the clutch,

$$T_f = n\mu F_a r_m \quad ...(10.7)$$

Number of plates = 10 – 20 ...(10.8)

Thickness of core plate = 0.75 – 3.0 mm ...(10.9)

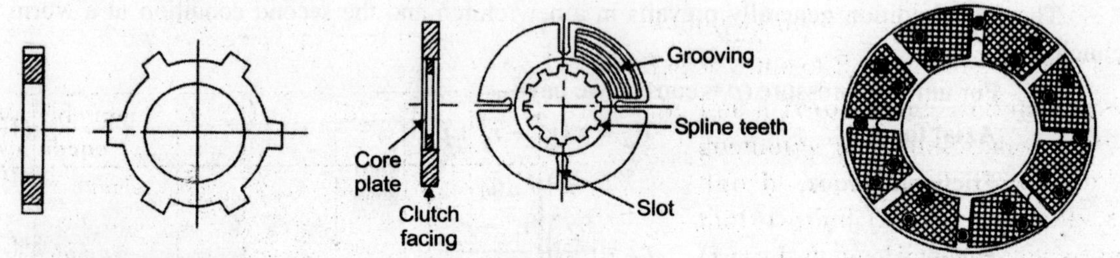

(a) Driving disc (b) Clutch disc (c) Slots on clutch facing

Fig. 10.3

The multiplate clutches work wet in only medium. The face of the wet clutch plate may be provided with radial slots, radial grooves or spiral grooves for proper circulation of oil through the clutch plate. The driving disc and clutch disc are shown in Fig. 10.3.

10.1.6 Dry and Wet Clutches

In case of single plate clutches, the contacting surfaces are either one or two. Due to large surface area available, heating is not a problem and the clutch is of the dry type. In case of multiplate clutches, the work done during engagement and disengagement is converted into heat. Heat dissipation is a serious problem in multiplate clutches because of large number of friction surfaces. Cooling oil is used to dissipate this heat. Therefore, these clutches are wet clutches.

10.1.7 Cone Clutch

A cone clutch has an inside conical surface or face on the driver which exactly fits into the outside conical surface of the driven member. The driven member may be shifted aloong the shaft by a forked lever provided to engaged or disengage by shifting the two conical surfaces. The torque is transmitted from one shaft to the other due to frictional resistance. A cone clutch is shown in Fig. 10.4.

Let p_n = intensity of normal pressure between contacting surfaces

r_1, r_2 = external and internal radii of friction surfaces respectively

r_m = mean radius of friction surface

μ = coefficient of friction

α = semi-cone angle of cone

b = width of the friction surface

(a) For uniform pressure, we have

$$F_a = \pi(r_1^2 - r_2^2)\, p_n \quad ...(10.10)$$

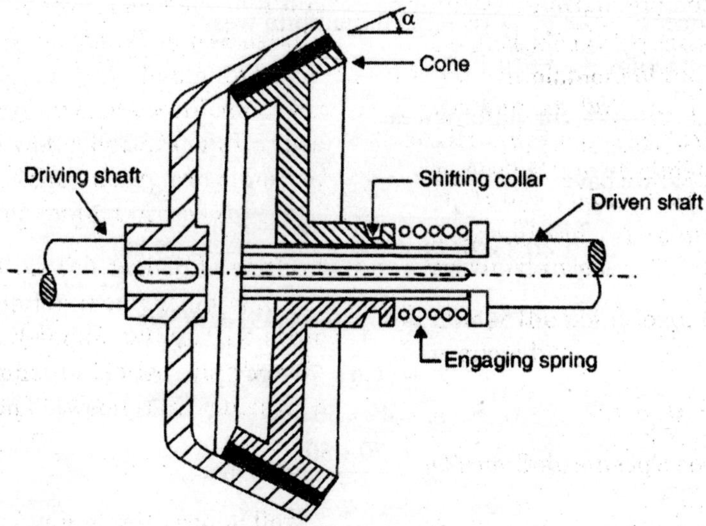

Fig. 10.4 Cone clutch.

Frictional torque, $$T_f = \mu F_a r_m \operatorname{cosec} \alpha \qquad ...(10.11)$$

where $$r_m = \left(\frac{2}{3}\right)\left[\frac{r_1^3 - r_2^3}{r_1^2 - r_2^2}\right]$$

(b) For uniform wear, we have

Total frictional torque,

$$T_f = \mu F_a r_m \operatorname{cosec} \alpha \qquad ...(10.12)$$

where $$r_m = \frac{r_1 + r_2}{2}$$

Axial force required for engaging the clutch,

$$F_e = F_n (\sin \alpha + 0.25 \mu \cos \alpha) \qquad ...(10.13)$$

■ **Example 10.1** Design a single plate clutch of a vehicle to transit 20 kW at a speed of 3000 rpm, allowing for 25% overload. The axial thrust is to be provided by 6 springs of about 20 mm mean coil diameter. The clutch may be made with 6 splines.

Solution : Given : $P = 20$ kW, $N = 3000$ rpm, Overload = 25%, Number of springs = 6, $D_m = 20$ mm, Number of splines = 6, $n = 2$

1. Preliminary decisions :

   We assume moulded asbestos friction material for which in dry condition,

   $$\mu = 0.2 - 0.5 \text{ and } p = 0.34 - 0.98 \text{ MPa}$$

   We take $\mu = 0.3$, $p = 0.4$ MPa and $\frac{r_2}{r_1} = 0.7$

   Shear stress in spring wire = 400 Mpa.

2. Torque transmitted,

   $$T = \frac{1.25 \times 10^3 \times P \times 60}{2 \pi N} = \frac{1.25 \times 10^3 \times 20 \times 60}{2 \times \pi \times 3000} = 79.58 \text{ N.m}$$

3. Axial force, $F_a = 2\pi\, p_a\, r_2\, (r_1 - r_2)$ for uniform wear
4. Frictional torque, $T_f = n\,\mu\, F_a r_m$

   where $r_m = \dfrac{r_1 + r_2}{2}$, for uniform wear
5. For $T = T_f$, we have

$$2 \times 0.3 \times 2\pi \times 0.4 \times r_2 \times \frac{(r_1^2 - r_2^2)}{2} = 79580$$

$$0.754 \times 0.7\, r_1 \times r_1^2\, (1 - 0.49) = 79580$$

$$r_1 = 66.6 \approx 70 \text{ mm}$$

$$r_2 = 49 \approx 50 \text{ mm}$$

$$r_m = \frac{70 + 50}{2} = 60 \text{ mm}$$

6. Mean peripheral speed,

$$v_m = \frac{2\pi\, r_m\, N}{60} = \frac{2\pi \times 60 \times 3000}{60 \times 10^3} = 18.85 \text{ m/s}$$

7. Axial force, $F_a = 2\pi\, p_a\, r_2\, (r_1 - r_2)$

$$= 2\pi \times 0.40 \times 50 \times (70 - 50) = 2513 \text{ N}$$

8. Spring design,

   Thrust to be exerted by the spring,

$$F = \frac{F_a}{\text{number of springs}}$$

$$= \frac{2513}{6} = 418.88 \simeq 420 \text{ N}$$

Let spring index, $C = 5$

Wahl's factor, $K_w = \dfrac{4C - 1}{4C - 4} + \dfrac{0.615}{C}$

$$= \frac{4 \times 5 - 1}{4 \times 5 - 4} + \frac{0.615}{5} = 1.31$$

Wire diameter, $d = \left[\dfrac{8FCK_w}{\pi\, \tau_d}\right]^{1/2}$

$$= \left[\frac{8 \times 420 \times 5 \times 1.31}{\pi \times 400}\right]^{1/2} = 4.18 \approx 4.25 \text{ mm}$$

Mean coil diameter,

$$D_m = 5 \times 4.25 = 21.25 \text{ mm} \approx 20 \text{ mm}$$

Outside spring diameter = $D + d = 25.5$ mm

Number of active coils = 6 (say)

Total coils, $i_t = i + 2 = 8$ for squared and ground ends

Maximum deflection, $\delta_{max} = \dfrac{8FC^3 i}{Gd}$

$$= \frac{8 \times 420 \times 5^3 \times 6}{78.45 \times 10^3 \times 4.25} = 7.56 \text{ mm}$$

Free length of spring,

$$l_f = i_t\, d + 1.15\,\delta_{max}$$
$$= 8 \times 4.25 + 1.15 \times 7.56 = 42.7 \text{ mm}$$

9. Splines of clutch shaft.

Torque transmission capacity of a splined shaft,

$$T = 0.25 \text{ phli } (d_1 + d) = 0.5 \text{ phli } (d + h)$$
$$b = 0.250\, d,\ h = 0.1\, d \text{ when loaded}$$

where d = minor diameter of splined shaft

d_1 = major diameter of splined shaft

h = thickness of spline

i = number of splines

l = length of spline

p = allowable bearing pressure for splines

= 7 MPa, for splines to slide when loaded

$$T_{shaft} = \left(\frac{\pi}{16}\right) d^3\, \tau$$

$$d = \left[\frac{16 \times 79580}{\pi \times 25}\right]^{1/3}$$

$$= 25.3 \approx 26 \text{ mm}$$

$$d_1 = d + 2h = 26 + 2 \times 0.1 \times 26 = 31.2 \approx 32 \text{ mm}$$
$$b = 0.250 \times 26 = 6.5 \text{ mm}$$
$$h = 0.1 \times 26 = 2.6 \approx 3 \text{ mm}$$
$$T = 0.25 \times 7 \times 3 \times 1 \times 6 \times (32 + 26) = 79580$$
$$l = 43.6 \approx 50 \text{ mm}$$

■ **Example 10.2** Design a suitable clutch for a 150 cc scooter to develop 5.8 HP at 5500 rpm. At maximum torque it develops 4 HP at 3200 rpm. Space restricts the outer diameter of the clutch to 150 mm.

Solution : Given : $P_{max} = 5.8$ HP, $N = 5500$ rpm, $P = 4$ HP at $N = 3200$ rpm, $r_1 = 75$ mm

1. Maximum torque, $T_{max} = \dfrac{4 \times 0.746 \times 1000 \times 60}{2\pi \times 3200} = 24.905$ N.m

2. Design torque, $T_d = 1.5\, T_{max} = 13.357$ N.m

3. Torque capacity of clutch assuming uniform wear,

$$T_f = n\,\mu\, F_a\, r_m$$

Assume $p = 0.08$ MPa, $\mu = 0.1$ for wet clutch,

$$\frac{r_2}{r_1} = 0.6, r_1 = 75 \text{ mm}, r_2 = 45 \text{ mm}$$

$$r_m = \frac{75+45}{2} = 60 \text{ mm}$$

$$F_a = 2\pi\, p_a\, r_2\, (r_1 - r_2)$$

$$= 2\pi \times 0.8 \times 45\,(75-45)$$

$$= 678.6 \text{ N}$$

4. Number of plates,

$$n \times 0.1 \times 678.6 \times 60 = 13357$$

$$n = 3.28 \approx 4$$

$$= n_1 + n_2 - 1$$

$$n_1 + n_2 = 5$$

$$n_1 = 3, n_2 = 2$$

5. Actual axial force,

$$F_a = \frac{T_d}{n\,\mu\, r_m} = \frac{13.357 \times 10^3}{4 \times 0.1 \times 60} = 556.54 \text{ N}$$

Actual pressure,

$$p = \frac{F_a}{2\pi\, r_2\,(r_1 - r_2)}$$

$$= \frac{556.54}{2\pi \times (45 \times 30)} = 0.0656 \text{ MPa}$$

6. Shaft diameter,

$$d_s = \left[\frac{16\,T}{\pi\,\tau}\right]^{1/3}$$

$$= \left[\frac{16 \times 13357}{\pi \times 25}\right]^{1/3} = 13.96 \approx 14 \text{ mm}$$

7. Splines,

$$b = 0.250\, d,\; h = 0.1\, d \text{ when loaded}$$

where $d = d_s$ minor diameter of splined shaft

$d_1 = d + 2h$, major diameter of splined shaft

h = thickness of spline

i = number of splines

l = length of spline

p = allowable bearing pressure for splines

= 7 MPa, for splines to slide when loaded

$$h = 0.1 \times 14 = 1.4 \approx 2 \text{ mm}$$
$$d_1 = d + 2h = 14 + 2 \times 2 = 18 \text{ mm}$$
$$b = 0.250 \times 14 = 3.5 \text{ mm}$$
$$T = 0.25 \text{ phli } (d_1 + d) = 0.5 \text{ phli } (d + h)$$
$$= 0.25 \times 7 \times 2 \times 1 \times 6 \times (18 + 14) = 13357$$
$$l = 1987 \approx 40 \text{ mm allowing for movement of hub.}$$

8. Design of spring
Thrust to be exerted by the spring,

$$F = \frac{F_a}{\text{number of springs}}$$
$$= \frac{556.54}{3} = 185.5 \text{ N}$$

Let spring index, $C = 6$

Wahl's factor, $$K_w = \frac{4C - 1}{4C - 4} + \frac{0.615}{C}$$
$$= \frac{4 \times 6 - 1}{4 \times 6 - 4} + \frac{0.615}{6} = 1.2525$$

Wire diameter, $$d = \left[\frac{8\,FCK_w}{\pi \tau_d}\right]^{1/2}$$
$$= \left[\frac{8 \times 185.5 \times 6 \times 1.2525}{\pi \times 400}\right]^{1/2} = 2.98 \approx 3 \text{ mm}$$

Mean coil diameter, $D = 6 \times 3 = 18$ mm
Outside spring diameter $= D + d = 21$ mm
Number of active coils = 5 (say)
Total coils, $i_t = i + 2 = 7$ for squared and ground ends
Maximum deflection,

$$\delta_{max} = \frac{8\,FC^3\,i}{Gd} = \frac{8 \times 185.5 \times 6^3 \times 5}{724.45 \times 10^3 \times 3} = 6.81 \text{ mm}$$

Free length of spring,

$$l_f = i_t\,d + 1.15\,\delta_{max} = 3 \times 3 = 1.15 \times 6.81 = 224.8 \text{ mm}$$

■ **Example 10.3** Design a cone clutch with leather facing and cast iron mating surface to transmit 15 kW at 750 rpm. The mean diameter is limited to 250 mm and the slope of the face is to be 15°. The clutch is to work under oily conditions.

Solution : Given : $P = 15$ kW, $N = 750$ rpm, $D_m = 250$ mm, $\alpha = 15°$

1. Preliminary decisions,
For leather on cast iron in oily conditions, coefficient of friction,

$$\mu = 0.15, \; p = (0.07 - 0.084) \text{ MPa}$$

We take, $$p = 0.08 \text{ MPa}$$

2. Torque transmitted,

$$T_f = \frac{10^3 \, P \times 60}{2\pi N} = \frac{10^3 \times 15 \times 10^3 \times 60}{2\pi \times 750} = 191 \text{ N.m}$$

3. Shaft diameter

Taking permissible shear stress in shaft material,

$$\tau = 40 \text{ MPa}$$

Shaft diameter, $d_s = \left[\frac{16\, T_f}{\pi \tau}\right]^{1/3}$

$$= \left[\frac{16 \times 191 \times 10^3}{\pi \times 40}\right]^{1/3} = 34.9 \approx 40 \text{ mm (adopt)}$$

4. Torque transmitted, $T_f = \frac{\mu F_n D_m}{2}$

Normal force, $F_n = \frac{191 \times 10^3}{0.15 \times 125} = 10{,}187 \text{ N}$

5. Axial force, $F_a = F_n (\sin \alpha + 0.25 \mu \cos a)$

$$= 10{,}187 (\sin 15° + 0.25 \times 0.15 \times \cos 15°)$$

$$= 3005.6 \text{ N}$$

6. Width of cone face, $b = \frac{F_n}{\pi D_m p}$

$$= \frac{10{,}187}{\pi \times 250 \times 0.08} = 162 \text{ mm}$$

7. Inside cone diameter, $D_i = D_m - b \sin \alpha$

$$= 250 - 162 \sin 15° = 208 \text{ mm}$$

8. Outside cone diameter, $D_0 = D_m + b \sin \alpha$

$$= 250 + 162 \sin 15° = 292 \text{ mm}$$

9. Helical spring design.

Let spring index, $C = 6$ and allowable shear stress in spring wire,

$$\tau_d = 500 \text{ MPa}$$

Wahl's correction factor,

$$K_w = \frac{4C - 1}{4C - 4} + \frac{0.615}{C}$$

$$= \frac{4 \times 6 - 1}{4 \times 6 - 4} + \frac{0.615}{6}$$

$$= 1.2525$$

Wire diameter, $d = \left[\frac{8\, F_a C K_w}{\pi \tau_d}\right]^{1/2}$

$$= \left[\frac{8 \times 3005.6 \times 6 \times 1.2525}{\pi \times 500}\right]^{1/2}$$

$$= 10.72 \approx 11 \text{ mm}$$

Mean coil diameter, $D = 6 \times 11 = 66$ mm

Let number of active turns, $i = 6$

Maximum deflection, $\delta_{max} = \left[\frac{8 F_a i D^3}{Gd^4}\right]$

$$= \frac{8 \times 3005.6 \times 6 \times 66^3}{78.45 \times 10^3 \times 11^4}$$

$$= 36 \text{ mm}$$

Free length of squared and ground ends,

$$l_f = (i+2)\, d + 1.15\, \delta_{max}$$

$$= (6+2) \times 11 + 1.15 \times 36 = 129.4 \text{ mm}$$

■ **Example 10.4** A single plate (one side) clutch can transmit 718 N.m torque. Calculate the inner diameter of friction lining using uniform pressure theory using following parameters :

$$D_0 = 300 \text{ mm}, \mu = 0.25, \text{ and } p_{max} = 835 \text{ kPa}.$$

Check the capacity of clutch using uniform wear assumption.

Solution : Given : $T_f = 718$ N.m, $D_0 = 300$ mm, $\mu = 0.25$, $p_{max} = 825$ kPa or 0.825 MPa, $n = 1$.

(i) Uniform pressue,

Axial force, $F_a = \left(\frac{\pi}{4}\right)(D_o^2 - D_i^2)\, p$

Torque transmitted, $M_t = \frac{n \mu F_a D_m}{2}$

Mean diameter, $D_m = \left(\frac{2}{3}\right)\left[\frac{D_o^3 - D_i^3}{D_o^2 - D_i^2}\right]$

$$M_t = 1 \times \mu \times \left(\frac{\pi}{4}\right)(D_o^2 - D_i^2)\, p \times \left(\frac{1}{3}\right)\left[\frac{D_o^3 - D_i^3}{D_o^2 - D_i^2}\right]$$

$$= \mu \times \left(\frac{\pi}{12}\right) p \times (D_o^3 - D_i^3)$$

$$718 \times 10^3 = \left(\frac{0.25 \times \pi}{12}\right) \times 0.825 \times (300^3 - D_i^3)$$

$$D_i = 239.3 \approx 240 \text{ mm}$$

(ii) Uniform wear,

Axial force, $F_a = \left(\frac{\pi}{2}\right) D_i\, (D_o - D_i)\, p$

Torque transmitted, $$M_t = \frac{n \mu F_a D_m}{2}$$

Mean diameter, $$D_m = \left(\frac{1}{2}\right)[D_o + D_i]$$

$$M_t = 1 \times \mu \times \left(\frac{\pi}{2}\right) D_i (D_o - D_i)\, p \times \left(\frac{1}{4}\right)(D_o + D_i)$$

$$= \mu \times \left(\frac{\pi}{8}\right) p \times D_i (D_o^2 - D_i^2)$$

$$718 \times 10^3 = \left(\frac{0.25 \times \pi}{8}\right) \times 0.825 \times D_i (300^2 - D_i^2)$$

$$D_i (300^2 - D_i^2) = 8864834$$

We find the value of D_i by hit and trial method.

D_i, mm	240	235	230	225
$D_i (300^2 - D_i^2)$	1088834	692709	331834	5451

We take $D_i = 225$ mm

10.1.8 Centrifugal Clutch

The centrifugal clutch shown in Fig. 10.5 consists of a number of shoes on the inside of a rim of a pulley. The outer surface of the shoes are covered with a friction materi^{a}l. These shoes can move radially in guides and are held against the spider on the driving shaft by means of springs. The springs exert a radially inward force which is assumed constant. The mass of the shoe, when revolving causes it to exact a radially outward centrifugal force. The magnitude of this centrifugal force depends upon the speed of rotation of shoe. When the centrifugal force exceeds the spring force, the shoe moves outwards and comes into contact with the driven member and presses against it. This pressing force is equal to the difference of centrifugal force and spring force. As the speed of rotation increases the shoe presses harder and enables more torque to be transmitted. It is used in small automobiles like mopeds.

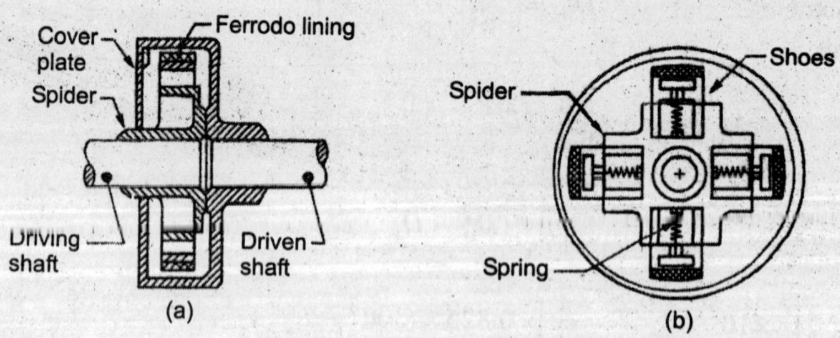

Fig. 10.5 Centrifugal clutch.

1. Mass of the Shoes

Let

m = mass of each shoe

n = number of shoes

r = distance of centre of gravity of shoe from the centre of spider

R = inside radius of pulley rim

N = rotational speed of pulley, rpm

Angular speed of pulley, $$w = \frac{2\pi N}{60} \text{ rad/s}$$

w_1 = angular speed at which engagement begins to occur, rad/s $= \frac{3}{4} w$

μ = coefficient of friction between the shoe and rim

Centrifugal force, $$F_c = mw^2 r$$

Inward force on each shoe exerted by the spring,

$$F_s = m w_1^2 r = m\left(\frac{3}{4} w\right)^2 r = \frac{9}{16} mw^2 r = \frac{9}{16} F_c$$

Net outward radial force on the rim $= F_c - F_r = F_c - \frac{9}{16} F_c = \frac{7}{16} F_c$

Frictional force acting tangentially on each shoe (Fig. 10.6),

$$F = \mu (F_c - F_s)$$

Frictional torque acting on each shoe

$$= F \times R = \mu (F_c - F_s) R$$

Total frictional torque,

$$T = n\mu (F_c - F_s) R = n\mu \times \frac{7}{16} F_c R$$

$$= \frac{7}{16} n\mu R \times m r w^2 \qquad \text{...(10.14)}$$

Calculate mass m of shoe from Eq. (10.14).

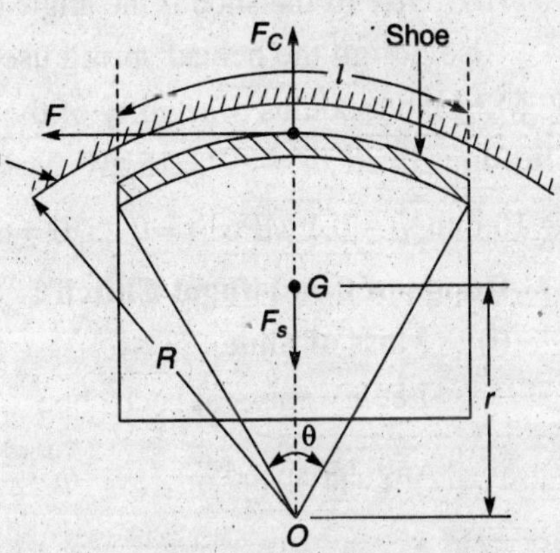

Fig. 10.6 Force acting on the shoe of a centrifugal clutch.

2. Size of the Shoes

Let

l – contact length of the shoes

b = width of the shoes

θ = angle subtended by the shoe at the centre of spider, radius $\simeq 60°$

p = intensity of pressure exerted on the shoe

$= 0.1 \text{ N/mm}^2$

Now $$\theta = \frac{l}{R}$$

or $$l = R\theta = \frac{\pi}{3} R \left(\text{for } \theta = \frac{\pi}{3}\right) \qquad \text{...(10.15)}$$

Area of contact of shoe, $A = l \times b$

Force with which shoe presses the rim $= pA = p \times lb$

Now $$lbp = F_c - F_r \quad \text{...(10.16)}$$

Calculate l from Eq. (10.16)

3. Dimensions of Spring

Load on the spring, $$F_s = \frac{9}{16} m w^2 r \quad \text{...(10.17)}$$

The spring can be designed under force F_s.

Example 10.5 A centrifugal clutch with four shoes is used to transmit 15 kW at 720 rpm. The speed at which engagement begins is 80% of full speed. The inside diameter of the drum is 325 mm and outside diameter of the drum is 350 mm. The radial distance of C.G. of each shoe from the axis of shaft is 120 mm. The radial decrease between the shoe and the rim is 10 mm. The pressure exerted on the shoe during engagement is 0.1 MPa. The coefficient of friction between shoe lining and the drum is taken as 0.25.

Determine :

(i) Mass of each shoe.

(ii) Size of the shoe if the angle subtended by shoe at centre is 60°.

Also design the helical spring used in the shoes taking the safe shear stress of spring material as 600 MPa and modulus of rigidity of the material as 80 kN/mm^2. Use the spring index of 6.

Solution : Given : $n = 4$, $P = 15$ kW, $N = 720$ rpm, $N_e = 0.8$ N, $D_i = 325$ mm, $D_o = 350$ mm, $r = 120$ mm, $C = 10$ mm, $p = 0.1$ MPa, $\mu = 0.25$, $\theta = 60°$, $\tau = 600$ MPa, $G = 80$ kN/mm^2, $C = 6$.

Design of Centrifugal Clutch :

(i) Mass of Shoe

Let m = mass of each shoe

Angular speed, $$\omega = \frac{2\pi N}{60} = \frac{2\pi \times 720}{60} = 75.4 \text{ rad/s}$$

Speed at engagement of clutch,

$$\omega_e = 0.8 \times 75.4 = 60.32 \text{ rad/s}$$

Radial distance of shoe C.G. from shaft axis, $r = 0.12$ m

Centrifugal force acting on each shoe,

$$F_c = mr\omega^2$$

$$= m \times 0.12 \times (75.4)^2 = 682.22 \text{ mN}$$

Centrifugal force at engagement of clutch,

$$F_e = mr\omega_e^2$$

$$= m \times 0.12 \times (60.32)^2 = 436.62 \text{ mN}$$

Torque transmitted at running speed, $$T = 10^3 \times P \times \frac{60}{2\pi N}$$

$$= 10^3 \times 15 \times \frac{60}{2\pi \times 720} = 198.94 \text{ N.m}$$

Also, $$T = \mu (F_c - F_e) Rn$$

where R = inside drum radius $= \frac{325}{2} = 162.5$ mm

n = number of shoes = 4

μ = co-efficient of friction between shoe lining and drum = 0.25

$$198.94 = 0.25\,(682.22 - 436.62)\, m \times 0.1625 \times 4$$

$$m = 4.985 \text{ kg}$$

(ii) Width of Shoes

Let l = contact length of shoes, mm

b = width of shoes, mm

Now, $$l = R\theta = \frac{162.5 \times \pi}{3} = 170.17 \text{ mm}$$

Area of contact of shoe, $A = lb = 170.17\, b \text{ mm}^2$

Force exerted by shoe on the drum $= Ap = 170.17\, b \times 0.1$

$$= 17.017\, b \text{ N}$$

$$= F_c - F_e$$

$$= (682.22 - 436.62)\, 4.985 = 1224.316 \text{ N}$$

$$b = 71.95 \approx 72 \text{ mm}$$

(iii) Design of Spring

Load on the spring

$$= F_e = 436.62 \times 4.985 = 2176.55 \text{ N}$$

Whal correction factor, $$K_w = \frac{4C - 1}{4C - 4} + \frac{0.615}{C}$$

$$= \frac{4 \times 6 - 1}{4 \times 6 - 4} + \frac{0.615}{6} = 1.2525$$

Shear stress in spring wire, $$\tau = \frac{K_w \times 8F_e C}{(\pi d^2)}$$

$$600 = \frac{1.2525 \times 8 \times 2176.55 \times 6}{\pi d^2}$$

$$d^2 = 69.42$$

$$d = 8.33 \text{ mm}$$

$$D = 6d = 50 \text{ mm}$$

Deflection of spring, $$\delta = \frac{8F_e C^3 n}{Gd}$$

$$10 = \frac{8 \times 2176.55 \times 6^3 \times n}{80 \times 10^3 \times 8.33}$$

Number of turns, $$n = 1.77 \approx 2$$

10.2 BRAKES

A brake is a device to retard or stop the motion of a machine. It absorbs either the kinetic or potential energy of a moving object. The absorbed energy is converted into heat energy, which has to be dissipated.

10.2.1 Types of Mechanical Brakes

The various types of mechanical brakes are :

1. Single block or shoe brake
2. Pivoted block or shoe brake
3. Double block or shoe brake
4. Simple band brake
5. Differential band brake
6. Band and block brake
7. Internal expanding brake

We shall study single block brake, simple band brake and differential and brake.

10.2.2 Factors on which Capacity of a Brake Depends

Capacity of a brake depends upon the following factors :

1. Unit pressure between braking surfaces.
2. Coefficient of friction between contacting surfaces.
3. Velocity of drum.
4. Projected area of friction surfaces.
5. Ability of brake to dissipate heat generated.

10.2.3 Single Block Brake

A single block brake is shown in Fig. 10.7.

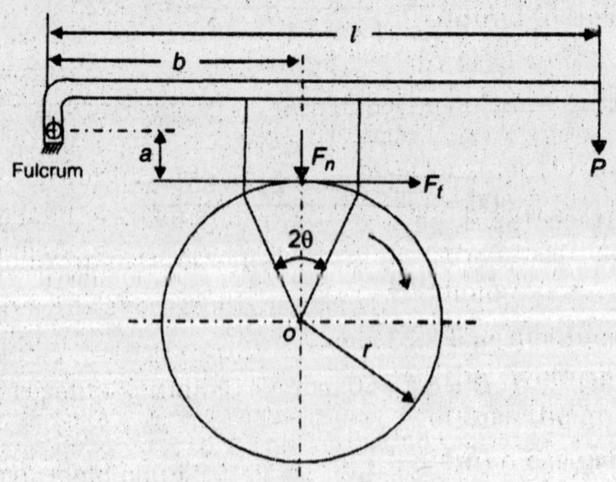

Fig. 10.7 Single block brake.

Let

P = force applied at the end of the lever

F_n = normal reaction between wheel and block

r = radius of wheel

2θ = angle of contact between block and wheel

μ = coefficient of friction

F_t = tangential braking force

$F_t = \mu F_n$

Braing torque, $T_{br} = F_t \,.\, r = \mu\, F_n \,.\, r$

Let the line of action of F_t be at a distance 'a' below the fulcrum, and the wheel rotates clockwise, as shown in Fig. 10.7. The forces acting on the lever are shown in its free body diagram. Taking moments about the fulcrum, we have

$$-F_n \times b + F_t \times a = P \times l$$

$$-F_n \times b + \mu \times F_n \times a = Pl$$

$$F_n = Pl\,(b + a\mu) \quad \text{...(10.18)}$$

Braking torque, $$T_{br} = \frac{\mu\, Pl\, r}{b + a\mu} \quad \text{...(10.19)}$$

Similar results can be obtained when F_t is above the fulcrum and wheel rotates clockwise or anti-clockwise.

Self-energizing brake : When the moment of frictional force adds to the moment of the applied force, then frictional force helps to apply the brake. Such a brake is called self-energizing brake.

Self-locking brake : When the frictional force is great enough to apply the brake without the help of external force, then the brake is said to be of the self-locking type.

10.2.4 Simple Band Brake

A simple band brake is shown in Fig. 10.8.

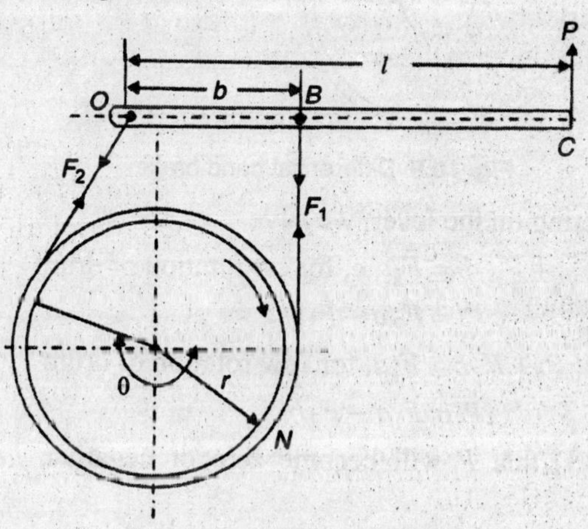

Fig. 10.8 Simple band brake.

Let

F_1, F_2 = tension in band on the tight and slack side respectively

θ = angle or wrap

μ = coefficient of friction between band and drum

r = radius of drum

Now
$$\frac{F_1}{F_2} = \exp(\mu\theta)$$

Braking torque,
$$T_{br} = (F_1 - F_2)\, r \qquad \text{...(10.20)}$$

Taking moments about the fulcrum, we have
$$Pl = F_1 \,.\, b \text{ for cw rotation of drum}$$
$$= F_2 \,.\, b \text{ for ccw rotation of drum}$$

Let

w = width of band

t = thickness of band

σ_t = permissible tensile strength of band material
$$F_1 = \sigma_t\, wt$$

10.2.5 Differential Band Brake

Consider the differential band brake shown in Fig. 10.9.

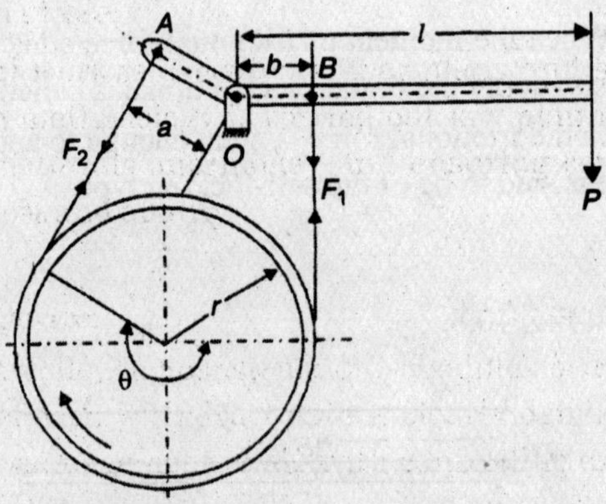

Fig. 10.9 Differential band brake.

Considering the forces acting on the lever, we have
$$P \,.\, l + F_1 \,.\, b = F_2 \,.\, a, \text{ for cw rotation of drum}$$
$$Pl = F_2 a - F_1 b \qquad \text{...(10.22)}$$
and
$$Pl + F_2 b = F_1 a, \text{ for ccw rotation of drum}$$
$$Pl = F_1 a - F_2 b \qquad \text{...(10.23)}$$

We find from Eq. (10.13) that P will become zero or negative, *i.e.*, the brake will become self-locking, when,
$$F_2 a - F_1 b \le 0$$
or
$$\frac{F_2}{F_1} \le \frac{b}{a} \text{ for cw rotation of drum}$$
and
$$\frac{F_1}{F_2} \le \frac{b}{a} \text{ a for ccw rotation of drum}$$

■ **Example 10.6** Design a differential band brake for a winch lifting a load of 50 kN by means of a rope wound round a barrel of 450 mm diameter.

Solution : Given : $W = 50$ kN, $D_b = 450$ mm

1. Preliminary decisions

   Let angle of wrap, $\theta = 240°$, coefficient of friction, $\mu = 0.3$

   Diameter of rope drum $D > D_b = 600$ mm (say)

   The brake is shown in Fig. 10.10.

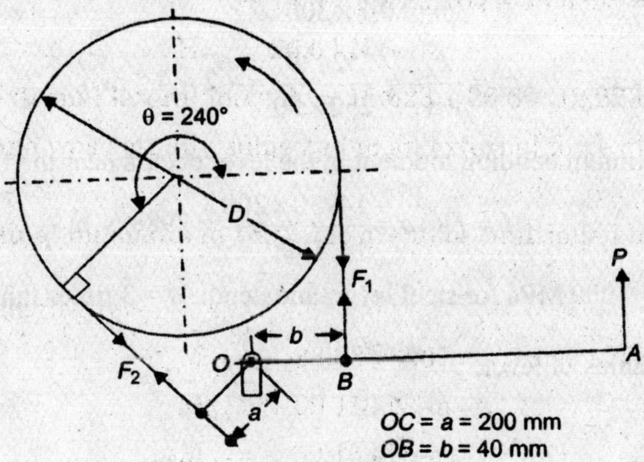

Fig. 10.10

2. Ratio of tensions,

$$\frac{F_1}{F_2} = \exp(\mu\theta)$$

$$= \exp\left(\frac{0.3 \times \pi \times 240}{180}\right)$$

$$= 3.51$$

3. Braking torque,

$$T_{br} = (F_1 - F_2)\, D/2$$

$$= \frac{WD_b}{2}$$

$$F_1 - F_2 = \frac{50 \times 10^3 \times 450}{600} = 37{,}500 \text{ N}$$

$$F_2 = \frac{37{,}500}{2.51} = 14940 \text{ N}$$

$$F_1 = 52439 \text{ N}$$

4. Applied effort at end of lever,

$$P = F_2 \frac{[a - b\exp(\mu\theta)]}{l}$$

To avoid self-locking, let

$$a = 1.5 \times b\exp(\mu\theta) \approx 5b$$

For $a = 200$ mm, $b = \dfrac{200}{5} = 40$ mm

Let $l = 1000$ mm

$$P = \frac{14940\,(200 - 40 \times 3.51)}{1000} \approx 600 \text{ N}$$

5. Band width and thickness,

Band width, $w = \dfrac{F_1}{pr} = \dfrac{52439}{0.7 \times 300} = 249 \approx 250$ mm

Band thickness, $t = \dfrac{F_1}{w\sigma_t} = \dfrac{52439}{250 \times 50} = 4.2 \approx 4.5$ mm

6. Lever : Maximum bending moment in the lever occurs near the fulcrum,

$$M = F_2 \,.\, a = \frac{14930 \times 200}{1000} = 2988 \text{ N.m}$$

Taking $\sigma_t = 1000$ MPa for steel lever, and depth, $d = 3$ times thickness, h

Section modulus of lever, $z = \dfrac{h \times (3h)^2}{6} = \dfrac{3h^3}{2}$

$$z = \frac{M}{\sigma_t}$$

$$\frac{3h^3}{2} = \frac{2988 \times 10^3}{100}$$

$$h = 27.1 \approx 28 \text{ mm}$$

$$d = 84 \text{ mm}$$

7. Pins :' Fulcrum reaction,

$$R = [(52439)^2 + (14940 \cos 60°)^2 + (14940 \sin 60°)^2]^{1/2} = 61290 \text{ N}$$

Pin at *O* and *A*

$$\frac{l_p}{d_p} = 2 \text{ and } p_b = 10 \text{ MPa}$$

$$2d_p \times d_p \times 10 = 61290$$

$$d_p = 55 \text{ mm and } l_p = 110 \text{ mm}$$

Boss diameter, $d_p = 1.5\, d_p \approx 85$ mm

Pin at *B*

$$2d_p \times d_p \times 10 = 14940$$

$$d_p = 28 \text{ mm and } l_p = 56 \text{ mm}$$

Boss diameter, $d_p = 45$ mm

8. Check of the lever at the fulcrum,

$$\text{Section modulus, } z = \frac{\left[\dfrac{100\,(85^3 - 60^3)}{12}\right]}{\left(\dfrac{85}{2}\right)}$$

$$= 85870 \text{ mm}^3 > 29880 \text{ mm}^3$$

Hence safe.

■ **Example 10.7** A band brake acts on the 3/4th of circumference of a drum of 450 mm diameter which is keyed to the shaft. The band brake provides a braking torque of 225 N.m. One end of the band is attached to a fulcrum pin of the lever and the other end to a pin 100 mm from the fulcrum. If the operating force is applied at 500 mm from the fulcrum and the coefficient of friction is 0.25, find the operating force when the drum rotates in anti-clockwise direction.

If the brake lever and the pins are to be made of mild steel having permissible stress in tension and crushing as 70 MPa and for shear 56 MPa, design the shaft, key, lever and pins. The bearing pressure between pin and lever arm may be taken as 8 MPa.

Solution : Given : D = 450 mm, T_{br} = 225 N.m, OB = 100 mm, L = 500 mm, $\mu = 0.25$, $\tau = 56$ MPa, $\sigma_t - \sigma_c = 70$ MPa

(a) The band brake is shown in Fig. 10.11. The operating force P will act upwards to apply the brake when the drum rotates ccw.

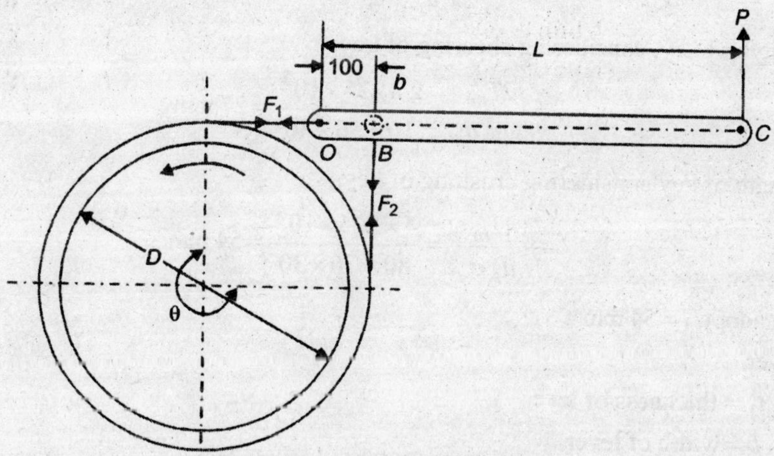

Fig. 10.11

1. Angle of wrap, $\theta = \dfrac{3 \times 360}{4} = 270°$ or $\dfrac{3\pi}{2}$ rad

2. Ratio of tensions,

$$\frac{F_1}{F_2} = \exp(\mu\theta) = \exp\left[\frac{0.25 \times 3\pi}{2}\right] = 3.25$$

3. Braking torque, $T_{br} = (F_1 - F_2)\,D/2$

$$F_1 - F_2 = \frac{2 \times 225 \times 10^3}{450} = 1000 \text{ N}$$

$$F_2 = \frac{1000}{2.25} = 444.45 \text{ N}$$

$$F_1 = 1444.45 \text{ N}$$

4. Taking moments about fulcrum, we get

$$P = \frac{F_2 \times OB}{500} = \frac{444.45 \times 100}{500} = 88.9 \text{ N}$$

5. Shaft design,

Braking torque, $$T_{br} = \left(\frac{\pi}{16}\right) d_s^3\ \tau$$

Shaft diameter, $$d_s = \left[\frac{225 \times 10^3 \times 16}{\pi \times 56}\right]^{1/3} = 27.35 \approx 30 \text{ mm}$$

6. Key size.

Width of key, $$w = \frac{d_s}{3} = \frac{30}{3} = 10 \text{ mm}$$

Thickness of key, $$t = \frac{d_s}{4} = \frac{30}{4} = 7.5 \approx 8 \text{ mm}$$

Length of key, considering shearing of key,

$$L = \frac{2T_{br}}{w\,\tau\,d_s} = \frac{2 \times 225 \times 10^3}{10 \times 56 \times 30} = 27 \text{ mm}$$

Length of key considering crushing of key,

$$l = \frac{4T_{br}}{t\,d_s\,\sigma_c} = \frac{4 \times 225 \times 10^3}{80 \times 70 \times 30} = 54 \text{ mm}$$

We adopt, $l = 54$ mm

7. Lever.

Let t_1 = thickness of lever

b = width of lever

Bending moment of the lever,

$$M = PL = 88.9 \times 500 = 44450 \text{ N.mm}$$

Section modulus of lever,

$$z = \frac{t_1 b^2}{6} = \frac{b^3}{12} \text{ for } b = 2t_1$$

$$z = \frac{M}{\sigma_t}$$

$$\frac{b^3}{12} = \frac{44450}{70}$$

$$b = 20 \text{ mm}$$

$$t_1 = 10 \text{ mm}$$

8. Pins.

Let d_p = diameter of pins

$l_p = 1.25\, d_p$ = length of pins

$p_b = 8$ MPa

Considering bearing of pins, we have

$$F_1 = d_p\, l_p\, p_b$$

$$1444.45 = 1.25\, d_p^2 \times 8$$

$$d_p = 12 \text{ mm}$$

$$l_p = 15 \text{ mm}$$

Check of pins in double shear,

$$1444.45 = 2 \times \frac{\pi}{4} \times 12^2 \times \tau_p$$

$\tau_p = 6.4$ MPa, which is quite safe.

Check of pins in bending,

Maximum bending moment,

$$M_{\max} = \frac{5F_1 l_p}{24} = \frac{5 \times 144.45 \times 15}{24} = 4515 \text{ N.mm}$$

Section modulus, $z_p = \frac{\pi d_p^3}{32} = \frac{\pi \times 12^3}{32} = 170 \text{ mm}^3$

Bending stress $= \frac{M_{\max}}{z_p} = \frac{4514}{170} = 26.5 \text{ MPa} < 70 \text{ MPa}$

9. Eye hole in lever.

Thickness of each eye, $t_e = \frac{l_p}{2} = \frac{15}{2} = 7.5$ mm

Outer diameter of eye, $d_e = 2d_p = 2 \times 12 = 24$ mm

10. Bush size in the eye,

Thickness of bush $= 3$ mm

Diameter of hole in the lever $= d_p + 2 \times 3 = 12 + 6 = 18$ mm

Outer diameter of boss $= 2d_p = 24$ mm

Length of boss $= l_p = 15$ mm

11. Check of lever at the fulcrum in bending,

$$M = PL = 88.9 \times 500 = 44450 \text{ N.mm}$$

$$z = \frac{\left[\frac{15\,(24^3 - 18^3)}{12}\right]}{12} = 833 \text{ mm}^3$$

$$\sigma_b = \frac{44450}{833} = 53.3 < 70 \text{ MPa}$$

Hence safe.

10.2.6 Thermal Considerations in Brake Designing

The energy absorbed by the brake is transformed into heat, which increases the temperature of the rubbing surfaces. This heat must be dissipated to the surrounding air in order to avoid excessive temperature rise of the brake lining. When the temperature increases, the coefficient of friction decreases, which adversely affects the torque capacity of the brake. At high temperature, there is rapid wear of the friction lining which reduces the life of the lining. Therefore, the temperature rise should be kept within permissible range, as given in Table 10.1.

Table 10.1 : Permissible temperature and intensity of pressure limits for friction materials.

Material	*Permissible temperature, °C*	*Permissible intensity of pressure, p, MPa*	*Coefficient of friction, μ*
1. Cast iron on cast iron	300	1.00	0.15 – 0.20
2. Wood on cast iron	60	0.35	0.25 – 0.30
3. Leather on cast iron	60	0.25	0.30 – 0.50
4. Woven - asbestos on metal	250	0.65	0.35 – 0.40
5. Moulded - asbestos on metal	250	1.00	0.40 – 0.45
6. Sintered metal on metal	300	2.75	0.20 – 0.40

The energy absorbed (or heat generated) and the rate of wear of the brake lining at a particular speed are dependent on the normal pressure between the braking surfaces, therefore it is an important factor in the design of brakes. The permissible normal pressure between the braking surfaces depends upon the material of the brake lining, coefficient of friction and the maximum rate at which the energy is to be absorbed.

The energy absorbed (or heat generated) is given by :

$$E = H_g = \mu\, pAv \text{ watts}$$

where μ = coefficient of friction

p = normal pressure between the braking surfaces, MPa

A = projected area of contacting surfaces, m^2

v = peripheral speed of brake drum, m/s

Also heat generated,

$$H_g = E_k + E_p$$

where E_k = total kinetic energy absorbed

E_p = total potential energy absorbed

Heat dissipated,

$$H_d = C\,(t_1 - t_2)\,A_r$$

where C = coefficient of heat transfer, $W/(m^2\,.\,°C)$

$\simeq 29.5$ to $44\ W/(m^2\,.\,°C)$ for $\Delta t = 40 - 200°C$

t_1 = temperature of exposed radiating surface, °C

t_2 = temperature of surrounding air, °C

A_r = area of radiating surface, m^2

It is very difficult to precisely calculate temperature rise. The expressions given above for heat dissipation are quite approximate. It has been found that 10 to 25 percent of the heat generated is immediately dissipated to the surrounding air while the remaining heat is absorbed by the brake drum, causing its temperature to rise. The rise in temperature of brake drum is given by,

$$\Delta t = \frac{H_g}{mC}$$

where Δt = temperature rise of brake drum assembly, °C

m = mass of the brake drum assembly, kg

C = specific heat of brake drum material, J/(kg.°C)

H_g = heat generated by the brake, J

In brakes, it is very difficult to precisely calculate the temperature rise. In preliminary design analysis, the product $p.v$ is considered in place of temperature rise. When the coefficient of friction is constant, the rate of heat generated is proportional to the product $p.v$, where p = intensity of normal pressure in MPa (N/mm^2), and v is the rubbing velocity in m/min. The recommended values of product $p.v$ are given in Table 10.2.

Table 10.2 : Recommended values of $p.v$

Application	$p.v$
1. Intermittent applications, comparatively long period of rest and poor dissipation of heat.	115
2. Continuous application and poor dissipation of heat.	58
3. Continuous application and good dissipation of heat.	175
4. Vehicle brake.	125

■ **Example 10.8** A flywheel of 120 kg mass and 400 mm radius of gyration is rotating at 600 rpm. It is brought to rest by a brake. The mass of the brake drum assembly is 6 kg. The brake drum is mass of cast iron FG260 (C = 460 J/kg.°C). Assuming that the total heat generated is absorbed by the brake drum, calculate its temperature rise.

Solution : Given : M = 120 kg, K = 0.4 m, N = 600 rpm, m = 6 kg, C = 460 J/(kg °C)

$$\omega_1 = \frac{2\pi N}{60} = \frac{2\pi \times 600}{60} = 62.83\ \text{rad/s}$$

$$\omega_2 = 0$$

$$I = MK^2 = 120 \times (0.4)^2 = 19.2\ \text{kg.m}^2$$

$$\Delta KE = \frac{1}{2} I (\omega_1^2 - \omega_2^2) = \frac{1}{2} \times 19.2 [(62.83)^2 - 0] = 37897 \text{ J}$$

$$\Delta t = \frac{H_g}{mC} = \frac{\Delta K_E}{mC} = \frac{37897}{6 \times 460} = 13.73°\text{C}$$

■ **Example 10.9** A vehicle of mass 1000 kg is moving down the hill at a slope of 1 : 6 at 60 km/h. It is to be stopped in a distance of 40 m. The diameter of the tyre is 0.6 m. Determine the average braking torque to be applied to stop the vehicle, neglecting all the frictional energy except for the brake. If the mass of the brake drum is 20 kg, determine the average temperaure rise of the drum. The specific heat for cast iron may be taken as 520 J/kg.°C.

Solution : Given : M = 1000 kg, slope = 1 : 6, v = 60 km/h, s = 40 m, d = 0.6 m, m = 20 kg, C = 520 J/kg.°C

$$v = \frac{60 \times 10^3}{3600} = 16.67 \text{ m/s}$$

$$E_K = \frac{1}{2} mv^2 = \frac{1}{2} \times 1000 \times (16.67)^2 = 138{,}889 \text{ N.m}$$

$$E_P = Mgs \times \text{slope} = 1000 \times 9.81 \times 40 \times \frac{1}{6} = 65{,}400 \text{ N.m}$$

$$E = E_K + E_P = 138{,}889 + 65{,}400 = 204{,}289 \text{ N.m}$$

Tangential braking force,

$$F_t = \frac{E}{s} = \frac{204{,}289}{40} = 5107.2 \text{ N}$$

Average braking force required to stop the vehicle,

$$T_b = F_t \times \frac{d}{2} = 5107.2 \times \frac{0.6}{2} = 1532.2 \text{ N.m}$$

Now,
$$E = m \times C \times \Delta t$$

$$\Delta t = \frac{204{,}289}{20 \times 520} = 19.6°\text{C}$$

SOLVED QUESTIONS FROM UNIVERSITY EXAMINATIONS

■ **Example 10.10** Describe a centrifugal clutch. How this clutch is designed? Where does it find applications? [MDU, Dec., 2007]

Solution : Refer to art. 10.1.8.

■ **Example 10.11** Explain various types of brakes indicating the applications. What is self energizing condition brakes? Discuss thermal considerations in brake design. [MDU, Dec., 2007]

Solution : Refer to art. 10.2.1.

■ **Example 10.12** With the help of a neat sketch explain the working of single plate clutch. Give the design procedure. [MDU, May, 2008]

Solution : A single plate clutch is shown in Fig. 10.12. It consists of a clutch plate whose both sides are lined with a frictional material. It is mounted on the hub which is free to move axially along the splines of the driven shaft. The pressure plate is mounted inside the clutch body which is bolted to the flywheel. Both the pressure plate and flywheel rotate with the engine crankshaft. The pressure plate pushes the clutch plate towards the flywheel by a set of springs arranged radially inside the body. The release levers are carried on pivots suspended from the case of the body. When the clutch pedal is pressed the forked shaft moves forward on which bearing is mounted. The release levers are arranged in such a manner so that the pressure plate moves away from the flywheel by the inward movement of a thrust bearing. When the clutch pedal is pressed the clutch disengages otherwise remains in engaged position.

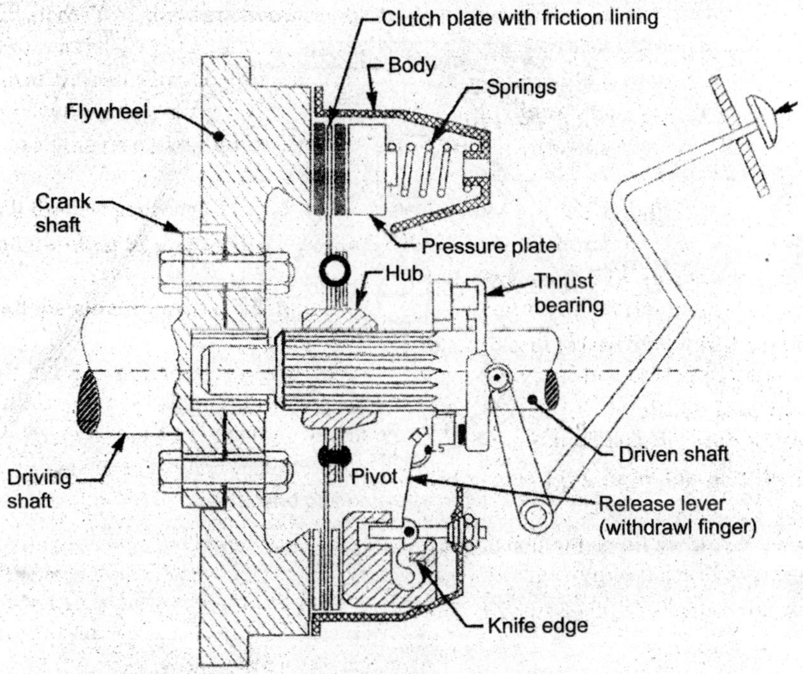

Fig. 10.12 Single disc or plate clutch.

Design procedure :

1. For the given power to be transmitted and speed, calculate the frictional torque to be transmitted.

$$P = \frac{2\pi N T_f}{60 \times 10^3}$$

2. For uniform wear consideration, axial force is given by,

$$F_a = 2\pi r_2 (r_1 - r_2) p_a$$

3. Mean radius, $r_m = \frac{1}{2}(r_1 + r_2)$ for uniform wear

4. $T_f = n \mu F_a r_m$, n = number of frictional sides of plate.
5. Calculate the r_1 and r_2 radii of clutch plate.
6. Design the spring.

■ **Example 10.13** With the help of a neat sketch, describe the principle of operation of an internal expanding shoe brake. Derive the expression for the braking torque. [MDU, May, 2008]

Solution : The internal expanding shoe brake shown in Fig. 10.13 consists of two shoes S_1 and S_2. One end of the shoes pivots about the fixed fulcrum O_1 and O_2 and the other end faces the cam. The outer surface of the rim is lined with some frictional material. When brakes are applied, cam rotates which pushes the shoes outwards to press the brake lining against the rim of the drum. The shoes are held in off position by the spring S, which is in compression. When the cam rotates, the spring is in tension and the brakes are applied.

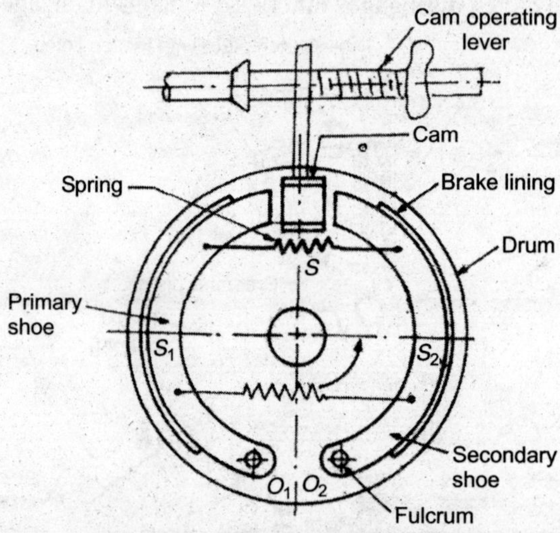

Fig. 10.13 Internal expanding brake.

Expression for braking torque : Consider the forces acting on the brake when are drum rotates in ccw direction as shown in Fig. 10.14.

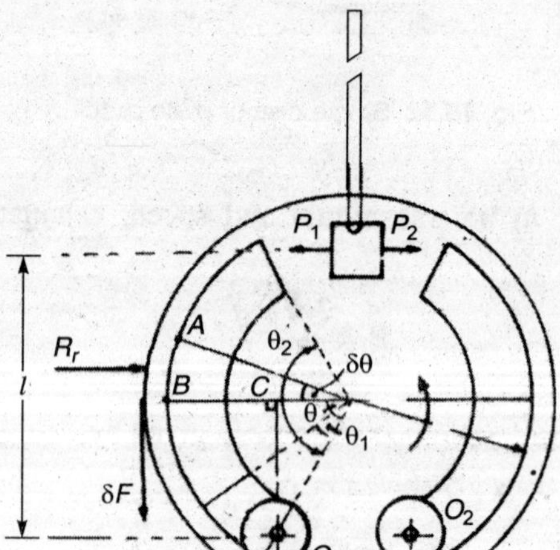

Fig. 10.14 Forces on the brake.

Let

p_1 = maximum intensity of normal pressure

p_n = normal pressure

r = internal radius of the drum

b = width of brake lining

T_b = braking torque

P_1 = force exerted by the cam on the leading or primary shoe

P_2 = force exerted by the cam on the trailing or secondary shoe

R_n = normal force

F = friction force

μ = Coefficient of friction between shoe and drum

M_n = moment of normal forces

M_f = moment of frictional force

Consider a small element AB of brake lining subtending an angle $\delta\theta$ at the centre of the drum. Join O_1 to O. The pressure distribution on the shoe is assumed to be nearly uniform, however, the shoe wears out more at the free end. The rate of wear of the shoe lining varies directly as the perpendicular distance from O_1 to B, *i.e.*, O_1C.

From Fig. 10.14, we get

$$O_1C = OO_1 \sin\theta \quad \text{...(i)}$$

Normal pressure at B

$$P_n \propto \sin\theta$$

or

$$P_n = p_1 \sin\theta$$

Area of the element = $b.r.\delta\theta$.

Normal force acting on the element

$$\delta R_n = \text{Normal pressure} \times \text{Area of element}$$

$$= P_n \times b.r.\delta\theta = p_1 \sin\theta \, b.r.\delta\theta$$

We know that friction force on the element is given by

$$\delta F = \mu.\delta R_n = \mu \, p_1 \sin\theta.b.r.\delta\theta \quad \text{...(ii)}$$

Braking torque on the element about O

$$\delta T_b = \delta F.r = \mu \, p_1 \sin\theta.b.r\delta\theta.r$$

$$= \mu p_1 \sin\theta.br^2 \, \delta\theta$$

Total braking torque for whole shoe about O

$$T_b = \mu \, p_1 \, br^2 \int_{\theta_1}^{\theta_2} \sin\theta.d\theta \qquad \because \int \sin\theta \, d\theta = -\cos\theta$$

$$= \mu \, p_1 \, br^2 (\cos\theta_1 - \cos\theta_2)$$

Moment of normal force δR_n of element about O_1

$$\delta M_n = \delta R_n \times O_1C \qquad \text{(Substituting } O_1C \text{ from equation (i))}$$

$$= p_1 \sin\theta.b.r\delta\theta \times (OO_1 \sin\theta)$$

$$= OO_1.p_1 \sin^2\theta.b.r.\delta\theta$$

So, total moment of the normal force about O_1

$$M_n = OO_1 \cdot p_1 \cdot b \cdot r \int_{\theta_1}^{\theta_2} \sin^2\theta \, d\theta \qquad \left(\because \sin^2\theta = \frac{(1-\cos 2\theta)}{2}\right)$$

$$= OO_1 \; p_1 \; br \int_{\theta_1}^{\theta_2} \left(\frac{1-\cos 2\theta}{2}\right) d\theta$$

$$= \frac{OO_1 \; p_1 \; br}{2} \left[\theta - \frac{\sin 2\theta}{2}\right]_{\theta_1}^{\theta_2}$$

$$= \frac{1}{2} p_1 \; br \; OO_1 \left[(\theta_2 - \theta_1) + \frac{1}{2}(\sin 2\theta_1 - \sin 2\theta_2)\right]$$

Moment of frictional force δF about O_1

$$\delta M_f = \delta F \, . BC \qquad (\because BC = OB - OC = r - OO_1 . \cos\theta)$$

$$= \mu \; p_1 \; br . \sin\theta \, (r - OO_1 . \cos\theta) \, \delta\theta$$

(Substituting the value of δF from equation (ii))

$$= \mu \; p_1 \; br \left(r \sin\theta - \frac{OO_1 \sin 2\theta}{2}\right) . \delta\theta$$

Total moment of frictional force

$$M_f = \mu \; p_1 \; br \int_{\theta_1}^{\theta_2} \left(r \sin\theta - \frac{OO_1 \sin 2\theta}{2}\right) d\theta$$

$$= \mu \; p_1 \; br \left[-r\cos\theta + \frac{OO_1}{4} \cos 2\theta\right]_{\theta_1}^{\theta_2}$$

$$= \mu \; p_1 \; br \left[r(\cos\theta_1 - \cos\theta_2) + \frac{OO_1}{4}(\cos 2\theta_2 - \cos 2\theta_1)\right]$$

For leading shoe, take moments about O_1

$$P_1 . l = M_n - M_f \qquad \ldots\text{(iii)}$$

and for trailing shoe, take moments about O_2

$$P_2 . l = M_n + M_f \qquad \ldots\text{(iv)}$$

If the moment of the friction forces M_f about O_1 is greater than the moment of normal forces about O_1, *i.e.*, $M_f > M_n$ the brake will be self-locking for leading shoe.

Self-locking is useful when the motion only in one direction is allowed such as in conveyers, hoists, etc. In case of a lift if it is moving upwards or downwards and the power is off, it remains suspended (or hangs) at that particular location because of self-locking. Thus self-locking acts as a safety device.

■ **Example 10.14** Compare dry and wet clutch design. [MDU, Dec., 2008]

Solution : Refer to art. 10.1.6.

■ **Example 10.15** A multiple-disc clutch is to be made up of 6 steel and 5 bronze discs, with an inner radius of 25 mm. Determine the outer diameter and the axial pressure to transmit a torsional moment of 18 N.m. [MDU, Dec., 2008]

Solution : Given : $n_1 = 6, n_2 = 5, r_2 = 25$ mm, $T = 18$ N.m

Number of pairs of contact surfaces,

$$n = n_1 + n_2 - 1 = 6 + 5 - 1 = 10$$

$$T = n \mu F_a r_m$$

Assume $p = 0.1$ MPa and $\mu = 0.1$,

$$F_a = 2\pi p_a r_2 (r_1 - r_2)$$

$$= 2\pi \times 0.1 \times 25 (r_1 - 25) = 5\pi (r_1 - 25) \text{ N}$$

$$r_m = \frac{1}{2}(r_1 + r_2) = \frac{1}{2}(r_1 + 25) \text{ mm}$$

$$18 \times 10^3 = 10 \times 0.1 \times 5\pi (r_1 - 25) \times \frac{1}{2}(r_1 + 25)$$

$$2291.83 = r_1^2 - 625$$

$$r_1 = 54 \text{ mm}$$

Outer diameter, $D_1 = 2r_1 = 2 \times 54 = 108$ mm

Axial force, $F_a = 5\pi (54 - 25) = 455.5$ N

■ **Example 10.16** Is self locking in brakes desirable? If so, under what conditions?

[MDU, Dec., 2008, May, 2009]

Solution : When the frictional force is great enough to apply the brake without the help of external force, then the brake is said to be of the self-locking type.

Self-locking is useful when the motion in one direction only is allowed such as in conveyors, hoists, etc. In case of a life if it is moving upwards or downwards and the power goes off, it remains suspended (or hanged) at that particular location because of self-locking. Thus self-locking acts as a safety device.

■ **Example 10.17** A single block brake with torque capacity of 15 N.m is shown in Fig. 10.15. The coefficient of friction is 0.3 and the maximum pressure on the brake lining is 1 N/mm^2. The width of the block is equal to its length.

Calculate :

(i) The actuating force (ii) The dimensions of the block (iii) The resultant hinge-pin reaction, and (iv) The rate of heat generated, if the brake drum rotates at 50 rpm. [MDU, Dec., 2008, May, 2009]

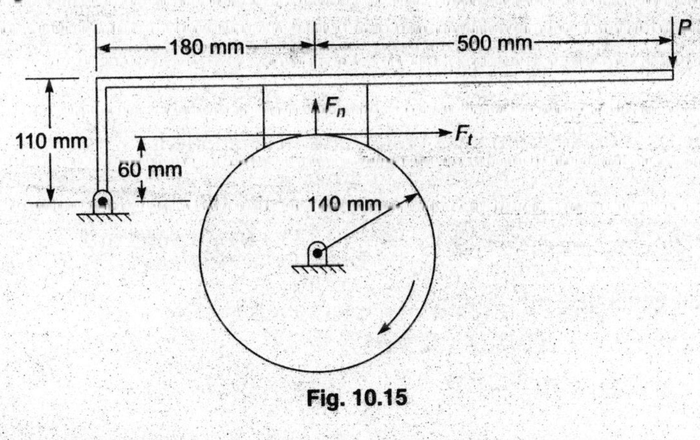

Fig. 10.15

Solution : $T_{br} = 15$ N.m, $\mu = 0.3$, $p = 1$ N/mm^2, $w = r$, $N = 50$ rpm, $b = 180$ mm, $l = 680$ mm, $a = 60$ mm, $r = 140$ mm

(i) Let F_n = normal reaction between block and brake drum

F_t = tangential braking force

$= \mu F_x$

Braking torque, $$T_{br} = F_t \times r = \mu F_n \times i$$

$$15 \times 10^3 = 0.3 F_n \times 140$$

$$F_n = 357.14 \text{ N}$$

$$F_t = 0.3 \times 357.14 = 107.14 \text{ N}$$

Taking moments about the fulcrum, we have

$$F_n \times b = F_t \times a + P \times l$$

$$357.14 \times 180 = 107.14 \times 60 + P \times 680$$

Actuating force, $$P = 85.08 \text{ N}$$

(ii) Area of shoe = width × length = $w \times h = w^2$

Maximum pressure on brake lining,

$$p = \frac{F_n}{w^2}$$

$$1 = \frac{357.14}{w^2}$$

$$w = 18.898 \simeq 19 \text{ mm}$$

$$h = 19 \text{ mm}$$

(iii) Considering the equilibrium of forces acting on the brake in horizontal and vertical directions, we have

$$R_x = 107.14 \text{ N}$$

$$R_y = F_n - P = 357.14 - 85.08 = 272.06 \text{ N}$$

Resultant hinge-pin reaction, $R = \sqrt{F_x^2 + R_y^2}$

$$= \sqrt{(107.14)^2 + (272.06)^2} = 292.4 \text{ N}$$

(iv) $$\omega = \frac{2\pi N}{60} = \frac{2\pi \times 50}{60} = 5.236 \text{ rad/s}$$

Initial velocity of brake drum,

$$v_i = \omega r = 5.236 \times 0.140 = 0.733 \text{ m/s}$$

Final velocity of brake drum after brake have been applied,

$$v_f = 0$$

Average velocity, $v_{avg} = \frac{1}{2}(v_i + v_f) = \frac{1}{2}(0.733 + 0) = 0.3665$ m/s

Rate of heat generated $= F_t \times v_{avg}$

$$= 107.14 \times 0.3665$$

$$= 39.27 \text{ W}$$

■ **Example 10.18** A multi-disk clutch, steel on bronze, is to transmit 4 kW at 750 rpm. The inner and outer diameters of the contacting surfaces are 44 mm and 72 mm respectively. The coefficient of friction is 0.1 when the clutch operates in oil. The average pressure is 0.35 N/mm^2 maximum. Determine :

(i) Total disks of steel and bronze required.

(ii) Axial force required. [MDU, May, 2009]

Solution : Given : $P = 4$ kW, $N = 750$ rpm, $r_1 = 36$ mm, $r_2 = 22$ mm, $\mu = 0.1$, $p = 0.35$ N/mm^2

(i)
$$P = \frac{2\pi N T_f}{60 \times 10^3}$$

$$T_f = \frac{60 \times 10^3 \times 4}{2\pi \times 750} = 50.93 \text{ N.m}$$

$$r_m = \frac{1}{2}(r_1 + r_2) = \frac{1}{2}(36 + 22) = 29 \text{ mm for uniform wear theory.}$$

$$T_f = n \mu F_a r_m$$

$$50.93 \times 10^3 = n \times 0.1 \times 677.3 \times 29$$

$$n = 25.9 \simeq 26$$

Now
$$n = n_1 + n_2 - 1$$
$$26 = n_1 + n_2 - 1$$
or
$$n_1 + n_2 = 27$$
$$n_1 = 14, n_2 = 13$$

Number of steel plates, $n_1 = 14$

Number of bronze plates, $n_2 = 13$

(ii) Axial force, $F_a = 2\pi r_2 (r_1 - r_2)\, p$ for uniform wear theory

$$= 2\pi \times 22\,(36 - 22) \times 0.35 = 677.3 \text{ N}$$

■ **Example 10.19** Distinguish between uniform pressure theory and uniform wear theory. [MDU, Dec., 2009]

Solution : Uniform pressure theory : In uniform pressure theory the pressure on the clutch plate surface remains uniform at all radii. Therefore,

$$p \propto r \text{ or } \frac{p}{r} = \text{const.}$$

Such a situation arises in relatively new clutches.

Uniform wear theory : In uniform wear theory, the pressure on the clutch plate surface varies inversly with the plate radius.

$$p \propto \frac{1}{r} \text{ or } pr = \text{const.}$$

Such a situation arises in relatively worn out clutches.

■ **Example 10.20** A multiple-disk wet clutch is to be designed for transmitting a torque of 85 N.m. Space restrictions limit the outside disc diameter to 100 mm. Design values of molded friction material and steel disks to be used, when $\mu = 0.06$, $p_{max} = 1.4$ MPa. Determine approximate values for the disk inside diameter and the total number of disks. [MDU, Dec., 2009]

Solution : Given : $T_f = 85$ N.m, $r_1 = 50$ mm, $n = 0.06$, $p_{max} = 1.4$ MPa

The uniform wear theory is applicable for worn out wet friction clutches. According to this theory the intensity of pressure is inversely proportional to the radius of clutch plate. Therefore, maximum pressure occurs at the inside radius of friction plate surface.

$$\therefore \quad F_a = 2\pi r_2 (r_1 - r_2)\, p_{max}$$

$$= 2\pi r_2 (50 - r_2) \times 1.4 = 8.796\, r_2 (50 - r_2) \text{ N}$$

Frictional torque, $T_f = n \mu F_a r_m$

where
$$r_m = \frac{1}{2}(r_1 + r_2) = \frac{1}{2}(50 + r_2) \text{ mm}$$

$$85 \times 10^3 = n \times 0.06 \times \frac{1}{2}(50 + r_2) \times 8.796\, r_2 (50 - r_2)$$

$$= 0.26388 \times n = (2500 - r_2^2)\, r_2$$

or
$$n\, r_2 (2500 - r_2^2) = 322116$$

Let
$$\frac{r_1}{r_2} = 2 \text{ or } r_2 = \frac{r_1}{2} = \frac{50}{2} = 25 \text{ mm}$$

$$\therefore \quad n \times 25 (2500 - 625) = 322116$$

$n = 6.87 \simeq 8$, as pairs of contacting surfaces have to be even.

$$n = n_1 + n_2 - 1$$

$$8 = n_1 + n_2 - 1$$

or
$$n_1 + n_2 = 9$$

$$n_1 = 5,\ n_2 = 4$$

Inside diameter of disk, $D_2 = 2r_2 = 50$ mm

Total number of disks = 9.

■ **Example 10.21** A short shoe external drum brake shown in Fig. 10.16, has 40 mm width, coefficient of friction 0.3 and $p_{max} = 600$ kPa. Determine the limiting force that can be applied to the lever, braking torque and radial force on the lever pivot. Also determine whether the brake is self-energizing or de-energizing for the direction of rotation involved. [MDU, Dec., 2009]

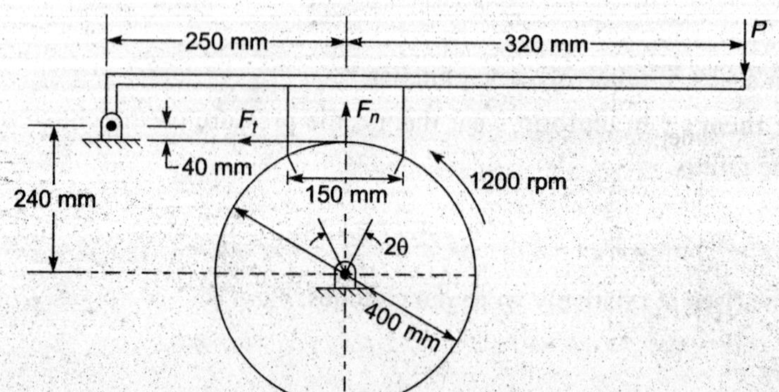

Fig. 10.16

Solution : Given : width of shoe, $h = 40$ mm

length of shoe $w = 150$ mm

Drum radius, $r = 200$ mm

$$b = 250 \text{ mm}, l = 570 \text{ mm}, a = 40 \text{ mm}, N = 1200 \text{ rpm (ccw)}$$

$$p_{max} = 0.6 \text{ MPa}, \mu = 0.3$$

$$\sin\theta = \frac{\omega}{2r} = \frac{150}{2 \times 200} = 0.375$$

$$\theta = 22°$$

Angle of contact $2\theta = 44°$

Area of shoe,

$$A = h \times 2r \sin\theta = 40 \times 2 \times 200 \sin 22° = 6000 \text{ mm}^2$$

$$p = \frac{F_n}{A}$$

$$F_n = pA = 0.6 \times 6000 = 3600 \text{ N}$$

$$F_t = \mu F_n = 0.3 \times 3600 = 1080 \text{ N}$$

Taking moments about fulcrum, we have

$$F_n \times b = F_t \times a + P \times l$$

$$3600 \times 250 = 1080 \times 40 + P \times 570$$

$$P = 1503.16 \text{ N}$$

Braking torque, $T_{br} = F_t \times r = 1080 \times 0.2 = 216 \text{ N.m}$

We observe that the frictional force F_t increases the effect of applied load, therefore, the brake is self-energizing.

EXERCISES

1. Determine the main dimensions of a single plate friction clutch with both sides effective to transmit 30 kW at 3000 rev./min. The pressure intensity of the friction lining is not to exceed 125 kN/m^2. Surface speed at mean radius must not exceed 2000 m/min. The outside diameter of the friction surface is to be 1.3 times the inside diameter. The axial thrust is to be provided by six springs of about 300 mm mean diameter. Sketch the clutch designed by you.
2. A double plate lever release clutch is required for a motor having a maximum torque capacity of 150 N-m. The width of the facings is limited to one-fourth at their outside diameter. The allowable unit pressure is 0.25 MPa and the coefficient of friction is 0.35. There are 9 pressure springs having outside diameter of 300 mm. The movement of the pressure plate during release is to be 2.5 mm, and the effective pedal movement, 120 mm. An overload torque capacity of 100 per cent is to be provided for starting and shock loading. Determine the outside and inside facing diameters.
3. 10 kW is to be transmitted by a multiple friction clutch from a shaft running at 1500 rev./min. Design and draw the clutch. Assume all data suitably.

4. Design and sketch a friction clutch with two plates, both sides of each plate being effective. The clutch is to transmit 60 kW of 1500 rev./min.

5. A leather-faced cone clutch is to transmit 20 kW at 900 rev./min. The large end diameter of the clutch is 300 mm and the cone face angle is 10 degrees. The coefficient of friction is 0.25. Determine :
   (a) required face width of the clutch to limit the bearing pressure on the contact surface to a maximum of 0.15 MPa.
   (b) axial force necessary to engage the clutch.
   Assume uniform wear.

6. The rope drum of an elevator hoist is 600 mm in diameter and the speed of the elevator is 2.5 m/s. This drum is fitted with a brake drum 900 mm in diameter having four cast iron brake shoes each subtending an arc of 45 degrees on the brake drum. The elevator weighs 18 kN when loaded and the brake is to have sufficient capacity to stop the elevator in 3.5 m.
   Determine the principal dimensions of the brake and prepare its dimensioned sketch. How much heat is generated in stopping this elevator?

7. A simple brake band has a 750 mm drum fitted with a steel band 4 mm thick, lined with Ferodo, having a coefficient of friction of 0.3, when not sliding. The arc of contact is 240 degrees. The brake drum is attached to a 600 mm diameter hoisting drum that sustains a rope load of 10 kN. The operating forces has a moment arm of 1.5 m and the band is attached at 150 mm away from the pivot point.
   (a) Find the force required to just support the load.
   (b) What force will be required if the direction of rotation is reversed?
   (c) What width of steel band is required if the tensile stress is limited to 50 MN/m^2?
   (d) How will you convert thin as a self-locking brake?

8. A simple band brake operates on a drum 600 mm diameter that is running at 200 rev./min. The coefficient of friction is 0.25. The brake band has a contact of 270 degrees and one end is fastened to a fixed pin and the other end to the brake arm 125 mm from the fixed pin. The straight brake arm is 750 mm long and is placed perpendicular to the diameter that bisects the angle of contact. Determine :
   (a) the minimum pull necessary on the end of the brake arm to stop the wheel if 40 kW is being absorbed.
   (b) direction of rotation for the minimum pull.
   (c) width of band 2 mm thick if the maximum tensile stress in the band is not to exceed 50 N/mm^2.
   Draw a neat dimensioned sketch of the brake.

9. A mine hoist is equipped with a block brake having two 60 degree shoes each 300 mm wide and faced with asbestos blocks. The cast steel brake drum is 4 m in diameter by 350 mm wide and is attached to a rope drum 3 in diameter. The mine is 300 m deep and the weight of cage handled by the brake is 100 kN. This hoist operates continuously, lowering in 1.25 min, raising in 1.5 min, and resting while loading and unloading 15 s per round trip. Assume the radiating surface to be 20 per cent greater than the exposed surface of the brake drum and the coefficient of friction to be 0.3. Determine (a) the probable operating temperature of the brake, and (b) the probable variation in temperature during each cycle of operation.

10. A passenger car weighing 50 kN and moving on level ground at 20 km/h must be stopped in the shortest distance possible. The diameter of the steel wheels is 600 mm. Assuming a coefficient of friction of 0.35 between the wheels and the rails, and a friction area of the brake shoes of 20×10^3 mm^2 at each of the four wheels, determine the distance and the time needed to stop the car.

❑❑❑

2208

B.E. 5th Sem. (Mech. Engg.)

Examination – December, 2009

MECHANICAL MACHINE DESIGN - I

Paper : ME-303-E

Time : Three Hours] *[Maximum Marks : 100*

Before answering the questions, candidates should ensure that they have been supplied the correct and complete question paper. No complaint in this regard, will be entertained after examination.

Note : (i) Attempt any *five* questions. All questions carry equal marks.

(ii) Use of PSG design data book is permitted.

(iii) Any missing data may be suitable assumed.

1. (a) Explain the following with examples : 10

(i) Manufacturing process considerations

(ii) Standardization

(b) Graphically represent the clearance, transition and interference fits for hole and shaft base systems. What do you understand by allowance? 10

2. (a) Explain the importance of material selection decision for machine elements. 10 + 10

(b) Give a generalized definition of Factor of Safety. Explain the factors responsible for uncertainties in design necessitating a suitable factor of safety.

3. A bracket is supported by means of 4 rivets of same size, as shown in figure 1. Determine the diameter of the rivet if the maximum allowable shear stress is 100 MN/mm^2. 20

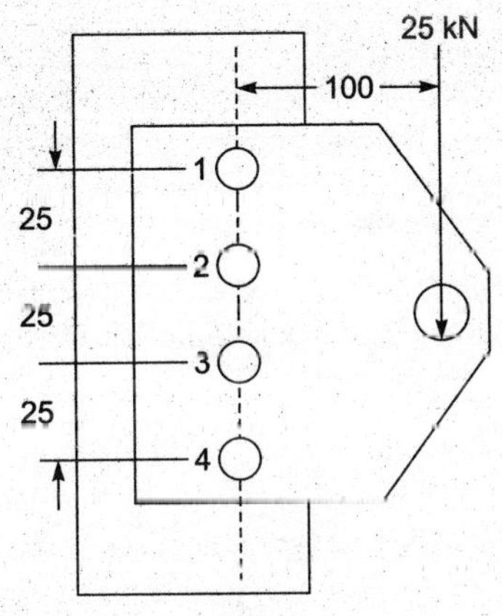

Fig. 1

4. Design and draw a cotter joint subjected to slow reversal of 30 kN load. Clearly state the design decisions taken. 20

5. A V-belt drive is to transmit 15 kW to a compressor. The motor speed is 750 rpm and the compressor pulley runs at 250 rpm. Give complete design of the drive. 20

6. Design of a pin type flexible coupling to connect two shafts of 50 mm diameter transmitting 45 kW at 1200 rpm. The bearing pressure between rubber bush and cast iron flange is 0.5 N/mm^2 and the allowable shear stress for the pin is 30 N/mm^2. Thicknesses of input and output flanges are taken as diameter of bush and half of shaft diameter respectively. Assume that there are six pins used in the coupling and the diameter of each pin is 20 mm. 20

7. (a) Distinguish between Uniform pressure theory and Uniform wear theory. 5 + 15

   (b) A multiple disk wet clutch is to be designed for transmitting a torque of 85 Nm. Space restriction limit the outside disk diameter to 100 mm. Design values of molded friction material and steel disks to be used are $\mu = 0.06$, $P_{max} = 1.4$ MPa. Determine appropriate values for the disk inside diameter and the total number of disks.

8. A short shoe external drum brake is shown in figure 2 has 40 mm width, coefficient of friction 0.3 and $P_{max} = 600$ kPa. Determine the limiting force that can be applied to the lever, braking torque and radial force on the lever pivot. Also determine whether the brake is self-energizing or de-energizing for the direction of rotation involved. 20

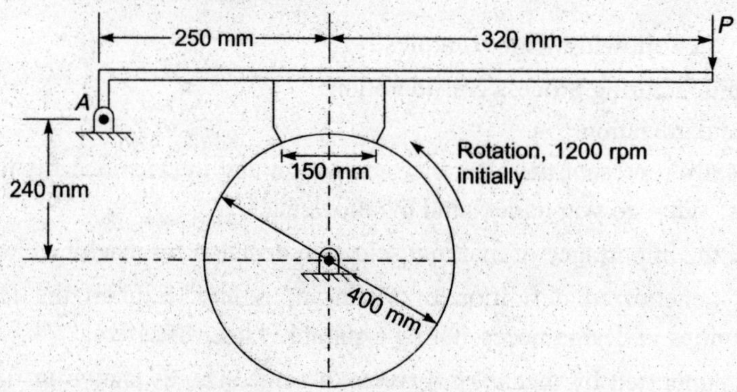

Fig. 2

❑❑❑

INDEX

A

Aesthetic considerations 10
Assembly considerations 43

B

Brain storming 4
Brakes 320
Bush type flange coupling 178

C

C.I. castings design 45
Cast irons 26
Centrifugal clutch 316
Chain drive 273
Clamp or compression coupling 171
Classification of materials 25
Clutches 305
Combined fluctuating stresses 90
Completely reversed stresses
Composite materials 39
Cone clutch 308
Couplings 168

D

Design for finite (limited) life 83
Design for infinite (unlimited) life 83
Design of welded joints 123
Design philosophy 1
Design process 5
Differential band brake 322
Disc clutch 306
Dry and wet clutches 308

E

Earle model 6
Eccentric loading on bolts 216
Eccentrically loaded riveted joints 115
Eccentrically loaded welded joints 128
Elastic constants 61
Endurance limit 77
Ergonomic considerations 11

F

Factor of safety 25, 61
Factors affecting endurance limit 80
Fatigue considerations 77
Fatigue failures 76
Fatigue life 77
Fatigue strength 77
Fatigue stress – concentration factor 78
Fits and tolerances 50
Flat belt drive 259
Fluctuating loading 73
Fluctuating stresses 75
Flywheel 183
Forged end flange coupling 173
Frames and housings 45

G

Gib and cotter joint 139
Good design requirements 16

H

Heat treatment methods 34

I

Interchangeability 13
Internal expanding shoe brake 332

K

Keys 162
Knuckle joint 143

L

Lozenge joint 111

M

Machine design 2
Manufacturing considerations 43
Materials selection 18
Mechanical engineering design 2
Mechanical joints 100
Modes of failure 63
Morphology of design 2
Multiple disc clutch 307

N

Non-ferrous metals 35
Non-metallic materials 37
Notch sensitivity 78

P

Plastics 38
Polygonal effect in chain 280
Polymers 38
Power screws 222
Preferred numbers 13
Principal strains 60

R

Rigid protected type flange coupling 174
Riveted boiler joints 105
Riveted joints 100
Ropes 283

S

Screw jack 235
Screw thread forms 210
Screwed joints 209
Shafts 247
Simple band brake 321
Single block brake 320
Sleeve or muff coupling 168
Special alloy steels 32
Spigot and socket cotter joint 136
Splines 165
Standardisation 12
Static loading 56
Steel castings design 47
Steel wire rope drums 288
Steels 28
Strength considerations 56
Stress – concentration 73
Stress – strain diagram 20
Stresses in machine members 57
Stresses in screw 227
Stresses in screwed joints 212
Structural riveted joint 111
Support and retainment 44

T

Theories of failure 64
Thermal considerations in brake designing 328

V

V-belts drive 269

W

Weighted factor method 40
Welded joints 121
Welding design 49
Wohler's curve 77
Working stress 61, 88

❑❑❑

Win Prizes!

ATTENTION READERS

We request you for your frank assessment regarding some of the aspects of the book given as under:

Mechanical Machine Design - I

By

Dr. Sadhu Singh

Please fill up the given space in neat capital letters.

(i) What topics of your syllabus that are important from you examination point of view are not covered in the books ?

(ii) What are the chapters wherein the treatment of the subject matter is not systematic or organised or updated ?

(iii) Have you come across misprints/mistakes/factual inaccuracies in the book? Please specify the chapters and the page numbers.

(iv) What other books on the same subject have you found/heard better that the present book? Please specify the terms of price and quality.

(v) Further suggestions and comments for the improvement of the book:

(PS: If need be please attach a separate sheet to write your views.)

B-63/0204 2021

Other Details

(i) Who recommended you the book ? (Please tick in the box near the option relevant to you.)

☐ Teachers ☐ Friends ☐ Bookseller

(ii) Name of the recommending teacher, his designation and address.

....

....

....

....

(iii) Name and address of the bookseller you purchased the book form :

....

....

....

....

(iv) Name and address of your institution (Please mention the University of Board, as the case may be)

....

....

....

....

(v) Your name and complete postal address :

....

....

....

....

(vi) Course you are enrolled in :

....

....

....

....

The best assessment will be awarded each month. The award will be in the form of our publications as decided by the Editorial Board.

Please mail the filled up coupon at your earliest to:

S.K. Kataria & Sons®
4760-61/23, Ansari Road, Daryaganj Delhi-110002
Ph. : 23243489, 23269324
Fax : 011-23243489
e-mail: katariabooks@yahoo.com
katson_sanjeev@yahoo.co.in
www.skkatariaandsons.com.